ARTEMISIA

Medicinal and Aromatic Plants – Industrial Profiles
Individual volumes in this series provide both industry and academia with in-depth coverage of one major medicinal or aromatic plant of industrial importance.

Edited by Dr Roland Hardman

ARTEMISIA

Edited by

Colin W. Wright
The School of Pharmacy
University of Bradford
UK

CRC Press
Taylor & Francis Group
Boca Raton London New York

CRC Press is an imprint of the
Taylor & Francis Group, an **informa** business

CRC Press
Taylor & Francis Group
6000 Broken Sound Parkway NW, Suite 300
Boca Raton, FL 33487-2742

First issued in paperback 2019

© 2002 by Taylor & Francis Group, LLC
CRC Press is an imprint of Taylor & Francis Group, an Informa business

ISBN-13: 978-0-415-27212-4 (hbk)
ISBN-13: 978-0-367-39682-4 (pbk)

British Library Cataloguing in Publication Data
A catalogue record for this book is available from the British Library

Library of Congress Cataloging in Publication Data
A catalog record for this book has been requested

Visit the Taylor & Francis Web site at
http://www.taylorandfrancis.com

and the CRC Press Web site at
http://www.crcpress.com

CONTENTS

PREFACE TO THE SERIES

There is increasing interest in industry, academia and the health sciences in medicinal and aromatic plants. In passing from plant production to the eventual product used by the public, many sciences are involved. This series brings together information which is currently scattered through an ever increasing number of journals. Each volume gives an in-depth look at one plant genus, about which an area specialist has assembled information ranging from the production of the plant to market trends and quality control.

Many industries are involved such as forestry, agriculture, chemical, food, flavour, beverage, pharmaceutical, cosmetic and fragrance. The plant raw materials are roots, rhizomes, bulbs, leaves, stems, barks, wood, flowers, fruits and seeds. These yield gums, resins, essential (volatile) oils, fixed oils, waxes, juices, extracts and spices for medicinal and aromatic purposes. All these commodities are traded worldwide. A dealer's market report for an item may say "Drought in the country of origin has forced up prices".

Natural products do not mean safe products and account of this has to be taken by the above industries, which are subject to regulation. For example, a number of plants which are approved for use in medicine must not be used in cosmetic products.

The assessment of "safe" to use starts with the harvested plant material which has to comply with an official monograph. This may require absence of, or prescribed limits of, radioactive material, heavy metals, aflatoxin, pesticide residue, as well as the required level of active principle. This analytical control is costly and tends to exclude small batches of plant material. Large scale contracted mechanised cultivation with designated seed or plantlets is now preferable.

Today, plant selection is not only for the yield of active principle, but for the plant's ability to overcome disease, climatic stress and the hazards caused by mankind. Such methods as *in vitro* fertilisation, meristem cultures, and somatic embryogenesis are used. The transfer of sections of DNA is giving rise to controversy in the case of some end-uses of the plant material.

Some suppliers of plant raw material are now able to certify that they are supplying organically-farmed medicinal plants, herbs and spices. The Economic Union directive (CVO/EU No 2092/91) details the specifications for the obligatory quality controls to be carried out at all stages of production and processing of organic products.

Fascinating plant folklore and ethnopharmacology leads to medicinal potential. Examples are the muscle relaxants based on the arrow poison, curare, from species of *Chondrodendron*, and the antimalarials derived from species of *Cinchona* and *Artemisia*. The methods of detection of pharmacological activity have become increasingly reliable and specific, frequently involving enzymes in bioassays and avoiding the use of laboratory animals. By using bioassay linked fractionation of crude plant juices or extracts, compounds can be specifically targeted which, for example, inhibit blood platelet aggregation, or have antitumour, or antiviral, or any other required activity. With the assistance of robotic devices, all the members of a genus may be readily screened. However, the plant material must be fully authenticated by a specialist.

The medicinal traditions of ancient civilisations such as those of China and India have a large armamentarium of plants in their pharmacopoeias which are used throughout South East Asia. A similar situation exists in Africa and South America. Thus, a very high percentage of the World's population relies on medicinal and aromatic plants for their medicine. Western medicine is also responding. Already in Germany all medical practitioners have to pass an examination in phytotherapy before being allowed to practise. It is noticeable that throughout Europe and the USA, medical, pharmacy and health related schools are increasingly offering training in phytotherapy.

Multinational pharmaceutical companies have become less enamoured of the single compound magic bullet cure. The high costs of such ventures and the endless competition from me too compounds from rival companies often discourage the attempt. Independent phytomedicine companies have been very strong in Germany. However, by the end of 1995, eleven (almost all) had been acquired by the multinational pharmaceutical firms, acknowledging the lay public's growing demand for phytomedicines in the Western World.

The business of dietary supplement in the Western World has expanded from the Health Store to the pharmacy. Alternative medicine includes plant based products. Appropriate measures to ensure the quality, safety and efficacy of these either already exist or are being answered by greater legislative control by such bodies as the Food and Drug Administration of the USA and the recently created European Agency for the Evaluation of Medicinal Products, based in London.

In the USA, the Dietary Supplement and Health Education Act of 1994 recognised the class of phytotherapeutic agents derived from medicinal and aromatic plants. Furthermore, under public pressure, the US Congress set up an Office of Alternative Medicine and this office in 1994 assisted the filing of several Investigational New Drug (IND) applications, required for clinical trials of some Chinese herbal preparations. The significance of these applications was that each Chinese preparation involved several plants and yet was handled as a single IND. A demonstration of the contribution to efficacy, of each ingredient of each plant, was not required. This was a major step forward towards more sensible regulations in regard to phytomedicines.

My thanks are due to the staff of Harwood Academic Publishers who have made this series possible and especially to the volume editors and their chapter contributors for the authoritative information.

Roland Hardman

PREFACE

The genus *Artemisia* comprises some 400 species including a number which are important for their medicinal properties. Chief among these is *A. annua*, the source of the antimalarial drug artemisinin. According to a legend, in the 5th century B.C. the city of Selinus in Sicily was delivered from a plague thought to be malaria and this event was commemorated by the minting of a coin. The obverse of the Selinus coin (figure 1a), shows a chariot driven by Artemis representing the women of the city who suffered severely from the fever accompanied by her brother Apollo, who, as the god of healing is firing arrows to destroy the plague (McGregor, 1996). The reverse of the Selinus coin (figure 1b), also includes Artemis and Apollo represented as a cockerel and a bull respectively while the central figure is the river god Selinus representing Empedocles, a student of Pythagoras who is reputed to have delivered the city from the fever by cleansing the surrounding marshes. Although this interpretation of the reverse of the coin is not generally accepted by modern scholars, the association of Artemis with *Artemisia* (= plant sacred to Artemis) and with the deliverance of Selinus from malaria is fascinating.

The herb *A. annua* has been used in China for thousands of years to treat fevers and the isolation of artemisinin in 1971, a highly potent antimalarial agent has proved to be of considerable importance as malaria parasites are increasingly developing resistance to established antimalarial drugs. Artemisinin is unusual in that the molecule contains a peroxide group which is rarely found in nature although another example is the explosive compound ascaridole, the major constituent of

1a   1b

Figure 1   The Selinus coin. The British Museum

chenopodium oil. This book reflects the importance of artemisinin in that it includes chapters which describe the traditional use, cultivation, genetics and phytochemistry of *A. annua*, the development of artemisinin and its derivatives, their mode of action and their clinical use in the treatment of malaria. In addition, a chapter discussing the regulation of artemisinin-like drugs as antimalarials is included as it is imperative that steps are taken to prevent or at least delay the development of malaria parasite resistance to these agents.

The remaining species of *Artemisia* described in this book have been chosen to illustrate the wide variety of species in this genus. *A. absinthium* became notorious as a result of the toxic effects of absinth liqueur while the culinary uses of *A. dracunculus* (tarragon) are well known. *A. herba-alba* is of particular interest on account of its traditional use as an antidiabetic and *A. vulgaris* has a place in European herbal medicine as an emmenagogue and appetite stimulant. The account of *A. ludoviciana* includes an interesting discussion of the ethnobotany of this Mexican species while the chapter on the Indian *A. pallens* highlights the importance of the latter as a source of oil used in perfumery. A number of other *Artemisia* species are mentioned in the comprehensive introduction to the genus which forms the first chapter of the book and in the second chapter which is concerned with the analysis and quality control of commercial *Artemisia* species.

It is hoped that this volume will stimulate further work on many of the hitherto little known species of *Artemisia* and also remind us of the potential of plant species to provide new therapeutic agents. Finally I would like to express my sincere thanks to each of the authors who have contributed to this volume for their hard work and patience.

REFERENCE

McGregor, I.A., 1996. Malaria. In: Cox, F.E.G. (Ed.), The Wellcome Trust History of Tropical Diseases, The Wellcome Trust, London, pp. 230–247.

CONTRIBUTORS

B.M. Beattie
Department of Primary Industry &
   Fisheries
Devonport
Tasmania 7310
Australia

R.S. Bhakuni
Central Institute of Medicinal and
   Aromatic Plants
P.O. Box CIMAP
Lucknow-226015
India

C. Darbellay
Mediplant
Centre des Fougéres
1964 Conthey
Switzerland

S. Deans
Aromatic and Medicinal Plants Group
Scottish Agricultural College
Auchincruive, KA6 5HW
Scotland

N. Delabays
Federal Agricultural Research Station
1960 Nyon
Switzerland

N. Galland
University of Lausanne
1015 Lausanne
Switzerland

G.N. Heazlewood
Department of Primary Industry &
   Fisheries
Devonport
Tasmania 7310
Australia

M. Heinrich
The School of Pharmacy
University of London
London, WCIN 1AX
UK

D.C. Jain
Central Institute of Medicinal and
   Aromatic Plants
P.O. Box CIMAP
Lucknow-226015
India

A.I. Kennedy
Aromatic and Medicinal Plants Group
Scottish Agricultural College
Auchincruive, KA6 5HW
Scotland

R.N. Kulkarni
Central Institute of Medicinal and
   Aromatic Plants
Allalasandra
Bangalore – 560 065
India

J.C. Laughlin
1/14A Sherburd Street
Kingston
Tasmania 7050
Australia

P.A. Linley
The School of Pharmacy
University of Bradford
West Yorkshire, BD7 1DP
UK

S. Looareesuwan
Faculty of Tropical Medicine
Mahidol University
Bangkok 10400
Thailand

M. Maffei
Departments of Morphophysiology and
    Plant Biology
University of Turin
1-10125 Turin
Italy

M. Mucciarelli
Departments of Morphophysiology and
    Plant Biology
University of Turin
1-10125 Turin
Italy

P. Phillips-Howard
Division of Control of Tropical
    Diseases
World Health Organization
Geneva
Switzerland

N. Pras
Groningen Research Institute of
    Pharmacy
Groningen University Institute of Drug
    Exploration
9713 AW Groningen
The Netherlands

P. Proksch
Institut für Pharmazeutische Biologie
Universitatsstr. 1., Gebaude 26.23
40225 Dusseldorf
Germany

R.P. Sharma
Central Institute of Medicinal and
    Aromatic Plants
P.O. Box CIMAP
Lucknow-226015
India

E. Simpson
Aromatic and Medicinal Plants Group
Scottish Agricultural College
Auchincruive, KA6 5HW
Scotland

D.C. Warhurst
London School of Hygiene and Tropical
    Medicine
University of London
London, WCIE 7HT
UK

P. Wilairatana
Faculty of Tropical Medicine
Mahidol University
Bangkok 10400
Thailand

H.J. Woerdenbag
Groningen Research Institute of
    Pharmacy
Groningen University Institute of Drug
    Exploration
9713 AW Groningen
The Netherlands

C.W. Wright
School of Pharmacy
University of Bradford
West Yorkshire, BD7 1DP
UK

H. Yu
East West Biotech Ltd.
Kingham
Oxon, OX7 6UP
UK

S. Zhong
East West Biotech Ltd.
Kingham
Oxon, OX7 6UP
UK

1. INTRODUCTION TO THE GENUS

MARCO MUCCIARELLI AND MASSIMO MAFFEI

Department of Morphophysiology and Plant Biology, University of Turin, Viale P.A. Mattioli 25, I-10125 Turin, Italy.

GENERAL FEATURES OF THE ASTERACEAE FAMILY

The genus *Artemisia* is one of the largest and most widely distributed of the nearly 100 genera in the tribe Anthemideae of the Asteraceae (Compositae).

Asteraceae is a natural family, with well established limits and a basic uniformity of floral structure, represented in all its members by the common possession of characters such as aggregation of the flowers into *capitula* and production of achenes (*cypselae*) as the typical fruits of the family. The *cypsela* structure, the presence or not of a *pappus*, and the *capitulum* itself, that seems to have evolved in order to simulate an individual flower with respect to pollinator attraction, are all morphological features which have an adaptive significance and whose taxonomic value has been demonstrated in only a few groups such as the Anthemideae and the Cardueae (Heywood *et al.*, 1977).

Asteraceae, in spite of their relatively uniform capitular and floral features, occupy a wide range of habitat types and are found in abundance on every continent except Antarctica.

On a global scale the family has *ca.* 23,000 species (Bremer 1994), almost 10% of the total Angiosperm flora of the world (Wilson 1986).

Ecological diversification must have begun very early in the development of the family and this ecological plasticity has been made possible largely through habitat diversification, which includes a wide range of forms, from highly reduced, small, montane annuals (less than 1 cm high) to relatively large tropical trees (up to 20 meters tall), most of which are soft-wooded. Slow-growing, cold-enduring, hard-wooded shrubs such as *Artemisia* occur along with fast-growing, cold-sensitive, herbaceous genera, over a wide range of latitudes.

The family is very distinctive in its chemical attributes; several classes of plant compounds are characteristic of this family, as we will describe later in this chapter. Many of these substances elaborated by the family are toxic or show other significant physiological activity. This may be one reason why plants of the Asteraceae are rarely used in human diets or for animal fodder, with few important exceptions such as lettuce (*Lactuca* sp.) and sunflower (*Helianthus annuus*), the latter being probably the most useful economic plant of the family (Heywood *et al.*, 1977).

The rich accumulation of essential oils and other terpenoids in certain Asteraceae is responsible for the use of various members such as tansy (*Tanacetum vulgare*) and wormwood (*Artemisia absinthium*) for flavouring foods or liqueurs. Terpenoids and certain phenolic compounds are also responsible for the value of many species of Asteraceae in pharmacy and medicine (Wagner 1977).

The subdivision of the family into 13 tribes by Bentham (1873) has been largely accepted, and the basic classification of the family as recognised today is little different from the original one. Bentham's tribal classification has stood the test of time, and with some modifications such as those introduced by Hoffmann (1894), and by Dalla Torre and Harms in their *Genera Siphonoganarum* (1907) still form the basis of most current work (Heywood *et al.*, 1977). More recent proposals for classification include the division of the family into two phyletic groups (Carlquist 1976), which often led to the creation of two or more subfamilies (Wagenitz 1976; Jeffrey 1978), and the recent classification of the Asteraceae according to Bremer (1994), in which the family is divided into 17 tribes split into two subfamilies (the Asteroideae and the Cichorioideae).

ANTHEMIDEAE – BIOLOGICAL AND SYSTEMATIC REVIEW

The Anthemideae is an important tribe in that it contains about 8% of the total genera and 13% of the species of the Asteraceae. The most distinctive characters of the tribe are the truncate-penicilliate styles with parallel stigmatic lines, the ecaudate or short-tailed anther appendages and the usually aromatic leaves (Heywood *et al.*, 1977).

Generally, the place of origin of Angiosperms, and specifically that of the Asteraceae, are the Andean regions of northern South America or what was once western Gondwanaland (Raven and Axelrod 1974). The late Cretaceous or early Eocene is suggested to be the time of probable origin of the Asteraceae, at approximately the same date as the first upheaval of the Andes (Small 1919).

Of the thirteen tribes adopted, only two, the Asteroideae and the Senecioideae, may be said to be cosmopolitan or nearly so. The Cichorieae, the Cynaroideae and the Anthemideae belong to the northern hemisphere with chief centres in the Mediterranean and central Asiatic regions, and a few have spread over North America and even down the Andes to extra-tropical South America (Turner 1977). The Eurasian concentration of the tribe Anthemideae was largely due to their peripheral development in north-eastern Lauratia with respect to their Gondwanaland centre as a consequence of the phenomenon of continental drift.

The original Asteraceae were shrubby or woody plants, now they are mostly herbaceous as seen in the more largely herbaceous tribes of the Anthemideae and Cichorieae, in which many cases of secondary reversion to the woody condition have occurred. An example of this secondary reversion is represented by the genus *Artemisia*, as stated by Stebbins (1977), who refers to it as one of the most specialized genera of a relatively specialized tribe, the Anthemideae. This tribe has an intermediate percentage (39%) of woody or partly wooded genera, but the three genera (*Hymenopappus*, *Hymenothrix* and *Leucampyx*) which link the Anthemideae to other tribes, particularly the largely herbaceous Helenieae, are all herbaceous. Furthermore, the anomalous stem anatomy observed in the shrubby species of

Artemisia would be expected in a secondary revertant from herbaceous forms (Diettert 1961; Moss 1940).

Geographical Patterns

The Anthemideae are essentially palaearctic in distribution and about 75% of the species occur in the northern hemisphere. They have many species and genera centred in middle Asia, South-West Asia, the western Mediterranean region and the Far East.

Some of the larger genera including *Artemisia* and *Achillea* are widely distributed in north temperate regions, but most genera have more discrete centres of distribution.

The 400 and more known species of *Artemisia* are mainly found in Asia, Europe and North America, but a few species occur in most temperate countries. They are mostly perennial herbs and shrubs dominating the vast steppe communities of Asia, the "sage-brush" communities of the New World and Karoo scrub of South Africa (Heywood and Humpries 1977). Koekemoer, (1996) has recently given a detailed account of the distribution of the Asteraceae in South Africa, stressing the huge distribution in this region of the family. Fifteen of the 17 tribes recognized by Bremer (1994) are present, accounting for a total of 24,731 taxa, among which the Anthemideae with 270 species (12%) and 35 genera, 60% of which are endemic, comprise a large component of southern African flora (Koekemoer 1996 and refs. cited therein) (Table 1).

Floral Biology

There are two pollination modes in the Anthemideae, anemophily (wind pollination) in *Artemisia* and in its immediate allies (Ainswort Davis 1908) such as *Crossostephium*, and entomophily (insect pollination) occurring in the remaining genera.

A number of pollen studies have been conducted in the Anthemideae, using both light and electron microscopy, and many of these studies have been concerned with the characterization and identification of *Artemisia* pollen (Skvarla *et al.*, 1977). Stix, (1960) recognized two basic pollen types in the Anthemideae, *Anthemis* and *Artemisia*, distinguishing them by spine presence in the former and absence or great reduction in the latter.

Floral modification characteristic of anemophilous species include inconspicuous often pendulous *capitula* and florets, loss of nectar production and dry *spineless* pollen, in contrast with the sticky *spiny* pollens of the entomophilous genera (Skvarla and Larson 1965).

Some other important features of *Artemisia* spp. pollen anatomical structure, such as the complex and interwoven branches of the basal grain *columellae* and the reduced grain dimensions, have been investigated by Vezey *et al.* (1993) and discussed at the tribal level.

Table 1   Number of Asteraceae species per tribe in the world (Bremer 1994) and in South Africa (Arnold 1994) (extracted from Koekemoer 1996).

Tribe	No. of species in the world	% of species in the world	No. of species in South Africa	% of species in South Africa
		Subfamily: Cichorioideae		
Mutisiae	970	4%	57	3%
Cardueae	2,500	11%	13	1%
Lactuceae	1,550	7%	53	2%
Vernonieae	1,300	6%	42	2%
Liabeae	160	1%	–	–
Arctoteae	200	1%	200	9%
		Subfamily: Asteroideae		
Inuleae	480	2%	48	2%
Plucheeae	220	1%	19	1%
Gnaphalieae	2,000	9%	543	24%
Calenduleae	110	0,5%	92	4%
Astereae	2,800	12%	276	12%
Anthemideae	1,740	8%	270	12%
Senecioneae	3,200	14%	570	25%
Helenieae	830	4%	7	0,3%
Heliantheae	2,500	11%	49	2%
Eupatorieae	2,400	11%	14	1%
TOTAL	22,960	–	2,253	–

Palinotaxonomic studies have proved to be valuable both in the systematic subdivision at the species level and in the phylogenetic evaluation of the genus *Artemisia* (Solomon 1983). From a morphological point of view, a palinological comparison has made it possible to differentiate *Artemisia campestris* ssp. *campestris* and *A. campestris* ssp. *borealis*, two very similar varieties from Piedmont, north-west Italy (Caramiello *et al.*, 1987; 1990; Caramiello and Fossa 1993–1994).

From the point of view of sex determination and expression, the most heterogeneity among the genera can be found in *Cotula* and *Artemisia*, which, except for gynodioecious forms (in which individual plants have either female or hermaphrodite flowers), have representatives in all the sex classes, including monoclinous (hermaphrodite), gynomonoecious (plants have both female and hermaphrodite flowers), monoecious (plants have both male and female flowers) and diclinous dioeceous (plants have male or female flowers) species.

Karyology

The most common base chromosome number of the Anthemideae is n = 9, but n = 8, 10, 13 and 17 also occur. Few cases of dysploidy have been observed (Mitsuoka and Ehrendorfer 1972). The best known examples of descending dysploidy are those found in the annual species of the *Pentzia globifera* group with n = 8, 6 and in *Artemisia* with n = 9, 8. Polyploidy occurs quite widely in the tribe and is especially marked in *Leucanthemum*, in *Cotula* (in the section *Leptinella*) and in *Artemisia*, where polyploidy is regarded as one of the main evolutionary mechanisms (Oliva and Vallès 1993). In *Artemisia* genus the diploid chromosome number goes from 14 to 110 (Heywood and Humphries 1977).

Division into Subtribes

Traditionally, the Anthemideae have been divided into two subtribes, the Anthemidineae Dumort. and the Chrysanthemineae Less. (Lessing, 1832), on the basis of the presence or absence of receptacular scales. For many genera this subtribal division is quite workable, but on the other hand the artificiality of the scale character has been demonstrated for many other groups and genera of the tribe. In her survey of the Asteraceae, Poljakov (1967) recognized six subtribes in the Anthemideae, based principally upon the informal groups of Bentham – the Anthemidineae, the Anthanasineae, the Chrysanthemineae, the Artemisineae, the Seriphidineae and the Oligosporineae. A relatively more recent analysis and grouping of the whole tribe is that of Reitbrecht (1974). On the basis of anatomical studies of the cypsela (the dry, single seeded fruit), Reitbrecht suggests that the generic assemblages of the Anthemideae can be arranged into seven provisional groups, among which the "*Artemisia gruppe*" includes *Artemisia*, *Crossostephium* and their allies.

The subdivision into subtribes however is unsatisfactory and has varied, as shown above, from a traditional two-subtribe split based on "artificial" characters to the informal recognition of five, six, or more generic assemblages.

So far the combined data from *cypsela* studies, phytochemical investigations, embryological analyses and cytological studies and breeding systems together with gross morphological studies have provided useful insights into the relationships of some of the principal genera, particularly the northern hemisphere *Anthemis* and *Chrysanthemum* complexes and parts of other genera such as *Achillea*, *Artemisia*, *Dendranthema* and *Cotula* (Heywood and Humphries 1977). Other revisions of the tribe arrangement have followed or are still in the course of definition; relevant references have been reported in following paragraphs.

Molecular Systematics of the Tribe

The Anthemideae Cass. is the seventh largest tribe in the Asteraceae, it is monophyletic and composed of 109 genera and 1740 species (Bremer and Humpries 1993). L.E. Watson, (1996) has undertaken a study of the restriction site of the chloroplast genome (cpDNA phylogeny) to evaluate phylogenetic relationships among the tribe. Watson's results support the monophylogeny of the subtribes, with a strong concordance between the molecular and morphological phylogenies. *Dendranthema* (DC.) Des Moulins, *Artemisia* L. and *Seriphidium* (Besser) Poljakov, form a monophyletic clade (branch), corresponding to the subtribe Artemisineae of Bremer and Humphries (1993).

ARTEMISIA – THE GENUS

Morphological and Taxonomic Aspects

As reported by Tutin and Persson (1976) the 400 species of *Artemisia* share the following common morphological characters:

Herbs or small shrubs, frequently aromatic. Leaves alternate. *Capitula* small, usually pendent, racemose, paniculate or capitate inflorescences, rarely solitary. Involucral bracts in few rows. Receptacle flat to hemispherical, without scales, sometimes hirsute. Florets all tubular. Achenes obvoid, subterete or compressed, smooth, finely striate or 2-ribbed; pappus absent or sometimes a small scarious ring.

They are mostly perennial herbs and shrubs dominating the vast steppe communities of Asia, the New World and South Africa. *Artemisia* is a highly evolved genus with a wide range of life forms, from tall shrubs to dwarf herbaceous alpine plants, occurring in a variety of habitats between Arctic alpine or montane environments to the dry deserts. Many species are not always well known and a world revision of the genus, from a systematic point of view is needed, starting from the big available bulk of regional, sectional and species group accounts (Heywood *et al.*, 1977 and refs. cited therein).

Asia seems to show the greatest concentration of species with 150 accessions for China (Hu 1965), 174 in the ex U.S.S.R. (Poljakov 1961a) and about 50 reported to occur in Japan (Kitamura 1939, 1940). Tutin and Persson (1976) recognize 57 species for the European region and 30 or so New World species are reasonably well documented (Clements and Hall 1923; Ward 1953; Beetle 1959, 1960).

The first rational and natural arrangement of the genus was given by Besser (1829), and a major part of his work was included in the text of De Candolle (1837) and Hooker (1840). Besser divided the genus into four sections based on fundamental differences in the floral structure (Table 2).

Phylogenetically, *Abrotanum* and *Absinthium* represent the more primitive sections, while *Dracunculus* and *Seriphidium* are the most advanced (Clements and Hall 1923).

Some taxonomists have elevated Besser's sections to the subgenus level (Kelsey and Shafizadeh 1979; Poljakov 1961b) and reduced the number of subgenera to three, by combining *Abrotanum* with *Absinthium* to form the subgenus *Artemisia*, which was further divided into three sections: *Artemisia*, *Abrotanum* and *Absinthium* (Poljakov 1961b).

The shrubby members of the subgenus *Seriphidium* endemic to North America have been recognized as being closely related, distinct from the Old World *Seriphidium* and grouped together in the section *Tridentatae* of the subgenus *Seriphidium* (Kelsey and Shafizadeh 1979). Many authors have not accepted this assignment and McArthur and Plummer (1978) have argued that the New World *Tridentatae* and the rest of the New World *Seriphidium* species should be considered as separate taxonomic entities. A similar more or less close relationship among different species of the section *Abrotanum*, was recognized by Kelsey and Shafizadeh (1979), who grouped them in the subsection *Vulgares*.

Table 2   Besser's division of *Artemisia* (extracted from Kelsey and Shafizadeh 1979).

Morphological characters	Section
1. Heads heterogamous, the marginal flowers pistillate	
2. Central flowers fertile, with normally developed achenes	
3. Receptacle not hairy	1. *Abrotanum*
3. Receptacle long hairy	2. *Absinthium*
2. Central flowers sterile, their achenes aborted	3. *Dracunculus*
1. Heads homogamous, marginal flowers absent	4. *Seriphidium*

Kelsey and Shafizadeh (1979) have given a detailed chemotaxonomic review of the sesquiterpene lactones isolated from species of the genus *Artemisia*. They tried to solve or at least to provide answers to the systematic problem of the species grouped by Keck in the subsection *Vulgares* to understand whether they are a distinct group in the subgenus *Abrotanum*, while considering also the problem of the origin and relationship of the subgenus *Seriphidium* of the section *Tridentatae*.

According to Kelsey and Shafizadeh (1979), there are no chemical characteristics supporting the subdivision of *Artemisia* species into the two subgenera *Abrotanum* and *Absinthium*, since both produce eudesmanolides and guaianolides, that are identical from a qualitative point of view and are biosynthetically related. Such a statement should confirm the demonstration that the presence of hairs on the floral receptacle is an artificial character at the genus level as well as at the subtribe level and suggests that the two subgenera could be combined into one (*Artemisia*) as formerly proposed by Poljakov (1961b).

The species belonging to the subsection *Vulgares* arose as a phylogenetically closely related group, having a chemotaxonomic profile similar to but not entirely distinct from other New World *Abrotanum* species.

The subgenus *Seriphidium*, which is composed of two geographical groups, one in the Old World and one in the New World (*Tridentatae*), seems to be polyphyletic, with a striking separation between the species belonging to the section *Tridentatae* of the Americas, which produce guaianolides, which appear very similar and structurally related those produced by the New World *Abrotanum* species but which are absent in the Old World *Seriphidium* species. Therefore the *Tridentatae* may have originated from ancestors in *Abrotanum* rather than from the Old World *Seriphidium* and should be recognized as a subgenus separated and distinct from the latter (Kelsey and Shafizadeh 1979).

An excellent study with respect to the solution of the unsatisfactory intrageneric division of the genus *Artemisia*, has been done by Belenovskaja (1996), in her recent work dealing with the taxonomic value of flavonoids at the subgeneric level. She has demonstrated that flavonoid data support subdivision of the genus into three subgenera: *Artemisia*, *Dracunculus* and *Seriphidium*. The occurrence of a large number of 6-methoxyflavonoids in *Artemisia* correspond well with the postulated advanced position of the entire genus in the Asteraceae (see below for further discussion).

Thereafter the distribution of different types of flavonoids reflects the best subdivision of subgenera into taxonomic groups or the so called Krascheninnikov's series (Krascheninnikov 1948, 1958). A link is suggested, through the comparative analysis of data, between the species of the section *Absinthium* and the *Tridentata* group of North American species of *Seriphidium*. Data confirmed the Krascheninnikov's division of the subgenus *Artemisia*, with its sections *Artemisia*, *Abrotanum* and *Absinthium*, in several taxonomic series.

The subgenus *Seriphidium*, subdivided in the Flora of USSR (Poljakov 1961b), into two sections, *Junceum* and *Seriphidium*, showed a flavonoid distribution which correlates well with the division of the subgenus into six sections and 16 subsections, as proposed by Filatova (1986).

All these arguments support the possibility of applying flavonoid and other chemical data to the solution of the different taxonomic problems, concerning the complex phylogenetical relationships at the intrageneric and interspecific level of the genus *Artemisia*.

Geological and Evolutionary History

The traditional criteria used for establishing the initial appearance of a taxon in a geologic record are 1) the identification must be certain; 2) the fossil must be contained within the rock; 3) it must be indigenous to the rock; and 4) the age of the rock must be known (Graham 1996).

The geological history of the Asteraceae is strongly linked to that of the genus *Artemisia* and of particular interest here is that the most convincing early fossils of the Asteraceae include *Artemisia* pollen of the late Oligocene of central Europe, *Artemisia* fruits and seeds from Poland (middle Miocene), and *Artemisia* pollen from eastern and western North America, of the late Miocene and late Oligocene, respectively.

If the oldest reports of *Asteraceae* are accepted, the family would be represented in the Eocene of central to southern South America, with the Mutisieae and the AHH complex (Astereae-Heliantheae-Helenieae *et al.*) as the oldest tribes, followed by the *Artemisia* type of the late European Oligocene (Graham 1996).

Until recently, general consensus placed the origin of the genus *Artemisia* in central Asia with subsequent migration to North America through the Bering Land Bridge (Clements and Hall 1923; McArthur and Plummer 1978; Stebbins 1974; McArthur 1979). However, the biological evidence points to Eurasia as the centre of origin.

Most of the species are present in the Old World compared with only about 50 in North America. The two genera *Chrysanthemum* and *Tanacetum*, the nearest and least specialised relatives of *Artemisia* as well as other genera in the tribe Anthemideae, grow predominantly in Eurasia and Africa (McArthur 1979). *Artemisia* species may have migrated from the Old World through most of the late Tertiary and periodically during the Pleistocene when the Bering Land Bridge was present (Hopkins 1967), thus providing sufficient time for the origin and dispersal of the present *Tridentatae* in North America. However, taking into account what is reported by Graham (1996) on the eastern and western coasts of North America, it is not possible to exclude the New World as a centre of evolution of the genus.

Pharmaceutical and Economic Uses

The biological and therapeutic applications of the plants of the Asteraceae are the result of popular tradition and of systematically conducted chemical and pharmacological research. In addition to drugs known since antiquity, from plants such as *Chamomilla*, *Cynara* and *Sylibum*, there are many other species in the family which have found therapeutic applications due to their antihepatotoxic, choleretic, spasmolytic, anthelmintic, antiphlogistic, antibiotic or antimicrobial activity. Wagner (1977) in his detailed review on the pharmaceutical properties of Asteraceae, has also pointed out the eminent role played in this regard by the genus *Artemisia*.

Antibiotic, fungistatic and insecticidal agents

In Wagner's work *Artemisia capillus* is referred to as a valuable source together with *Carlina acaulis*, of polyacetylenes as well as of capillin, which possesses significant antibiotic action against dermal mycoses (Shulte *et al.*, 1967).

The presence of santonin, a sesquiterpene lactone with antihelmintic activity, has been reported in *Artemisia cina* (Wagner 1977) and in *Artemisia coerulescens* L. ssp. *cretacea*. Water soluble extracts from *A. cina* possess a strong larvicidal activity against *Culex pipiens* L., an arbovirus human disease carrying mosquito (Aly and Badran 1995).

Tests on the antifeedant and insecticidal properties of *Artemisia absinthium* extracts, have been run in the field for the biological control of some important crop pests, e.g. against *Crocidolomia binotalis*, the cabbage webworm (Facknath and Kawol 1993). The volatile constituents of the green parts of *Artemisia princeps* var. *orientalis* have both phytotoxic and antimicrobial activity and also inhibit radicle elongation of receptor plants and markedly inhibit the growth of *Bacillus subtilis*, *Aspergillus nidulans*, *Fusarium solani* and *Pleurotus ostreatus* (Yun *et al.*, 1993).

The essential oils of several *Artemisia* species have been found to exert other biological activities (Janssen *et al.*, 1987). For example, studies undertaken at the Scottish Agricultural Center at Auchincruive focusing on the antimicrobial nature of volatile oils revealed that the oil of *A. afra* (also known as African wormwood), which contains a mixture of monoterpenes, is active against bacteria such as *Acinetobacter calcoaceticus*, *Klebsiella pneumoniae*, *Brevibacterium linens*, *Yersinia enterocolitica*, and many others including *Escherichia coli* (Deans and Svoboda 1990). The essential oil of *A. herba-alba* inhibited the asexual reproduction of *Aspergillus niger*, *Penicillium italicum* and *Zygorrhynchus* sp. (Tantaoui-Elaraki *et al.*, 1993). In some cases a direct relationship has been found between the presence of an oil constituent and its biological activity, as in the case of eugenol, *cis*-ocimene and α-phellandrene from the oil of *A. dracunculus*, which significantly inhibited the growth of most of the microorganisms mentioned above, including *Staphylococcus aureus* and *Proteus vulgaris* (Deans and Svoboda 1988).

Pharmacological properties and clinical experimentation

Scoparone (6,7-dimethoxycoumarin), a coumarin isolated from the well known hypolipidaemic Chinese herb, *Artemisia scoparia*, shown to possess vasodilator and antiproliferative activities, with free radical scavenging effects *in vitro*, has been tested *in vivo* using hyperlipidaemic diabetic rabbits and shown to have an antiatherogenic action, consisting of a reduction of plaque formation and of the plasma cholesterol level (Chen *et al.*, 1994). Extracts from the same species were shown to contain Ca^{2+} channel blocker-like constituents, capable of *in vivo* hypotensive and bradycardiac effects, confirming the spasmolytic properties ascribed to *A. scoparia* in Chinese folk medicine (Gilani *et al.*, 1994). Other spasmolytic effects, consisting of a strong smooth muscle relaxing activity have been ascribed to four flavonols isolated from *Artemisia abrotanum* L. (Bergendorff and Sterner 1995). Hypoglycaemic effects, similar to those of *A. scoparia*, with consistent reduction in blood glucose level and with minimal adverse side-effects, have been demonstrated for *Artemisia herba-alba* aqueous extracts, after oral administration (Alkhazraji *et al.*, 1993).

A sulphated polysaccharide purified from the crude fraction of the leaves of *Artemisia princeps* Pamp. was shown to be effective in the acceleration (6000 fold), of the formation of thrombin-HC II complex in human plasma, thus representing a novel pharmacological molecule, with a mode of action quite distinct from that of heparin (Hayakawa *et al.*, 1995).

Cytoprotective agents, such as dehydroleucodine of *Artemisia douglasiana*, have been found to prevent the formation of gastric lesions (Guardia *et al.*, 1994).

Cytotoxic agents

Costunolides (sesquiterpene lactones) with antitumoral activity have been isolated from *Artemisia balchanorum* (Herout and Sorm 1959). Several terpenoids and flavonoids, isolated from *Artemisia annua*, showed significant cytotoxic activity when tested *in vitro* on several human tumour cell lines; among them artemisinin and quercetagetin proved to be responsible for the action observed on five of the tumour lines tested (Zheng 1994). Immunomodulatory activity of solvent extracts of the air-dried aerial parts of *A. annua* has been demonstrated, through a study of their *in vitro* effects on human complement and T lymphocyte proliferation (Kroes *et al.*, 1995).

Antimalarial activity

Artemisia annua L. (annual wormwood) is a herb of Asiatic and Eastern European origin, that has also become naturalized in the United States of America and is the source of the traditional Chinese herbal medicine Qing Hao, which has been used for over 2000 years to alleviate fevers.

The species has received considerable attention because of its antimalarial properties. The plant activity has been established to be due to artemisinin (*qinghaosu*) a cadinane-type sesquiterpene lactone endoperoxide, present in the aerial parts (Klayman *et al.*,

1984), whose semisynthetic derivatives, artemether, arteether and artesunate are effective against multi-drug resistant malaria caused by *Plasmodium falciparum* and against the life-threatening complication, cerebral malaria (White 1994).

Currently, efforts are being made to make these drugs available world-wide and in order to meet the industrial demand for it, many research programmes have started, focusing on the selection and cloning of high-artemisinin yielding chemotypes of *A. annua* (Ghan *et al.*, 1995). A review of recent agricultural techniques to improve yields from cultivated plants is also available (Laughlin 1994); see also chapter 10.

At the present time, the total synthesis of artemisinin is not yet economic for large-scale production, and so alternative approaches for improving the economics of drug production are under study, through the extraction of artemisinin related sesquiterpenes from *A. annua* plants and their subsequent conversion into artemisinin semisynthetically (Gupta *et al.*, 1996). Among these artemisinic acid is the most promising.

Many researchers world-wide are involved in the elucidation of the chemical pathways of the biosynthesis of artemisinin and related compounds from ^{14}C precursors, both *in vivo* and in cell free systems (Sangwan *et al.*, 1993).

Delgado (1996), analysing the bioactive constituents of some Mexican medicinal Asteraceae, hypothesized the presence of an artemisinin analogue, presumably in minor amounts, in *Artemisia ludoviciana* Nutt. ssp. *mexicana*, following the observation that extracts of this species inhibited *Plasmodium berghei*, thus stressing once more the importance of world biodiversity preservation.

Current research programmes for the improvement of high-producing artemisinin varieties and clones may lead to the development of successful techniques for the *in vitro* culture of *A. annua* plants.

The search for high yielding artemisinin plants led to the selection of chemotypes of *A. annua* showing the highest percentages of artemisinin (0.42%), when plants were twelve to thirteen weeks old (Chan *et al.*, 1995), whereas the combined concentration of artemisinin and its intermediates artemisinic acid and artemisinin B was particularly high (1.35%) in some Indian *A. annua* (Gupta *et al.*, 1996).

Bitter substances and liqueurs

Although bitter substances occur widely in the Asteraceae, only *Artemisia absinthium* (wormwood) and *Cnicus benedictus* are important in pharmacy and the food industry. The bitter taste of *A. absinthium* is due to the guaianolide lactones, absinthin and anabsinthin. The pharmaceutical usefulness of these bitter drugs is due primarily to their stimulation of stomach secretions, which is the result of reflex nervous activity. Furthermore, release of gastrin causes an increase in stomach acidity (Wagner 1977); the great popularity of absinthe for the preparation of aperitifs (e.g. *vermouth*) is largely due to the aromatic substances that it also contains. Vermouth wines are prepared predominantly from *Artemisia pontica* L. Since absinthe liqueurs are prepared from *Oleum Absinthii*, they contain considerable quantities of thujone, which in large doses is toxic, and can lead to chronic poisoning. The biological action of some *Artemisia* oils has been directly experienced by humans. Thujone, a typical monoterpene of some *Artemisia* species, causes chronic poisoning so that preparations of the liqueur absinthe from root extracts of *A. absinthium* are banned in several countries (Wagner 1977). In this plant the total thujone content (α- + β-thujone derivatives) may reach concentrations up to 60% of the total oil, and for those reasons several attempts have been done to select for low thujone *A. absinthium* chemotypes (Lawrence 1992). The bitter substance absinthin from *A. absinthium* also has insecticidal and larvicidal properties (Javadi 1989).

Another aromatized alcoholic beverage, which has a long tradition in northern Italy is "*genepi*", a strong flavoured liquor extracted from *A. umbelliformis*, *A. genipi* and *A. petrosa* and containing α- and β-thujones as the characteristic compounds in GC profiles (Appendino *et al.*, 1985).

Spices and flavouring agents

Apart from absinthe, only *Artemisia dracunculus* (tarragon) and *Artemisia vulgaris* (mugwort) are of economic importance, owing to their aromatic smell and taste. They are used to season salads, mayonnaise, gravies, fish dishes, pickled gherkins, and in the preparation of tarragon vinegar.

A. dracunculus L. together with *A. pontica* L., *A. vallesiaca* All. and *A. mutellina* Vill., in order of economic importance, are actively grown in Piedmont, a northern Italian region with a long tradition of aromatic plant cultivation. *A. pontica* and *A. vallesiaca* are exclusively grown for the dry herb production, to be employed for drinks and alcoholic beverages flavouring, while *A. dracunculus* is almost totally cultivated for its essential oil, in the food and perfumery industry (Chialva 1985).

Ethnobotany and Paleoethnobotany

The Asteraceae have been used for food, medicine and crafts since very ancient times, but only when sufficient skill was developed in agricultural techniques were these plants introduced into cultivation. Starting from the available archaeological evidence, consisting mainly of carbonised fruits, and less commonly waterlogged plant remains or, more rarely, of well preserved mummified specimens, it has been possible to identify different centres of diversity for cultivated Asteraceae (Nuñez and De Castro 1996).

Among those concerning the genus *Artemisia*, we may note *Artemisia capillaris* Thunb. from the Chinese and Japanese centre, which has been cultivated occasionally as a medicinal plant and *Artemisia dracunculus* L. (tarragon) which is widely cultivated in central Asia as a condiment. The latter species consists mainly of two polyploid strains, the "Russian tarragon" a fertile decaploid and the "French tarragon", a sterile tetraploid vegetatively propagated. Hybrids between these strains have been described (Nuñez and De Castro 1996).

All over the Mediterranean area shrubby *Artemisia* L. species have been cultivated as garden plants on account of their medicinal properties. The use of wormwood in the Roman period is well documented by Dioscorides for Pontus and Cappadocia in Turkey (Berendes 1902; Gunther 1968) and is most likely represented from *Artemisia pontica* L. Achenes of *Artemisia* species with a distinct longitudinal striation on the surface were found in samples from refuse deposits in Syria, dated back to the Bronze age. The cultivation and use by Spanish Muslims of several *Artemisia* (*A. absinthium* L., *A. chamaemelifolia* Vill. and *A. arborescens* L.) and *Tanacetum* L. species as insecticide and insect deterrent is well documented in the Medieval period (Nuñez and De Castro 1996 and refs. cited therein).

Recipes confirming the use of *A. absinthium* in Italy during the Roman period, have been reported by Wittmack (1903). Mention of the latter and of other *Artemisia* species, and their use in Classical antiquity, can be found in Theophrastus and Hippocrates. It seems that wormwood was associated during Talmudic period with the use of gall to alleviate the sufferings of condemned and sick persons.

The cultivation and gathering of *Artemisia* species for medicinal and magical purposes was common among the natives of the Canary Islands, during the so called Guanche period. Achenes of *Artemisia absinthium* and *A. abrotanum* L. have been found in sites of the Medieval period as weed components of British cereal crops (Nuñez and De Castro 1996 and references therein). *Artemisia vulgaris* L. achenes dating back to the Neolithic and Iron age are a very common finding in France.

In Germany, Albertus Magnus recorded the properties of many species of Asteraceae in his book of medicinal plants, including *Artemisia absinthium*, *A. abrotanum* and *A. vulgaris* (Biewer 1992), and similar data has been reported from other European territories including Czechoslovakia, Hungary, Poland, Denmark, Finland and Sweden.

Shah, (1996) describes the genus *Artemisia* as one of the largest and most difficult taxa to understand under an ethnobotanical point of view. The medicinal use of *Artemisia* spp. was introduced into the Indian Himalayas by different cultural and ethnic groups who entered this region in the past coming from Mediterranean and Arabian regions. *Artemisia nilagarica* (C.B. Clarke) Pamp. (syn. *A. vulgaris* sensu Hook.f.) is the most common species found in earlier Indian literature. It was used as a decoction and infusion for the relief of nervous and spasmodic afflictions by Himalayan people. The plant was used also for magical purposes; it was traditionally kept at front doors and under pillows to discourage evil spirits and ghosts, and the aerial parts were used in festivals for worshipping or offered to the local divinity (Shah 1996 and references therein). The essential oil is also finding its place in the indigenous perfumery industry. The use of *Artemisia* as incense has perhaps evolved in this region, and it is still used for this purpose by placing the immature leaves and inflorescences as dried material on burning charcoal in a special bowl. The preference of the immature plant parts over the mature ones is probably due to the higher content of thujone and 1,8-cineole, whose psychoactive properties could help people to forget severe cold conditions and the other hardships of the region (Shah and Thakur 1992).

A detailed description of the ethnobotanical research of different regions of Catalonia, Spain, is given by Vallès *et al.* (1996), where the authors refer to the medical properties and uses of *A. absinthium* L., *A. campestris* L., *A. chamaemelifolia* Vill., *A. verlotiorum* Lamotte and of *A. vulgaris* L.

China is a country with an old ethnobotanical tradition; each of its 56 nationalities has accumulated a wealth of experience in the utilization of plants, including about 500 species of Asteraceae used as medicinal plants. About 100 species are used in Tibetan medicine and those are mainly from the genera *Artemisia* L., *Saussurea, Taraxacum, Inula, Ligularia* and *Cremanthodium* Benth. (Huang and Ling 1996).

In the dietetic and medical traditions of Chinese ethnobotany, some species are used for specific folk customs. For example, in the Dragon Boat Festival on 5th of May of the Lunar Calendar, people like to hang the mugwort, *Artemisia argyi* or *A. indica* and *Acorus calamus* L. at the door. The herbs are thought to ward off evil spirits or at least to drive the harmful insects away, owing to their essential oil content. Mongolian nationals often use the roots, stems and leaves of *Artemisia brachyloba* to make a special national tea, *banderol gan tea* instead of *Camellia* tea, which is also a good drug for curing indigestion. In China government officers, experts and the public pay close attention to the development of food resources, industrial materials, sustainable energy and medicines from plants. The effective constituents of more than 240 species of Asteraceae have been analysed for medicinal properties, and about 86 chemical constituents have been identified from some species of *Artemisia* and *Saussurea*.

The number of *Artemisia* species used in China as medicinal, food (wild edible plants), forage, aromatic, nectariferous and oil producing plants is impressive (Huang and Ling 1996).

CHEMICAL OVERVIEW AND CHEMOTAXONOMY

Several secondary metabolites characterise the chemical composition of the genus *Artemisia*. A survey of the literature indicates that almost all classes of compounds are present in the genus, with particular reference to terpenoids and flavonoids. However, wax constituents, polyacetylenes and, to a lesser extent, nitrogen containing molecules have also been found in several species. The wide array of molecules present in the genus and the distribution of plants in several different habitats provides the opportunity for the study of genotypic and phenotypic variations as well as chemotaxonomic relationships among species. In this section we will give an overview of the chemical variability and taxonomic relationships within the genus *Artemisia*, by considering the main classes of secondary metabolites mentioned above. Owing to the introductory nature of this chapter the discussion will be more like a journey in the field of secondary metabolites of *Artemisia*. Let us start from the most studied class of metabolites, the terpenoids.

Terpenoids

The terpenoids present in *Artemisia* species are representative of all classes of compounds, from monoterpenes up to triterpenes. Most of the species are characterized by the typical fragrance of lower terpenoids, such as monoterpenes and sesquiterpenes. These volatile molecules are present in the essential oils, which impart strong aromatic odours to the plants. Among the various compounds, lower terpenoids such as camphor, thujone, borneol and 1,8-cineole are the most representative (Fig. 1). Recently, analytical methods to determine the oil components have been improved by the use of capillary electrophoresis, whilst the separation of enantiomers has been achieved by the use of β-cyclodextrin coated chiral capillary columns (Ravid *et al.*, 1992). As for many aromatic plants, the oil content of *Artemisia* is affected by environmental factors. The monoterpenoid content of some *A. tridentata* ssp. *tridentata* plants varied seasonally with the highest content reached in July (4.18%) and the lowest during May (0.97%) (Cedarleaf *et al.*, 1983).

The huge variability in the essential oil composition within *Artemisia* can be summarized by taking into consideration the data published on the most important oil crops. The variability in the monoterpenoids of the genus ranges from the artemisyl derivatives (such as artemisia alcohol, artemisia ketone, artemisia acetate, artemisia triene and yomogi alcohol/ketone) to santolinyl and lavandulyl skeleton derivatives. Furthermore, the presence of many irregular monoterpenes seems to be a particular characteristic of the tribe Anthemideae (Fig. 1) (Greger 1977).

A. annua oil composition has been a matter of several investigations. The essential oils from the inflorescence of *A. annua* of Chinese as well as Dutch origin were particularly rich in the monoterpene artemisia ketone (63%; Liu *et al.*, 1988; Woerdenbag *et al.*, 1992, 1993), while in Mongolian chemotypes the phenols thymol and carvacrol were the main compounds (Satar 1986). The leaf oil composition of a population of *A. annua* grown in a greenhouse in our department showed a high content of the monoterpene hydrocarbons thujene (α+β), α-terpinene and limonene, followed by good percentages of the oxygenated monoterpenes camphor and α-terpineol (Table 3). A few sesquiterpenes were identified, α-guaiene being the most abundant. In the same table, the percentage of oil components is flanked by the corresponding coefficient of variation. This statistical parameter indicates the variability within the population for every single compound. As shown in Table 3, a very low variability was evident for the main monoterpene hydrocarbons (values below 15%) as well as for the monoterpene alcohol α-terpineol. On the other hand, a great variability was observed for the pinenes, γ-terpinene, *cis*-sabinol (values above 80%) as well as for borneol and the sesquiterpenes.

The chemical composition of *Artemisia* oils has been investigated in many other species for chemotaxonomic reasons. However, the continuous search for new active and/or flavouring molecules is the driving force for the improvement and development of extracting and purifying techniques. Camphor, 1,8-cineole, camphene, terpinen-4-ol and α-terpineol are the main oil components of *A. sieberi* (Weyerstahl *et al.*, 1993), *A. hololeuca*, *A. gelinii* and *A. pontica* (Bodrug *et al.*, 1987), whereas artemisia ketone made up 94% of the total oil of *A. alba* (Bodrug *et al.*, 1987).

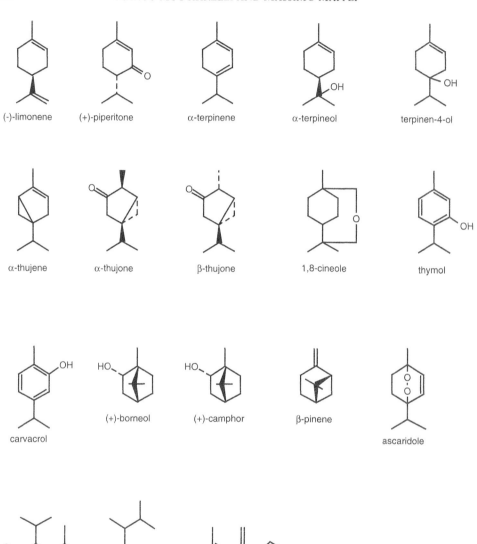

Figure 1 Structural formulae of some representative monoterpenoids found in the genus *Artemisia*.

However, the use of biotechnology, as in the case of micropropagation, does not always gives the same products as *in vivo* cultures. For example, micropropagation of *A. alba* afforded mainly isopinocamphone and camphor, but not artemisia ketone or α-thujone (Ronse and De Pooter 1990).

Table 3 Leaf oil chemical composition of a population of *Artemisia annua* grown in a greenhouse. C.V. = coefficient of variation.

Monoterpenes	%	C.V.
Hydrocarbons		
α-Pinene	2.63	91.63
α+β– Thujene	32.70	13.06
Camphene	1.31	48.09
β-Pinene	0.74	82.35
Sabinene	0.46	72.48
α-Terpinene	29.29	11.83
Limonene	3.00	11.97
γ-Terpinene	2.07	154.08
p-Cymene	0.14	39.77
Alcohols		
cis-Thuj-2-en-4-ol	1.13	72.11
Terpinen-4-ol	2.84	41.43
α-Terpineol	3.04	11.28
Borneol	0.24	31.69
cis Sabinol	0.11	86.59
Ketones		
Camphor	6.07	52.60
Sesquiterpenes		
α-Guaiene	1.59	60.53
Germacrene D	1.11	49.35
β-Caryophyllene	0.89	58.49

Davanone is the main compound of *A. thuscula* collected in the Canary islands (Bellomaria *et al.*, 1993), whereas β-thujone and chamazulene are the main constituents of *A. arborescens* essential oils of Italian origin (Biondi *et al.*, 1993). 1,8-Cineole, terpinen-4-ol and camphor were the main components of *A. vulgaris* oil, whereas thujone was the main constituent of *A. verlotiorum* (Carnat *et al.*, 1985). Limonene was found in high percentages in *A. frigida*, *A. santolinifolia*, *A. adamsi* and *A. pamirica*, whereas β-pinene was the main compound of *A. macrocephala* and *A. sphaerocephala* (Satar 1986). *Artemisia afra* oil was characteristic because of the presence of the monoterpenes α- and β-thujone, followed by 1,8-cineole, camphor and α-pinene (Graven *et al.*, 1992). The essential oils of *A. judaica* may be divided into two distinct chemotypes: an artemisyl-oil type and a piperitone-oil type (Putievsky *et al.*, 1992). In a study on the essential oils of some Indian *Artemisia* species, Thakur and Misra (1989) reported on the presence of azulenes in *A. absinthium*, high borneol and α-terpineol contents in *A. annua*, estragole and methyl chavicol in *A. dracunculus*, 1,4-cineole, sabinyl acetate and sabinol in *A. maritima*, α-thujone, camphor, β-eudesmol and 1,8-cineole in *A. vulgaris*, davanone and nerol in *A. pallens*, linalyl acetate, thujyl acetate and camphor in

A. roxburghiana, eugenol and agropinene in *A. scoparia* and β-himachalene and α-atlantone in *A. vestita*.

The use of the essential oils as chemotaxonomic markers has been applied to the classification of some *Artemisia* species growing spontaneously at high elevation. Table 4 shows the chemical composition of essential oils isolated from *A. abrotanum*, *A. absinthium*, *A. alba*, *A. annua*, *A. campestris*. ssp. *campestris*, *A. campestris* ssp. *borealis*, *A. chamaemelifolia*, *A. genipi*, *A. glacialis*, *A. petrosa* ssp. *eriantha*, *A. umbelliformis*, *A. vallesiaca*, *A. verlotiorum*, and *A. vulgaris*, growing spontaneously in the North-West Italian Alps (Mucciarelli *et al.*, 1995). Large data sets, like the one represented in Table 4, are usually hard to interpret exclusively on the basis of a direct comparison of compound percentage. In order to better characterize the relationship between the oil composition of different species, several authors adopt the use of multivariate statistical methods. In this context the most commonly used are cluster analysis, principal component analysis and discriminant analysis (Maffei *et al.*, 1993a). The cluster analysis of essential oils from the data matrix of Table 4 allowed us to devise a first group of species composed of *A. genipi*, *A. umbelliformis* and *A. petrosa*, which were statistically linked because of the high percentage of α-thujone and the low β-thujone levels. The remainder of the species were grouped in a cluster which was further subdivided into minor subclusters. *A. chamaemelifolia* was separated from the other plants by the high levels of *trans*-nerolidol, and carvacrol. *A. absinthium* was separated mainly by the presence of *trans*-chrysanthenyl acetate, while *A. abrotanum* was distinguished by the relatively high amount of 1,8-cineole. *A. campestris* ssp. *campestris* and *A. campestris* ssp. *borealis* showed a close statistical linkage due to the presence of high levels of α-pinene and β-pinene, whereas the remaining plants grouped in the last cluster were characterised by camphor (Fig. 2) (Mucciarelli *et al.*, 1995).

The economic importance of oil constituents of some *Artemisia* species prompted a variety of biotechnological investigations aimed at the production *in vitro* of the most valuable compounds (see below for further discussion).

The study of essential oils is not limited to chemical variations in the compound composition or to their flavouring properties. As stated above, there is also a great deal of interest in the biological action of oil constituents in both allelopathic and toxic mechanisms involved in the plants chemical defences. In *A. californica* monoterpenes such as camphor, 1,8-cineole, camphene, terpinen-4-ol and α-terpineol are responsible for allelopathic effects on neighbouring plants (Halligan 1975), while the oil from *A. princeps* var. *orientalis* suppresses seed germination and seedling elongation of several plants (Yun *et al.*, 1993).

Many other terpenoids have been studied in the genus *Artemisia*. Considering the sesquiterpenes, the aerial parts of *A. chrithmifolia* yielded a new cadinane derivative (Sanz *et al.*, 1991), whereas two new diasteromeric homoditerpene peroxides were isolated from *A. absinthium* (Reucker *et al.*, 1992). Besides the presence of germacranes, cadinanes and caryophyllene derivatives, which have been reported from the essential oils of several *Artemisia* species (Ahmad and Misra 1994; Mucciarelli *et al.*, 1995), tricyclic sesquiterpenes have been separated from the aerial parts of *A. chamaemelifolia* (Marco *et al.*, 1996).

Table 4 Percentage essential-oil composition (monoterpenes, sesquiterpenes, and phenols). From some *Artemisia* species growing spontaneously in North-West Italy.

Compounds	Artemisia species*													
	1	*2*	*3*	*4*	*5*	*6*	*7*	*8*	*9*	*10*	*11*	*12*	*13*	*14*
Monoterpenes (%)														
Hydrocarbons														
α-Pinene	2.9	16.5	0.9	tr	19.6	0.6	2.0	0.3	15.3	0.3	0.3	0.1	1.3	tr
α-Thujene	tr	tr	0.2	tr	tr	0.8	2.1	0.3	tr	0.8	0.1	tr	tr	0.1
α-Fenchene	tr	tr	tr	tr	tr	3.7	tr	tr	0.6	tr	0.6	tr	1.1	tr
Camphene	6.4	0.6	4.4	tr	2.4	tr	9.1	tr	0.5	tr	tr	7.4	1.2	tr
β-Pinene	1.8	10.7	1.6	tr	2.1	tr	2.1	0.7	9.8	1.3	0.3	0.2	tr	tr
Sabinene	tr	0.4	0.1	tr	1.1	tr	tr	1.6	tr	0.5	0.2	tr	tr	0.2
α-Terpinene	0.3	tr	0.2	0.6	0.4	tr	0.1	0.3	tr	0.3	tr	0.1	tr	0.2
Limonene	1.6	1.7	0.1	0.3	0.2	3.3	0.9	0.1	4.9	0.2	0.2	0.1	tr	0.1
γ-Terpinene	0.5	tr	0.1	0.5	0.7	tr	0.3	0.6	tr	0.5	0.1	0.1	tr	0.4
p-Cymene	1.5	0.9	0.4	7.9	1.0	2.7	0.6	1.0	2.2	0.6	0.5	0.6	1.0	0.4
Alcohols														
β-Sabinene hydrate	0.7	tr	0.1	tr	0.2	tr	tr	0.2	tr	tr	tr	0.2	tr	0.3
Yomogi alcohol	0.2	tr	0.1	tr	tr	tr	tr	tr	tr	tr	tr	tr	tr	tr
cis-Thuj-2-en-4-ol	0.8	tr	0.1	tr	0.3	tr	tr	0.4	tr	0.2	0.1	0.1	tr	0.2
Linalool	0.2	tr	0.1	tr	tr	tr	0.1	tr	0.8	0.1	tr	tr	1.7	0.1
Terpinen-4-ol	1.7	tr	0.7	2.2	2.0	tr	0.4	1.3	0.8	1.0	0.3	0.5	tr	1.4
Pinocarveol	0.8	2.7	2.0	tr	5.5	tr	0.2	tr	3.8	0.2	0.9	0.2	tr	0.1
trans-Verbenol	tr	1.1	0.7	tr	0.5	7.6	7.0	tr	2.4	tr	tr	0.9	tr	tr
α-Terpineol	0.9	0.3	0.2	0.7	1.8	tr	0.3	tr	1.0	0.3	0.2	0.1	tr	0.5
Borneol	0.8	1.3	tr	tr	1.0	17.6	0.4	tr	tr	0.1	0.2	27.7	2.5	tr
Myrtenol	0.3	tr	0.3	1.0	0.6	2.1	0.7	0.1	0.8	0.2	0.2	tr	tr	0.3
Esters														
trans-Chrysanthenyl ac.	tr	tr	tr	tr	tr	tr	tr	tr	tr	tr	tr	tr	21.6	tr
Bornyl acetate	tr	tr	4.1	3.3	tr	tr	0.7	tr	tr	0.1	tr	0.3	tr	tr
Sabinyl acetate	1.4	tr	tr	tr	tr	tr	0.1	1.6	tr	tr	0.6	tr	tr	0.2

Table 4 Percentage essential-oil composition (monoterpenes, sesquiterpenes, and phenols). From some *Artemisia* species growing spontaneously in North-West Italy. (*Continued*)

Compounds	Artemisia species*													
	1	2	3	4	5	6	7	8	9	10	11	12	13	14
Aldehydes														
Myrtenal	0.3	0.9	3.8	tr	0.5	tr	tr	0.1	1.8	0.2	tr	0.3	tr	0.3
Cuminaldehyde	1.0	4.7	13.5	tr	0.4	3.4	tr	0.2	2.2	0.9	0.7	tr	tr	0.3
Ketones														
α-Thujone	tr	tr	tr	0.4	tr	tr	tr	67.5	tr	79.8	69.7	0.2	tr	1.5
β-Thujone	tr	0.9	0.1	0.4	tr	tr	0.1	18.2	1.1	10.4	16.8	0.1	1.3	0.1
Camphor	31.7	2.7	39.3	tr	15.5	11.2	47.7	0.4	tr	tr	tr	40.5	17.1	0.1
Isopinocamphone	tr	1.8	10.2	tr	tr	tr	tr	tr	tr	tr	tr	tr	tr	tr
Verbenone	0.5	2.4	0.2	tr	0.6	tr	8.6	tr	tr	tr	tr	2.8	tr	tr
Carvone	3.9	tr	0.4	tr	0.3	tr	tr	tr	tr	tr	tr	0.5	tr	tr
Oxides														
2,3-Dihydro-1,8-cineole	0.1	0.3	tr	tr	0.1	1.1	tr	tr	5.2	tr	tr	0.1	tr	0.1
1,8-Cineole	15.3	19.2	0.4	34.7	23.3	10.6	3.9	2.0	tr	0.6	0.6	15.1	1.4	15.3
cis-Epoxy-ocimene	tr	tr	tr	tr	tr	tr	tr	tr	tr	tr	tr	tr	24.8	tr
β-Linalool oxide	tr	tr	0.1	tr	0.3	tr	tr	0.1	tr	tr	0.1	0.1	tr	tr
Ascaridole	tr	tr	tr	16.0	1.0	tr	0.1	tr	tr	tr	0.2	tr	tr	tr
Sesquiterpenes (%)														
Hydrocarbons														
α-Copaene	tr	0.3	0.3	tr	0.3	1.1	1.1	0.2	0.3	tr	tr	tr	tr	0.1
γ-Selinene	tr	tr	1.4	tr	3.5	tr	2.4	0.6	tr	tr	tr	0.2	tr	0.1
β-Caryophyllene	0.7	tr	0.6	tr	0.7	tr	4.3	0.5	tr	tr	tr	0.2	1.4	0.1
Alcohols														
Unknown Alcohol N°1	tr	5.3	tr	tr	tr	tr	tr	tr	tr	tr	tr	tr	tr	27.4
trans-Nerolidol	0.2	tr	tr	4.2	0.2	tr	tr	tr	1.1	0.5	tr	tr	tr	22.6
epi-Cubenol	tr	14.1	tr	tr	tr	tr	tr	tr	tr	tr	tr	tr	tr	tr
Spathulenol	9.3	18.7	1.7	4.0	tr	9.2	1.3	0.1	9.3	0.8	4.1	0.5	7.9	3.4
Unknown Alcohol N°2	tr	tr	tr	tr	tr	tr	tr	0.1	10.7	tr	tr	tr	tr	tr

Table 4 Percentage essential-oil composition (monoterpenes, sesquiterpenes, and phenols). From some *Artemisia* species growing spontaneously in North-West Italy. (*Continued*)

Compounds	Artemisia *species**													
	1	2	3	4	5	6	7	8	9	10	11	12	13	14
α-Bisabolol	tr	tr	tr	tr	tr	tr	tr	tr	1.2	tr	tr	tr	tr	3.0
(E)-α-Bergamotol	tr	2.2	tr	tr	tr	tr	tr	tr	tr	tr	tr	tr	tr	tr
α-Cadinol	tr	tr	tr	tr	tr	tr	tr	tr	2.7	tr	tr	tr	tr	tr
Oxides														
Caryophyllene oxide	6.4	5.7	1.1	tr	7.3	21.4	2.2	0.4	18.2	tr	0.3	0.2	10.0	0.5
Bisabolol oxide	tr	tr	tr	18.4	0.5	tr	0.3	tr	1.5	tr	tr	tr	4.5	3.0
Phenols (%)														
Carvacrol	1.6	1.5	0.7	3.1	4.1	tr	0.2	0.7	1.9	tr	0.3	tr	tr	16.3

* 1 = *A. glacialis*, **2** = *A. campestris* ssp. *campestris*, **3** = *A. alba*, **4** = *A. abrotanum*, **5** = *A. annua*, **6** = *A. verlotiorum*, 7 = *A. vulgaris*, 8 = *A. umbelliformis*, 9 = *A. campestris* ssp. *borealis*, 10 = *A. genipi*, 11 = *A. petrosa*, 12 = *A. vallesiaca*, 13 = *A. absinthium*, 14 = *A. chamaemelifolia*. From Mucciarelli *et al.* (1995), modified.

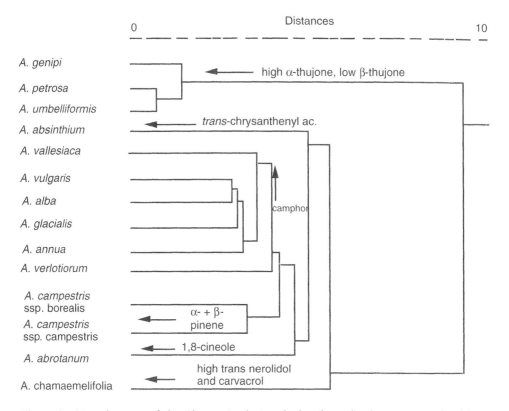

Figure 2   Tree diagram of the Cluster Analysis calculated on the data matrix of Table 4. Two main clusters are evident: one consisting of high α-thujone species and the other made of several minor subclusters. See comments on text. From Mucciarelli *et al.* 1995.

However, the most important group of sesquiterpenes found in the tribe Anthemideae is represented by the sesquiterpene lactones. These molecules have been extensively investigated in the genus *Artemisia* in chemotaxonomic and other studies. While germacranolides and guaianolides are the dominant compounds of the tribe, santanolides have been reported particularly in *Artemisia* (e.g. *A. cina*). However, a particular feature of *A. annua* is its biosynthetic capacity for the cyclization of a germacradiene precursor to a cadinanolide (Herz 1977), while several other germacranolides have also been found to occur in the genus (e.g. costunolide).

Dimeric guaianolides, such as absinthin and anabsinthin, are responsible for the bitter taste of extracts from *A. absinthium* and all these compounds are proazulenes or proazulenogenous. Azulenes are compounds of some pharmaceutical importance owing to their antiseptic and antibacterial properties and certain lactones are known to form azulenes (Herout and Sorm 1954; Novotny *et al.*, 1960). Among santanolides, douglanin and arglanin are present in *A. douglasia*, whereas artecalin and balchanin are found in *A. californica* and *A. balchanorum*, respectively. Several reports on new sesquiterpene lactones appear frequently in the bibliographic browsers. Here we will illustrate the wide variability of compounds present in the genus. Matricarin and hanphyllin were found in *A. argvi*, parishin B and C in *A. absinthium* (Ovezdurdyev *et al.*, 1987), six new eudesmanolides closely related to taurin were extracted from *A. santonicum* (Mericli *et al.*, 1988), two new eudesmanolides were obtained from *A. herba-alba* (Ahmed *et al.*, 1990), a new germacranolide and a novel germacranolide dimer were found in *A. barrelieri* (Marco *et al.*, 1991), whereas frigins A, B and C were isolated from *A. frigida* (Konovalova and Sheichenko 1991).

The aerial parts of *A. glabella*, a perennial plant widespread on the Kazakh dry steppe hills, contain the sesquiterpene lactones arglabin (Fig. 3), argolide and dihydroargolide (Adekenov *et al.*, 1995). New germacrane lactones have been isolated from *A. feddei* (Fig. 3), *A. herba-alba*, *A. ludoviciana* (Fig. 3) and *A. rutifolia*, whereas some new eudesmanolides were isolated from *A. gracilescens* (gracilin; Fig. 3), *A. argyi*, *A. nitrosa* (nitrosin; Fig. 3), *A. rutifolia*, *A. coerulescens*, *A. tournefortiana*, *A. fragrans*, *A. herba-alba*, *A. santolinifolia*, *A. splendens* and *A. xerophytica*, and new guaianolides were extracted from *A. feddei*, *A. frigida*, *A. argentea*, *A. argyi*, *A. douglasiana* (leucodin derivative; Fig. 3), *A. ludoviciana*, *A. mesatlantica* (mesantilantin A; Fig. 3), *A. rutifolia* and *A. xerophytica* (ligustrin derivative; Fig. 3) (Fraga 1992; 1993; 1994).

Quite recently two new eudesmanolides and two new guaianolides have been obtained from *A. lerchiana* (Todorova and Krasteva 1996), while a new eudesmenoic acid has been isolated from *A. phaeolepis* (Tan *et al.*, 1995). The extraction of air dried aerial parts of *A. eriopoda* afforded a sesquiterpene mixture containing three new eudesmane diols: $1\beta,6\beta$-dihydroxy-4(14)-eudesmene, 5α-hydroxyisopterocarpolone and 1-oxo-cryptomeridiol (Fig. 3) (Hu *et al.*, 1996a), whereas *A. pontica* was found to contain seven new 5-hydroxyeudesmanolides in addition to artemin (Fig. 3), 5-epi-artemin and 8-α-hydroxytaurin (Trendafilova *et al.*, 1996). In *A. herba-alba* a new eudesmanolide has been named herbalbin (1α-hydroxy-$3\alpha,4\alpha$-epoxyeudesm-$5\alpha,6\beta,7\alpha,11\beta$H-12,6-olide; Fig. 3) (Boriky *et al.*, 1996), whereas in *A. mongolica* the new eudesmane $6\alpha,8\alpha$-dihydroxyisocostic acid methyl ester has been isolated along with the eudesmane derivative ludovicin B (Hu *et al.*, 1996b). A new germacranolide: $4,5\beta$-epoxy-10α-hydroxy-1-en-3-one-*trans*-germacran-6,α12-olide has been isolated from aerial parts of *A. pallens* (Rojatkar *et al.*, 1996), whereas six sesquiterpene lactones with the uncommon rotundane skeleton have been extracted from the aerial parts of *A. pontica* (Fig. 3) (Todorova *et al.*, 1996).

Sesquiterpene lactones of the cadinanolide, germacranolide, guaianolide, eudesmanolide, secoeudesmanolide and helenanolide groups have been investigated by Bicchi and Rubiolo, (1996) in *A. umbelliformis* by HPLC-MS coupled through a particle beam interface. Several new germacranolides and guaianolides have been extracted from *A. reptans*, *A. turcomanica* and *A. deserti* (Marco *et al.*, 1993; Marco *et al.*, 1994a), whereas thirteen eudesmanolides and a 1,10-secoeudesmanolide have been obtained from the North African *A. hugueti* and *A. ifranensis* (Fig. 3) (Marco *et al.*, 1994b).

Artemisinin (Fig. 3), a sesquiterpene lactone that will be treated in more detail later in this book, is a typical santanolide of *A. annua*. A rapid, sensitive and specific method has been developed for the simultaneous determination of artemisinin and its bioprecursors by means of reversed-phase HPLC using electrochemical and UV detection in *A. annua* (Van den Berghe *et al.*, 1995).

Among diterpenes, phytene-1,2diol has been isolated from *A. annua* and the presumed biogenetic precursor of the diol is thought to be phytol (Brown 1994a). The triterpene fernerol was obtained from *A. vulgaris* (Hegnauer 1977), whereas a lanostane-type triterpenoid (9β-lanosta-5-ene-3α,27-diol 3α-palmitoleate) and two 13,14-seco-steroids (13,14-seco-cholest-7-ene-3,6α,27-triol 3,27-diocta-8′,6′-dienoate and 13,14-seco-cholest-5-ene-3β,27-methanoate 3β-hexadeca-11′,13′,15′-trien-1′-onate) have been isolated from the roots of *A. scoparia* (Sharma *et al.*, 1996).

Flavonoids and Coumarins

Flavonoids are widespread in the Anthemideae and their evaluation has proved to be a valuable tool for chemotaxonomic studies. However, according to Seeligmann, (1996) when considered at the familial level the distribution of the basic structures of flavonoids does not always reflect phylogenetic relationships between genera and is not in accordance with the degree of evolution.

In the Anthemideae, luteolin, apigenin and quercetin are the most common compounds (Fig. 4). More than 160 individual flavonoid components have been isolated in the genus *Artemisia* and about one third of them are derivatives of the flavones luteolin and apigenin (Belenovskaja 1996). Some derivatives of these compounds have also been isolated from *A. sacrorum*, like the isomeric compound genkwanin (Chandrashekar *et al.*, 1965) and from *A. pygmaea*, like rhamnazin (Fig. 4), the 7,3′-dimethyl ether of quercetin (Rodriguez *et al.*, 1972).

The rare compound 3,5-dihydroxy-6,7,8-trimethoxyflavone has been isolated from *A. klotzchiana* (Dominguez and Cardenas 1975). *A. pontica* at the flowering stage accumulated apigenin and luteolin methyl esters, whereas before flowering the plants produced mainly apigenin derivatives (Wollenweber and Valant-Vetschera 1996). Two new flavones were isolated from *A. giraldii* and their structures were identified as 4′,6,7-trihydroxy-3′,5′-dimethoxyflavone and 5′,-5-dihydroxy-3′,4′,8-trimethoxyflavone. These two compounds showed antibiotic activity against *Staphylococcus aureus*, *Sarcina lutea*, *Escherichia coli*, *Pseudomonas aeruginosa*, *Proteus* spp., *Aspergillus flavus* and *Trichoderma viride* (Zheng *et al.*, 1996).

The aerial parts of *A. stolonifera* afforded a new phenolic glycoside from natural sources, 2,4,6-trihydroxy acetophenone 2-O-β-D-glucopyranoside, coniferin and the acetophenone glycoside 2,4-dihydroxy-6-methoxy acetophenone 4-O-β-D-glucopyranoside. The latter compound was cytostatic to macrophages (Lee *et al.*, 1996). *Artemisia annua* leaf and stem extract contain apigenin, luteolin, kaempferol, quercetin, isorhamnetin, luteolin-7-methyl ether, isokampferide, quercetagetin 3-methyl ether, quercetagetin 4′-methyl ether, tomentin, astragalin, isoquercitrin, quercimeritrin and two chromenes (Fig. 4).

Some flavones extracted from *A. annua*, e.g. casticin, chrysoplenetin and cirsineol were found to enhance the antimalarial activity of artemisinin. In particular, casticin

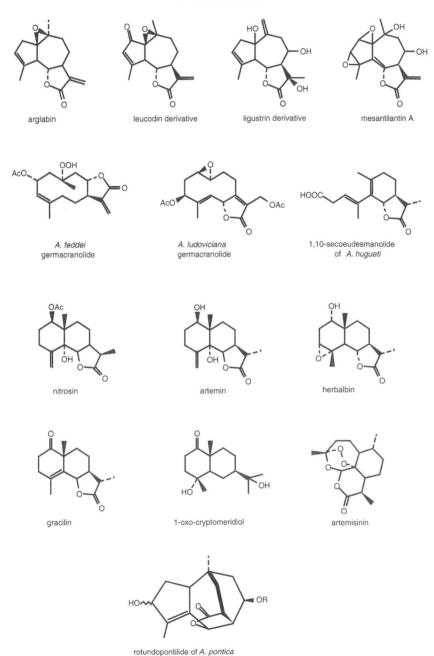

arglabin

leucodin derivative

ligustrin derivative

mesantilantin A

A. feddei
germacranolide

A. ludoviciana
germacranolide

1,10-secoeudesmanolide
of *A. hugueti*

nitrosin

artemin

herbalbin

gracilin

1-oxo-cryptomeridiol

artemisinin

rotundopontilide of *A. pontica*

Figure 3 Structural formulae of some representative sesquiterpene lactones found in the genus *Artemisia*.

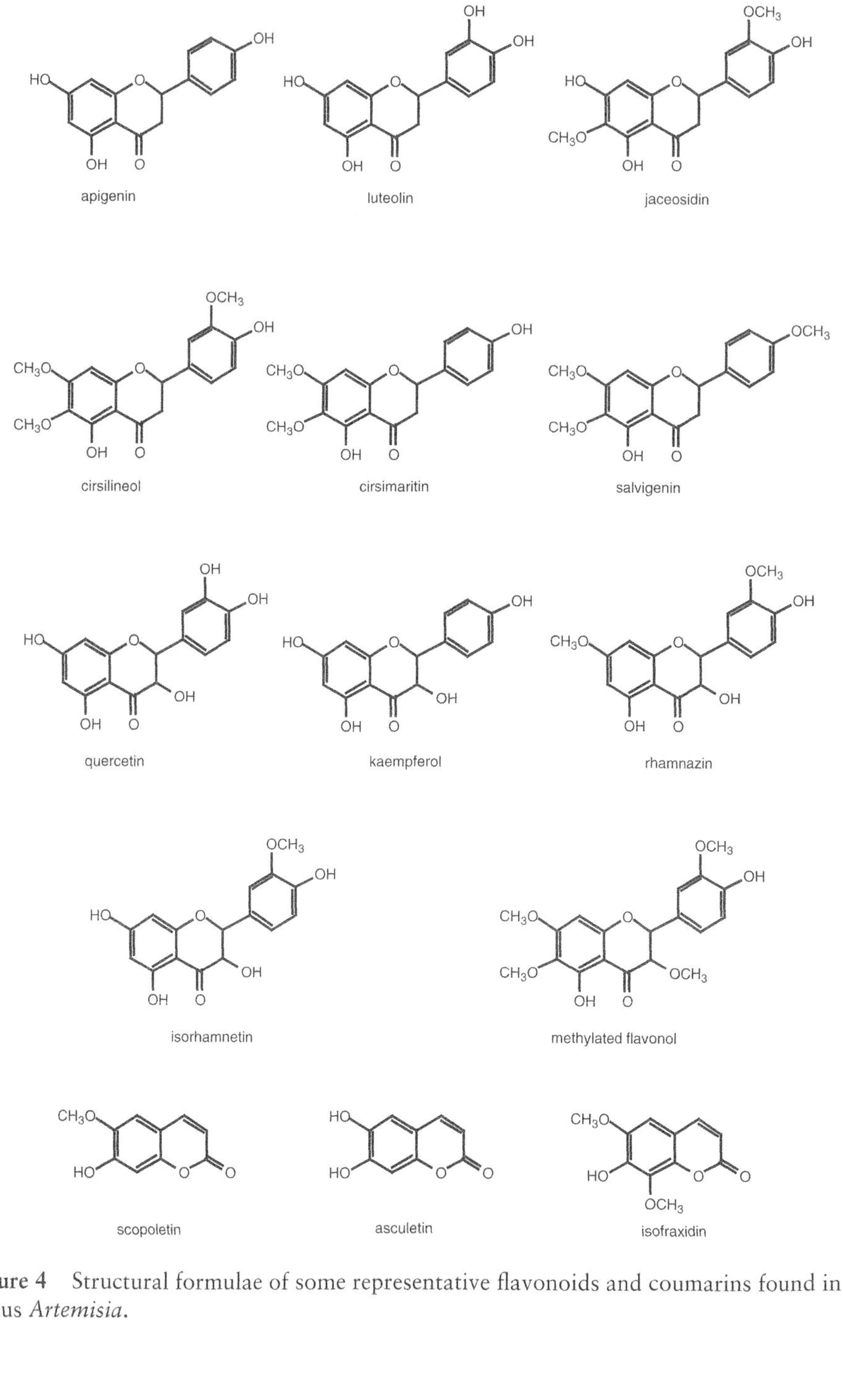

Figure 4 Structural formulae of some representative flavonoids and coumarins found in the genus *Artemisia*.

was found to be active in inhibiting some cytophysiological activities of the parasite (Yang *et al.*, 1995). In African species of *Artemisia* (reviewed by Saleh and Mosharrafa 1996), apigenin was found in aerial parts of *A. mesatlantica*, and its 7-methyl ester in *A. afra*. Acacetin, jacoesidin, cirsilineol, cirsimaritin (Fig. 4), hispidulin, isoschaftoside, lucenin-2 and vicenin-2 were all isolated from *A. judaica*, *A. monosperma* and *A. herba-alba*. Pectolinarigenin and salvigenin (Fig. 4), were extracted from *A. santolina*, luteolin from *A. afra*, chrysoeriol from *A. mesatlantica*, nepetin from *A. afra*, cirsiliol from *A. campestris*, eupatilin from *A. judaica*, tricin from *A. mesatlantica*, and an acerosin derivative from *A. herba-alba*. Aerial parts of *A. monosperma* contained acacetin-7-glycoside, whereas luteolin-3'-glycoside was found in *A. judaica*. Isovitexin was isolated from *A. herba-alba*, whereas neoschaftoside and neoisoschaftoside were extracted from *A. judaica*. The flavonols kumatakenin, axillarin and casticin were isolated from *A. campestris*, *A. afra* and *A. judaica*, respectively, whereas quercetin-5-glycoside, isorhamnetin 5-glycoside and patuletin-3-glycoside were extracted from *A. monosperma*, *A. judaica* and *A. herba-alba* (Saleh and Mosharrafa 1996).

The flavonoids mentioned above may occur both as O-glycosides and as methylated flavones and flavonols. Methylated flavones have been detected in *A. arctica* ssp. *saxicola* and *A. herba-alba*, whilst methylated flavonols have been identified in *A. arbuscula*, *A. tridentata*, *A. rothrockii*, *A. cana*, *A. absinthium*, *A. longiloba* and *A. arborescens* (Greger 1977).

Flavonoid components of 130 species of *Artemisia* have been used for the solution of taxonomic problems at the intrageneric level. By considering three sections (*Artemisia*, *Abrotanum* and *Absinthium*) and two subgenera (*Seriphidium* and *Dracunculus*), Belenovskaja (1996) was able to distinguish several groups of species according to the flavonoid composition: a) section *Artemisia*, two groups, one containing quercetin and isorhamnetin and the other 6-methoxyluteolin methylesters; b) section *Abrotanum*, several groups of species containing mainly common flavonols (quercetin, kaempferol, etc.), their O-glycosides, 6-methoxyluteolin, 6-methoxyquercetin derivatives and coumarins; c) section *Absinthium*, two groups, one with methylethers of flavonols and the other with methylethers of flavones; d) subgenus *Seriphidium*, quite similar patterns of distribution when compared to the genus *Artemisia*; e) subgenus *Dracunculus*, completely different from the other subgenera of *Artemisia* with flavanones with an unusual type of distribution.

Several species of *Artemisia* were analysed for their leaf exudate flavone and flavonol aglycone content and the majority of these compounds were 6-methoxylated, with additional substitutions at the 7-, 3'- and 4' position along with some coumarin derivatives (Valant-Vetschera and Wollenweber 1995). Finally, four flavonols with spasmolytic activity were isolated from *A. abrotanum* (Bergendorff and Sterner 1995), whereas several other flavonoids have been extracted from *A. austriaca* (Cubukcu and Melikoglu 1995) and from other *Artemisia* species (Al-Hazim and Basha 1991).

Hydroxycoumarins are typical constituents of the genus *Artemisia* representing a valuable chemotaxonomic character. Structurally complex coumarins such as scopoletin 7-dimethylallylether and methylene ethers of daphnetin and fraxetin have been isolated in A. *dracunculoides* (Herz *et al.*, 1970), whereas in A. *afra* root extracts contained isofraxidin while flowers contained scopoletin (Fig. 4) (Bohlmann and Zdero 1972). Esculetin (Fig. 4), isoscopoletin and their 7-O-glycosides esculin and methyl-esculin were isolated from A. *tridentata* ssp. *vaseyana* (Shafizadeh and Melnikoff 1970), while p-hydroxyacetophenone, herniarin, scoparone, scopoletin, umbelliferone and dehydrofalcari-3,8-diol were extracted from A. *marschalliana* (Ozhatay and Cubucku 1990). Tissue cultures of A. *vulgaris* produced significant amounts of the 6,7,8-trioxygenated coumarin isofraxidin, whereas A. *laciniata* was found to contain the chiral glycol of the trioxygenated isofraxetin (Murray 1995).

Leaf Wax Alkanes

So far, the evaluation of more than 550 species belonging to 11 plant families has proved that plant surface wax alkanes can be considered good chemotaxonomic markers at the familial, subfamilial and tribal level (Maffei 1994; 1996a; 1996b; 1996c; Maffei *et al.*, 1993b; 1997). This is mainly due to the universality of the occurrence of these molecules, species variation in composition, simplicity of sampling, and availability of rapid analytical tools.

The leaf-wax alkane profiles of some *Artemisia* species revealed the presence of odd and even carbon number alkanes as well as branched molecules (Maffei 1996c) (Table 5). The highest total alkane content was found for A. *siversiana* and A. *absinthium*, whereas the lowest contents were present in A. *abrotanum* and A. *pontica*. The alkane carbon chain-length ranged from C_{23} to C_{33}. The main components in all species were C_{29} and C_{31}, with A. *pontica* containing the highest percentage of the former and A. *vulgaris* of the latter compound. A. *alba* also contained high percentages of the branched alkane $2MC_{28}$. The common occurrence in Angiosperms of C_{29} and C_{31} has been documented by a large number of reports and the presence of *iso*-alkanes has been demonstrated in several plant waxes (Maffei 1994 and refs. cited therein).

IN VITRO CELL AND TISSUE CULTURES

Undifferentiated and differentiated *in vitro* tissue culture techniques have been recently developed, concerning the most economic valuable *Artemisia* species. The main efforts have been devoted to the *in vitro* selection of highly yielding clones and cell lines, producing secondary metabolites, with pharmacological and industrial applications. In this regard major attention has been paid to *Artemisia annua* L., owing to its antiplasmodial properties and moreover for its value as an aromatic plant, employed in fragrances, perfumery and cosmetic production.

Micropropagation techniques have been applied to A. *annua* in order to obtain *in vitro* multiple shoot-cultures from which desirable clones may be selected and conserved (Elhag *et al.*, 1992). Much interest has also been devoted to the close relationship which exists between *in vitro* plant morphology and artemisinin

Table 5 Leaf wax alkane composition (area percentages and mean values) and content of some *Artemisia* species

Taxon	Content µg g⁻¹	Percentage of Alkanes											
		C_{23}	C_{24}	C_{25}	C_{26}	C_{27}	C_{28}	$2MC_{28}$	C_{29}	C_{30}	C_{31}	C_{32}	C_{33}
Artemisia abrotanum	78	2.3	0.9	3.3	0.6	4.6	1.6	3.1	13.2	1.4	10.9	0.1	1.0
A. absinthium	621	0.1	0.1	0.3	0.2	0.8	0.6	0.2	24.3	2.8	27.4	1.6	4.8
A. alba	230	0.7	0.4	2.2	0.4	3.0	1.2	12.0	35.1	1.7	9.0	0.2	0.9
A. austriaca	171	0.9	0.2	2.2	0.3	5.2	1.8	2.6	34.6	1.9	22.1	1.1	1.4
A. coerulescens	461	0.1	0.1	0.5	0.2	2.6	1.8	2.0	28.9	4.4	24.4	5.1	1.3
A. dracunculus	164	0.1	0.1	0.3	0.2	0.8	0.4	2.9	23.2	2.8	20.6	0.1	0.2
A. ludoviciana	465	0.3	0.1	0.5	0.2	1.1	1.7	1.2	25.5	3.4	55.2	1.3	2.6
A. molinieri	130	1.5	0.4	1.8	0.4	3.3	0.4	1.8	10.0	0.2	7.0	1.9	0.3
A. persica	439	0.2	0.2	0.8	0.5	2.6	0.9	1.4	26.1	3.4	28.4	4.2	2.3
A. pontica	73	0.1	0.1	3.3	0.8	6.8	1.8	4.3	46.1	2.1	23.6	0.1	2.4
A. siversiana	792	0.7	0.4	1.2	0.8	2.5	2.0	1.1	31.4	3.0	28.5	1.7	2.9
A. verlotorum	345	0.1	0.3	1.4	0.1	3.2	0.1	4.5	5.7	1.5	44.1	1.4	3.9
A. vulgaris	321	0.4	0.3	1.1	0.4	1.3	0.7	0.9	21.5	2.9	62.3	1.7	4.3

production, i.e. the correlation of the biosynthesis of artemisinin and related compounds with flowering which is typical of mature plants (Gulati *et al.*, 1996). Ferreira *et al.* (1995a; 1995b) by studying *A. annua* in greenhouse, tissue culture and open field, found that artemisinin content was 4- to 11-fold higher in flowers than in leaves and that favourable results could be expected by intercrossing selected high artemisinin-yielding clones which had been grown in the greenhouse with prolonged lighting.

Valuable metabolites have also been obtained in undifferentiated callus and cell suspension cultures of *A. annua*; including the coumarin, scopoletin (Brown 1994b), and artemisinin (0.78 mg/g DW) from callus grown on MS medium, supplemented with 2,4-dichlorophenoxy-acetic acid and naphthalen-acetic acid. However, sometimes suspension cultures failed to accumulate any of the terpenoids found in the parent plant (Brown 1994b; Paniego and Giulietti 1994). Other than callus and multiple shoot cultures producing terpenoids *in vitro*, protocols for plantlet cultures in bioreactors have been developed for different *A. annua* geographical varieties, thus resulting in the production of camphor, 1,8-cineole and β-caryophyllene (Fulzele *et al.*, 1995).

There are several reports on the biosynthetic pathway leading to artemisinin as well as biotechnological approaches for the production of this sesquiterpene lactone *in vitro* (Elhag *et al.*, 1992; Sangwan *et al.*, 1993; Brown 1994c; Chen and Xu 1996; Gulati *et al.*, 1996). Artemisinin production has been obtained in transformed roots of *A. annua* after infection with *Agrobacterium rhizogenes*, in percentages suggesting a feasible commercial production of the active compound (Weathers *et al.*, 1994). Transformed shoot and root cultures of *A. annua* have also been established by infection with *Agrobacterium tumefaciens* (Vergauwe *et al.*, 1996) and were shown to produce artemisinin during prolonged subculturing (Paniego and Giulietti 1996).

A protocol was developed to preserve the hairy roots of artemisinin producing *A. annua* plants by cryopreservation in liquid nitrogen (Teoh *et al.*, 1996), while Ferreira and Janick (1996) developed a method for the immunoquantitative analysis of artemisinin using polyclonal antibodies. Besides its antimalarial characteristic, artemisinin also exhibits potent phytotoxic properties by inhibiting germination and reducing growth of many weeds and crop plants (Dukes *et al.*, 1987).

Callus cultures (Nin *et al.*, 1996) and protoplast isolation (Xu and Jia 1996) have also been obtained from other *Artemisia* species such as *A. absinthium* and *A. sphaerocephala*.

Various environmental factors, such as artificial light quality, sucrose concentration and hormonal culture media supplements have been shown to be effective in promoting *in vitro* essential oil accumulation, as occurred in the test-tube plants of *Artemisia balchanorum* (Bavrina *et al.*, 1994). Growth medium-related oil production was also found for *Artemisia alba* (Turra), as well as for other species of the Belgian flora, where the aim of the research was to develop a conservation program based on micropropagation (Ronse and De Pooter 1990).

Micropropagation techniques have also been applied in the flavour industry, for the isolation and growth of *Artemisia* clones of the "*genepi*-group"

(*A. umbelliformis* Lam. and *A. genipi* Weber), having a high degree of variability in their essential oil GC patterns and organoleptic characteristics (Gautheret *et al.*, 1984).

BIOSYNTHESIS AND SITES OF SYNTHESIS OF SOME *ARTEMISIA* SECONDARY METABOLITES

As we described in the previous section, the main secondary products of the genus *Artemisia* are terpenoids and flavonoids. The biochemical pathways and site of synthesis of most of these molecules have been investigated during the last two decades and this section will review some of the anatomical and biochemical features of metabolite production. We will start from the sites of synthesis and then give a short review of the biochemical pathways leading to the production of the most common secondary products.

Sites of Synthesis

The lipophilic substances produced by the genus *Artemisia* are synthesised and accumulated in secretory structures that can be generally classified into two categories: glandular trichomes and secretory ducts. Glandular trichomes are epidermal structures whilst ducts are composed of secretory tissue present below the epidermal layers occurring in many organs of the whole body of the plant. The biology and chemistry of glandular trichomes has been reviewed by several authors (Fahn 1979; 1988; Duffey 1986; Duke 1994; Rodriguez *et al.*, 1984); however, only a limited number of species of the genus *Artemisia* have been studied.

In general, glandular trichomes are made from a series of differentiated cells with different functional properties. With regard to terpenoid production, the essential oil accumulates between the cuticle and the cellulosic wall of the secretory cells, whose position is always at the top of the structure. Below secretory cells are the so called stalk cells, whose function is to connect secretory cells to the plant body. The base of the trichome is made by a foot (basal) cell, which is a modified epidermal cell that upholds the entire trichome structure. The number of secretory, stalk and foot cells may vary from plant to plant. In the genus *Artemisia*, the ontogeny of glandular trichomes is described as the appearance of a protodermal cell that undergoes anticlinal and periclinal division to yield a couple of basal and stalk cells, and three couples of secretory cells (Ascensão and Pais 1982).

During the early stages of trichome development, all cells are cytologically active and the secretion of oil components and oleoresins is transferred to the subcuticular layer that starts enlarging until rupture occurs. After this period the trichome degenerates beginning from the secretory cells down to the stalk cells. In *A. campestris* ssp. *maritima*, the glandular trichomes are embedded in a dense layer of oleoresins which cover the leaf surfaces. The removal of the oleoresins reveals the presence of numerous biseriate trichomes which are mostly present on adaxial surfaces (Ascensão and Pais 1987).

An ultrastructural study on *A. annua* carefully describes the structure of the biseriate 10-celled glandular trichomes (Duke and Paul 1993). Three apical cell pairs represent the secretory head, the first two top pairs contain amoeboid plastids without thylakoids, whereas in the two cell pairs below, ameboid chloroplasts show the presence of photosynthetic thylakoids. Stalk cell plastids containing starch grains and thylakoids were occasionally present in basal cells. An osmophilic gradient from the stalk to the uppermost secretory cells indicates the path of secretion and the related secretory activity of the cells (Fig. 5). The same situation was found for the glandular trichomes of *A. coerulescens* (Maffei, unpublished results).

In *A. umbelliformis*, besides essential oils, the subcuticular space accumulates some sesquiterpene lactones, like cytotoxic hydroperoxyeudesmanolides, that can be localized histochemically (Cappelletti *et al.*, 1986), whereas in *A. nitida* the glandular trichome histochemistry reveals the presence of terpenoids as well as steroids and catecol tannins (Corsi and Nencioni 1995). Sesquiterpene lactones have been localized in the glandular trichomes of *A. nova* (Kelsey and Shafizadeh 1980), whereas in *A. annua* the combined used of scanning electron microscopy and HPLC-EC of the flowering branches provided corroborating evidence that the biseriate trichomes that cover the epidermises are the site of sequestration of artemisinin (Ferreira and Janick 1995). Finally, the isolation and purification of glandular secretory cells has been obtained in *A. tridentata* ssp. *vaseyana* by a Percoll density gradient centrifugation (Slone and Kelsey 1985).

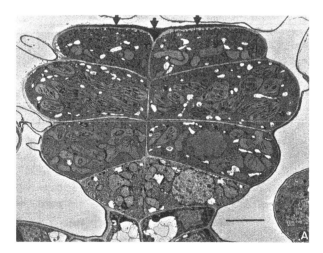

Figure 5   Transmission electron microscopy of a 10-celled glandular trichome of *Artemisia annua* showing a dense material (arrows) near the upper cell walls of the two apical secretory cells. Many ameboid plastids are evident in the three couples of secretory cells which appear more osmophilic than the lower foot and stalk cells. Scale bar = 5 μ. From Duke and Paul 1993.

Another important secretory structure present in the genus *Artemisia* is represented by resin ducts. The composition of the secreted substances in these structures varies from terpenes to polysaccharides to several other products. In *A. campestris* ssp. *maritima*, the resin ducts are distributed in the cortical parenchyma near the vascular bundles, running longitudinally in petioles and stems. The duct lumen is surrounded by secretory cells and contains oleoresins which stain with histochemical reactions typical of lipophilic substances such as terpenoids. The capability of secreting oleoresins inside the ducts appears to be related to the presence of a great amount of endoplasmic reticulum. The latter is found in association with plastids with few thylakoids contained in the epithelial and sub-epithelial cells that surround the duct cavity (Ascensão and Pais 1988). The same results have been obtained from *A. nitida*, where the oleoresin content of the ducts is similar to that of the glandular trichomes (Corsi and Nencioni 1995).

Biochemical Pathways

As discussed in the previous section, the essential oil composition of *Artemisia* species is made up of a mixture of mono- and sesquiterpenoids whose biochemical pathways have been studied in this genus as well as in many other aromatic plants. We will consider here only the biosynthetic pathways of the most characteristic *Artemisia* products starting from the monoterpenes thujone and camphor.

The condensation of two molecules of dimethylallyl pyrophosphate derived from isoprene *via* the the mevalonate pathway, gives rise to geranyl pyrophosphate (GPP), which is considered to be the common precursor of all monoterpenes. The cyclization of GPP by the catalytic action of GPP:(+)-sabinene cyclase yields sabinene. In *A. absinthium* sabinene has been found to act as a substrate for conversion to (+)-sabinyl acetate and (+)-3-thujone and microsomal preparations were shown to catalize the NADPH and O_2 dependent hydroxylation of (+)-sabinene to (+)-*cis*-sabinol. These results indicate that the synthesis of thujone requires a cytochrome P-450 dependent mixed function monooxygenase. In some other *Artemisia* species the NADPH-dependent stereoselective reduction of (+)-sabinone to (+)-3-thujone has been demonstrated (Fig. 6) (Croteau 1987 and refs. cited therein).

Another important monoterpene which often occurs in *Artemisia* essential oils is camphor. The formation of camphor involves the conversion of GPP to (+)-bornyl pyrophosphate which, after hydrolysis by a specific bornyl pyrophosphatase is converted to the bicyclic monoterpene alcohol (+)-borneol. An NAD-dependent dehydrogenase oxidizes borneol to the ketone camphor (Fig. 7) (Croteau 1992).

Artemisinin biosynthesis has been the subject of many studies owing to the importance of the therapeutic applications of the compound. The information available on the synthesis of artemisinin comes from both *in vivo* and *in vitro* studies. The biosynthesis of the cadinane skeleton of artemisinic acid has been studied in *A. annua*. The common precursor of all sesquiterpenoids, farnesyl pyrophosphate, is cyclised forming the bicyclic structure leading to the formation of artemisinic acid (Dewick 1995a), the probable precursor of arteannuin. The hydroxylation of the latter compound gives rise to a cadinanolide which is subsequently transformed to

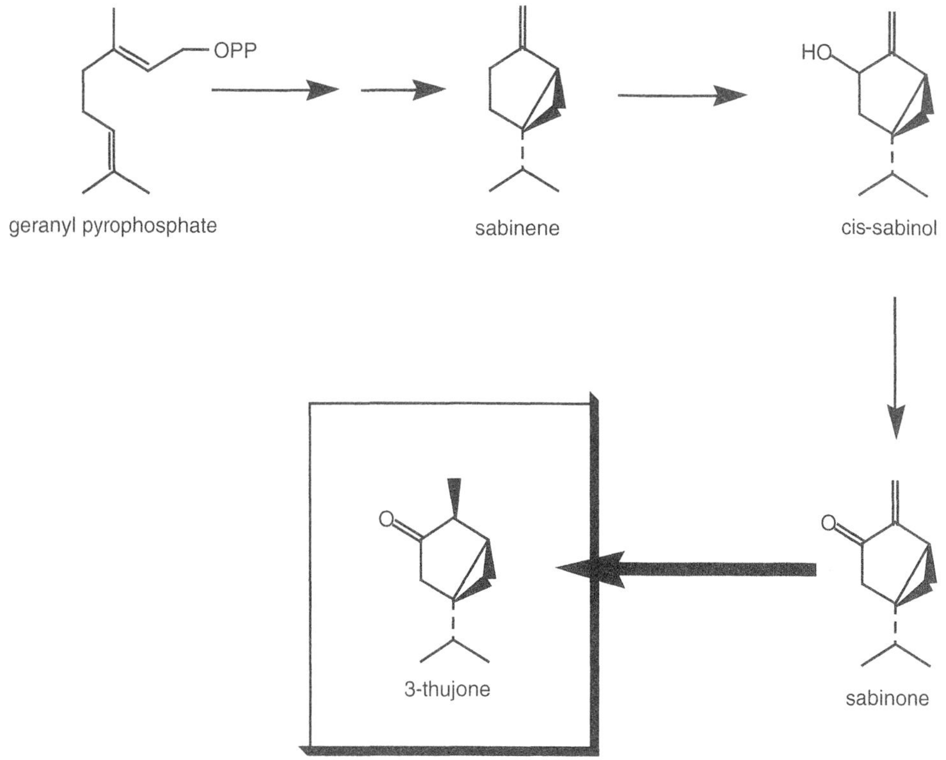
OPP
geranyl pyrophosphate
sabinene
HO
cis-sabinol
O
3-thujone
O
sabinone

Figure 6 Biosynthetic pathway of thujone

artemisene, the precursor of artemisinin (Fig. 8) (Brown 1994a). A fuller discussion of artemisinin biosynthesis is given in chapter 12.

Coumarins are a group of lactones formed through the shikimic acid pathway *via* phenylalanine and *trans*-cinnamic acid. Coumarins are active compounds able to inhibit various enzymes such as β-amylases, invertases and phenolases and to elicit the activity of peroxidases. The biosynthesis of the coumarin esculetin starts from

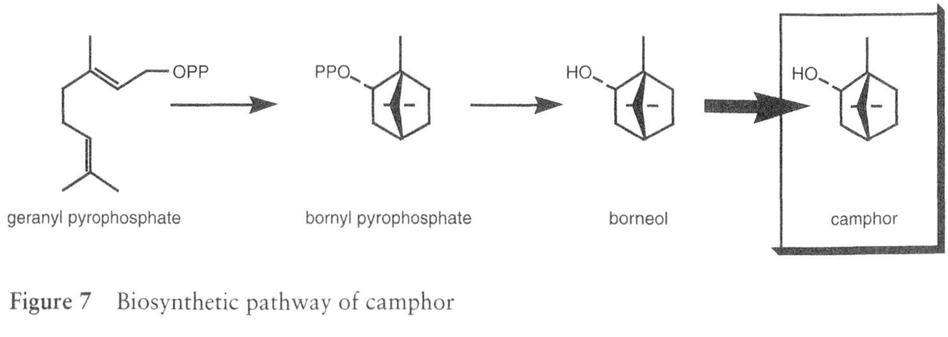
OPP
geranyl pyrophosphate
PPO
bornyl pyrophosphate
HO
borneol
HO
camphor

Figure 7 Biosynthetic pathway of camphor

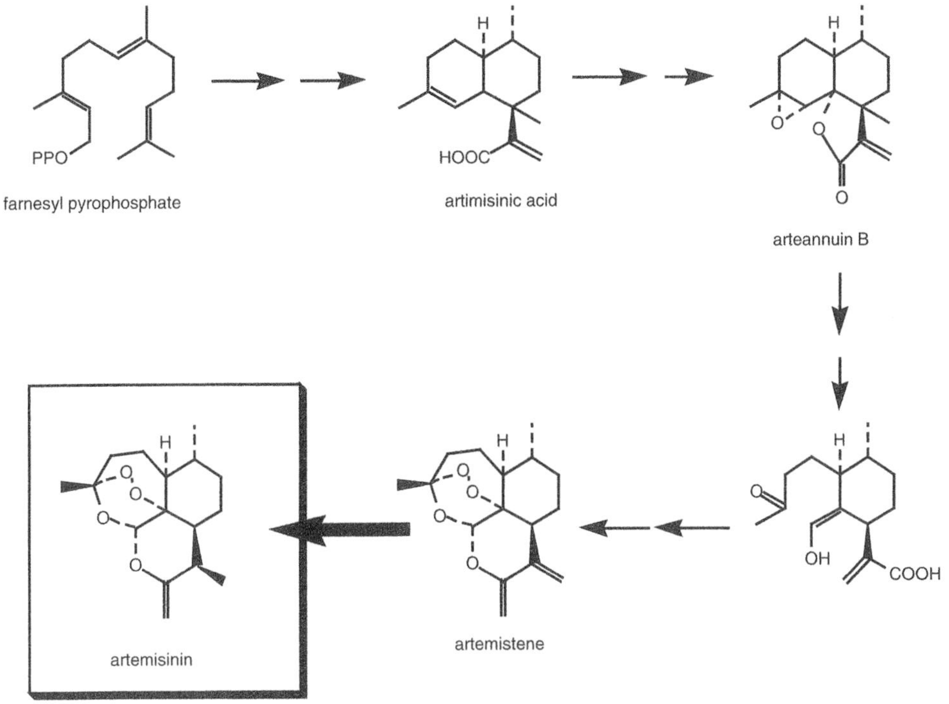

farnesyl pyrophosphate

artimisinic acid

arteannuin B

artemistene

artemisinin

Figure 8   Biosynthetic pathway of artemisinin

phenylalanine which is deaminated by phenylalanine ammonia lyase (PAL) to yield *trans*-cinnamic acid which is hydroxylated at position 4 to give *p*-coumaric acid. Hydroxylation of the latter molecule at position 2 gives the 2-hydroxy-*p*-coumarate which is then glycosylated by an ADP-dependent glucosyl transferase to 4-hydroxy-O-coumaryl glycoside. Cyclization of the molecule yields umbelliferone (a 7-hydroxylated coumarin), whose hydroxylation in the 6 position gives rise to esculetin (Fig. 9), (Brown 1990). The presence of the hydroxyl group in the 7 position has been detected in more than 500 coumarins, whereas only a dozen coumarins lack this hydroxylation.

   p-Coumaric acid is also the precursor of the base molecule for the synthesis of all flavonoids. Through the action of chalcone synthase, three units of acetyl-CoA are added to *p*-coumarate to yield chalcone, which is cyclized by chalcone isomerase to naringenin. This basic molecule is the substrate for several hydroxylases leading to the various flavonoid compounds as illustrated in Fig. 10 (for general reviews see Dewick 1995b; Holton and Cornish 1995; Hahlbrock 1981).

   The biochemistry of several other compounds has been investigated in the genus *Artemisia*; however the description of all the research done is beyond the scope of this introductory chapter. Owing to the great potential of *Artemisia* in producing valuable secondary products, research in the fields of the biochemistry and molecu-

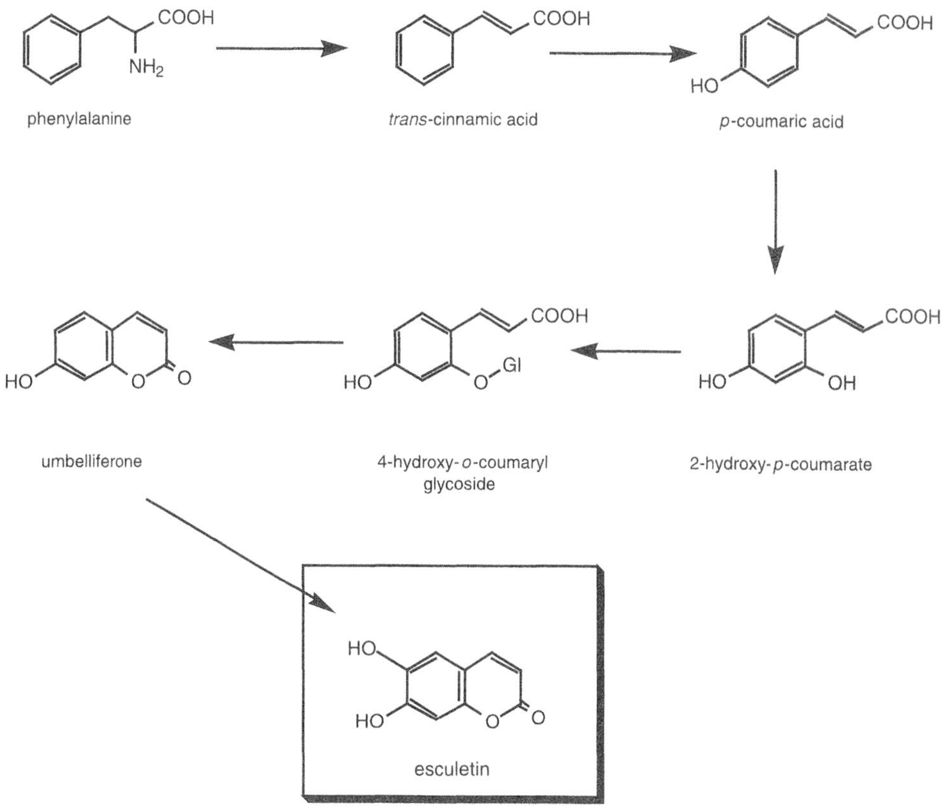

Figure 9   Biosynthetic pathway of esculetin

lar biology of the main pathways appears to be a basic and fundamental task for the understanding of the mechanism of product formation and the genetic basis for gene expression. In common with many other medicinal and aromatic plants, *Artemisia* metabolites are under environmental, nutritional and physiological control. Only a detailed knowledge of the regulation and synthesis of these compounds will enable the complete exploitation of the huge potential of the *Artemisia* genome. The invaluable contribution of all researchers in the continual search for new molecules and active compounds will also contribute to a thorough understanding of secondary plant metabolites.

ECOLOGICAL ASPECTS AND ENVIRONMENTAL MANAGEMENT

The floral landscapes of western North America, going from southern Canada to northern Mexico, through the eleven western United States are mainly dominated by the so called "sage-brush communities" (Beetle 1960). The latter are represented by *Artemisia* species, in a group considered to consist of between ten and twenty

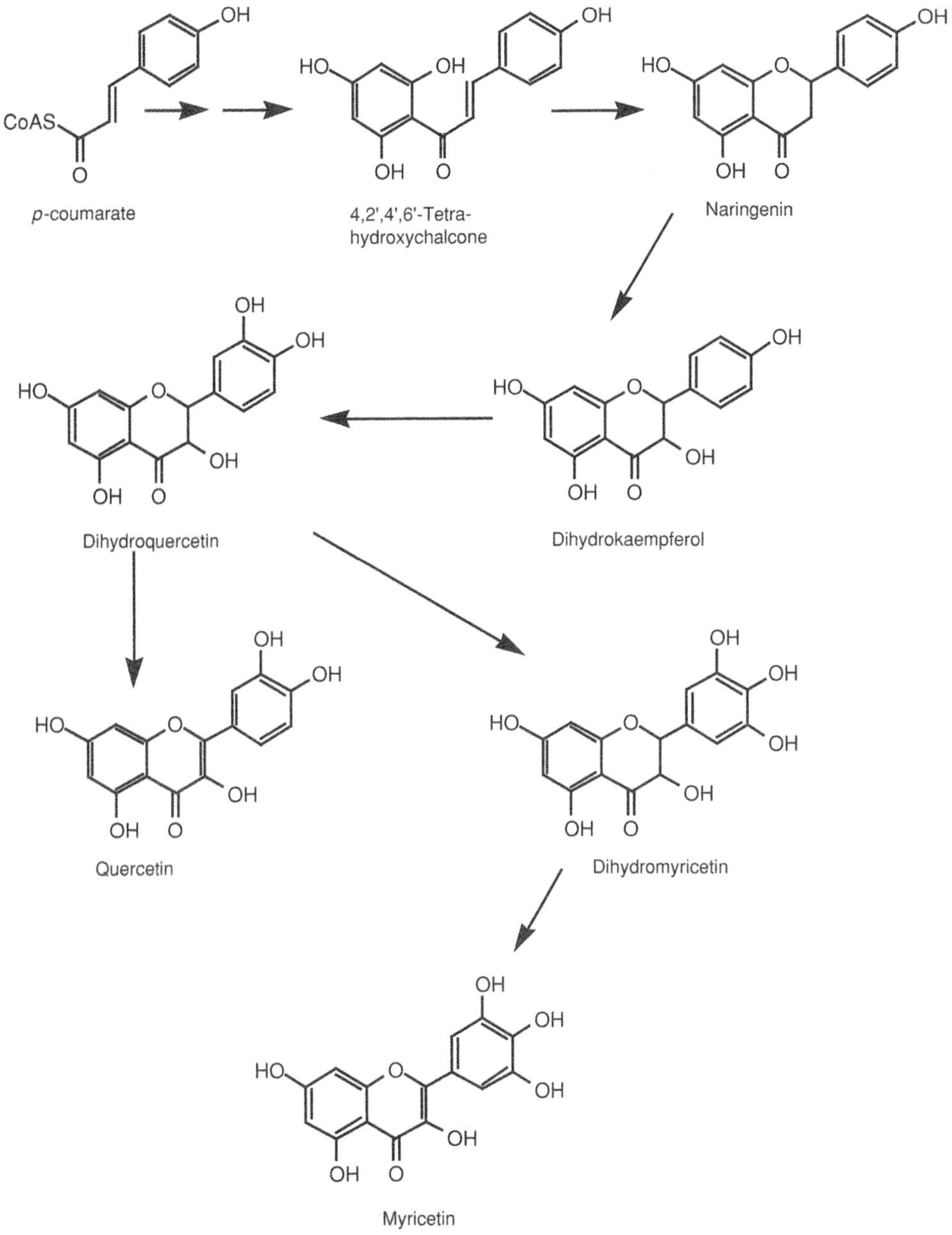

Figure 10   Biosynthetic pathway of some flavonoids

entities, of the subgenus or section *Tridentatae* (Shultz 1984). The sagebrush zone can be generally characterized as having an annual precipitation from 20 to 35 cm predominantly occurring in the winter. Sagebrush species dominate vast expanses of semi-arid rangelands, where the main renewal process in plant communities is wildfire, thus for these communities achene and seed-bed ecology is vital for the renewal process that leads to continued sagebrush dominance (Young and Mayeux 1996).

Three subspecies of big sagebrush are known: basin big sagebrush *A. tridentata* Nutt. ssp. *tridentata* Nutt., mountain big sagebrush *A. tridentata* Nutt. ssp. *vaseyana.*, Beetle and Wyoming big sagebrush *A. tridentata* Nutt. ssp. *wyomingensis* Beetle and Young.

An extremely efficient reproductive growth system that is active during the summer drought is present in these species and they benefit greatly by their highly efficient use of water at a time when competing plants are dormant (Young and Mayeux 1996). Big sagebrush plants can also carry on significant photosynthetic activity during cold periods, when other plants are completely dormant (Caldwell 1985).

Poorly timed, excessive, and continuous grazing of these lands by domestic livestock has favoured the big sagebrush and has virtually destroyed stands of perennial grasses over vast areas. Ecological problems have arisen from the sudden introduction of large numbers of domestic livestock, which caused a virtual purge of the native herbaceous layer and a dramatic increase in the density of the dominant shrubs (Young *et al.*, 1996). The native annual flora is now practically devoid of competitive annuals, especially grasses, and the failure to avoid the degradation of sagebrush ecosystems is partially due to the tremendous reproductive potential of sage brush, that allows the seedlings of these perennials to perform ecologically as annuals. In the USA., great effort is being devoted to better understand these complex plant communities from an ecological point of view and research programmes are being developed with the aims of limiting the damage caused by livestock and carrying out revegetation programmes using superior germplasms of sagebrush species with enriched protein content (Welch *et al.*, 1994).

Artemisia communities are also of importance in many other desert areas of the planet. Irina N. Safronova (1996) has described the distribution of 35 species of *Artemisia*, half of them belonging to the subgenus *Seriphidium*, in the Western Turan, the part of the Saharo-Gobi desert region which lies between the Caspian sea in the West and Aral sea and the Amu-Darya river in the East. It is limited by the mountains of Central Asia in the south, and in the north by the steppe zone at 48° N.

Artemisia species occupy about one third of the total area of West Turan, as dwarf semi-shrub dominants and codominants of the desert communities (Safronova 1996 and references therein). A few species have a wide ecological range of distribution, among which is *A. kemrudica* Krasch., an endemic western-southern Turan desert species, which is distributed both on loamy and sandy soils which are coarse textured and salty to various extents. Five species were found to be endemic to the Turan area. In contrast, some *Artemisia* species have a narrow ecological range: *A. camelorum* Krasch. and *A. pauciflora* Weber are halophytes (adapted to salty conditions), *A. lessingiana* Besser is a petrophyte (rock-growing) and *A. santolina* Schrenck is a psammophyte, typically found growing in the drift and hilly sands of central and southern deserts (Safronova 1996 and references therein).

Artemisia vulgaris is referred to by Rebele *et al.* (1993) as a good bioindicator plant, as it is one of the few species, able to grow in ecosystems around copper cave and industrial copper smelting complexes in Poland, where the upper soil layers are highly contaminated by copper, lead and other toxic pollutants such as sulphur dioxide.

CONCLUDING REMARKS

Owing to its high morphological variability, the genus *Artemisia* is still under discussion with regard to its systematics and phylogeny. Even though the morphological taxonomic approach has often proved to be useful in subdividing *Artemisia* into subgenera and sections leading to systematic models which correlate well with phylogenetic and paleobotanical studies, many problems still remain unresolved at the intrageneric, inter- and intra-specific level.

Today's use of chemical characteristics seems to be the most workable and natural taxonomic tool of investigation, having already been proven useful in many other instances. Furthermore, the impressive amount of therapeutic and pharmacological properties ascribed to *Artemisia* species, not only by their wide use in folk medicine but also as a result of biomedical research, prompts a world-wide chemical revision of this group of plants. Taking into account the great value of their chemical characteristics and related pharmacological activities, it is of primary importance to start research programmes utilising the great bulk of regional accounts on *Artemisia* species in order to search for valuable new varieties or chemotypes. From this viewpoint, a detailed study of the chemical properties of the genus with respect to the metabolic pathways of the biosynthesis of its more typical molecules is needed, not only for chemotaxonomic purposes, but also so that their industrial and therapeutic potential may be realised.

Cell genetics and the related tissue culture techniques, and notably those aimed at the genetic improvement of the genus, *via* protoplast hybridization or gene transformation are still limited in the family Asteraceae and should receive more attention in the future, especially as a result of the increasing importance of the antimalarial properties of *A. annua*.

In addition, we should not forget the great potential of the many species of *Artemisia* which have the ability to tolerate heat, drought and stress. These species may be suitable for use in regional agricultural renewal programmes (e.g. *Artemisia* spp. as aromatic alpine plants) as part of the environmental management of the arid regions of this planet.

REFERENCES

Adekenov, S.M., Rakhimova, B.B., Dzhazin, K.A., Alikov, V.B., Shaushekov, Z.K., Toregozhina, Zh.R. *et al.* (1995) Sesquiterpene lactones from *Artemisia glabella*. *Fitoterapia*, 66, 142–146.

Ahmad, A. and Misra, L.N. (1994) Terpenoids from *Artemisia annua* and constituents of its essential oil. *Phytochemistry*, 37, 183–186.

Ahmed, A.A., Aboul-El-Ela, M., Jakupovic, J., Seif El-Din, A.A. and Sabri, N. (1990) Eudesmanolides and other constituents from *Artemisia herba-alba*. *Phytochemistry*, 29, 3661–3663.

Ainswort Davis, J.R. (1908) *Knuth's Handbook of Flower Pollination*. Oxford, vol II.

Al-Hazim, H.M.G. and Basha, R.M.Y. (1991) Phenolic compounds from various *Artemisia* species. *J. Chem. Soc. Pak.*, 13, 277–289.

Alkhazraji, S.M., Alshamaony, L.A. and Twaij, H.A.A. (1993) Hypoglycaemic effect of *Artemisia herba alba*. 1. Effect of different parts and influence of the solvent on hypoglycaemic activity. *Journal of Etnopharmacology*, 40(3), 163–166.

Aly, M.Z.Y. and Badran, R.A.M. (1995) Mosquito control with extracts from plants of the Egyptian Eastern desert. *J. Herbs Spices & Med. Plants*, 3(4), 3–8.

Appendino, G., Belliardo, F., Bicchi, C., D'Amato, A., Frattini, C., Nano, G.M. *et al.* (1985) Distribuzione dei terpenoidi in liquori "genepi". In *Artemisie, ricerca ed applicazione*. Quaderno Agricolo, supplement 2, Federagrario, Torino, pp. 217–232.

Arnold, T.H. (1994) *Plants of southern Africa, Database*. Version 1.1, National Botanical Institute, Pretoria, South Africa.

Ascensão, L. and Pais, M.S.S. (1982) Secretory trichomes from *Artemisia chritmifolia*: some ultrastructural aspects. *Bull. Soc. Bot. Fr.*, 129, 83–87.

Ascensão, L. and Pais, M.S.S. (1987) Glandular trichomes of *Artemisia campestris* (ssp. *maritima*): ontogeny and histochemistry of the secretory product. *Bot. Gaz.*, 148, 221–227.

Ascensão, L. and Pais, M.S.S. (1988) Ultrastructure and histochemistry of secretory ducts in *Artemisia campestris* ssp. *maritima* (Compositae). *Nord. J. Bot.*, 8, 283–292.

Bavrina, T.V., Vorobev, A.S., Konstantinova, T.N. and Sergeeva, L.I. (1994) Growth and essential oil production in *Artemisia balchanorum in vitro*. *Russ. J. Plant Physiol.*, 41, 795–798.

Beetle, A.A. (1959) New names within the section *Tridentatae of Artemisia*. *Rhodora*, 61, 82–85.

Beetle, A.A. (1960) A study of sagebrush-The section *Tridentatae of Artemisia*. *Univ. Wyoming Agric. Exp. Stn. Bull.*, 386, 1–83.

Belenovskaja, L. (1996) *Artemisia*: the flavonoids and their systematic value. In D.J.N. Hind and H.J. Beentje (eds.), *Compositae: Systematics. Proceedings of the International Compositae Conference*, Royal Botanic Gardens, Kew, vol. I, chapter 18, pp. 253–259.

Bellomaria, B., Valentini, G. and Biondi, E. (1993) Essential oil composition of *Artemisia thuscula* Cav. from the Canary islands. *J. Essent. Oil Res.*, 5, 391–396.

Bentham, G. (1873) Notes on the classification, history and geographical distribution of Compositae. *J. Linn. Soc. (Bot.)*, 13, 335–577.

Berendes, J. (1902) *Des pedanios Dioskurides aus Anazarbos Arzeimittellehre in Fünf Büchern*. Ferdinand Enke, Stuttgart.

Bergendorff, O. and Sterner, O. (1995) Spasmolytic flavonols from *Artemisia abrotanum*. *Planta Med.*, 61, 370–371.

Besser, W.S. (1829) *Bull. Soc. Imp. Natl. Mosc.*, 1, 219.

Bicchi, C. and Rubiolo, P. (1996) High-Performance Liquid Chromatographic particle beam Mass Spectrometric analysis of sesquiterpene lactones with different carbon skeletons. *J. Chromatog.*, 727, 211–221.

Biewer, K. (1992) *Albertus Magnus. De vegetalibus*, VI, 2. Wissenschaftliche Verlagsgesellschaft mbH, Stuttgart.

Biondi, E., Valentini, G., Bellomaria, B. and Zuccarello, V. (1993) Composition of essential oil in *Artemisia arborescens* L. from Italy. *Acta Hortic.*, **344**, 290–304.

Bodrug, M.V., Dragalin, I.P. and Vlad, P.F. (1987) Introduction of some *Artemisia* species into Moldavia: characterization and chemical composition of their essential oils. *Izv. Akad. Nauk Mold. SSR, Ser. Biol. Khim. Nauk*, **2**, 14–16.

Bohlmann, F. and Zdero, C. (1972) Constituents of *Artemisia afra*. *Phytochemistry*, **11**, 2329–2330.

Boriki, D., Berrada, M., Talbi, M., Keravis, G. and Rouessac, F. (1996) Eudesmanolides from *Artemisia herba-alba*. *Phytochemistry*, **43**, 309–311.

Bremer, K. (1994) *Asteraceae: Cladistics and Classification*. Timber Press, Portland.

Bremer, K. and Humpries, B. (1993) Generic monograph of the Asteraceae – Anthemideae. *Bull. Nat. Hist. Mus.* (London), **23**, 71–177.

Brown, G.D. (1994a) Phytene-1,2-diol from *Artemisia annua*. *Phytochemistry*, **36**, 1553–1554.

Brown, G.D. (1994b) Secondary metabolism in tissue culture of *Artemisia annua*. *J. Nat. Prod.*, **57**, 975–977.

Brown, G.D. (1994c) Cadinanes from *Artemisia annua* that may be intermediates in the biosynthesis of artemisinin. *Phytochemistry*, **36**, 637–642.

Brown, S.A. (1990) Biosynthesis and distribution of coumarins in the plant, In *Cumarine, ricerca ed applicazioni*, Società Italiana di Fitochimica, pp. 15–37.

Caldwell, M.M. (1985) Cold desert. In B.R. Chabot & H.A. Mooney (eds.), *Physiological Ecology of North American Plant Communities*. Chapman and Hall, New York, pp. 198–212.

Cappelletti, E.M., Caniato, R. and Appendino, G. (1986) Localization of the cytotoxic hydroperoxyeudesmanolides in *Artemisia umbelliformis*. *Biochem. Syst. Ecol.*, **14**, 183–190.

Caramiello, R. and Fossa, V. (1993–1994) Flora palinologica italiana; schede morfopalino-logiche S216: *Artemisia borealis* Pallas. *Allionia*, **32**, 27–38.

Caramiello, R., Ferrando, R, Siniscalco, C. and Polini, V. (1987) Schede palinologiche di *Artemisia vulgaris* L., *Artemisia verlotorum* Lamotte, *Artemisia annua* L., su campioni freschi acetolizzati. *Aerobiologia*, **3**, 37–51.

Caramiello, R., Fossa, V., Siniscalco, C. and Potenza, A. (1990) Flora palinologica italiana-Schede di *Artemisia glacialis* L., *Artemisia genipi* Weber, *Artemisia umbelliformis* Lam. su campioni freschi ed acetolizzati (Schede n. S175, S176, S177). *Aerobiologia*, **6**, 221–238.

Carlquist, S. (1976) Tribal interrelationships and phylogeny of the Asteraceae. *Aliso*, **8**, 465–492.

Carnat, A.P., Gueugnot, J., Lamaison, J.L., Guillot, J. and Pourrat, H. (1985) The mugwort: *Artemisia vulgaris* L. and *Artemisia verlotiorum* Lamotte. *Ann. Pharm. Fr.*, **43**, 397–405.

Cedarleaf, J.D., Welch, B.L. and Brotherson, J.D. (1983) Seasonal variation of monoter-penoids in big sagebrush (*Artemisia tridentata*). *J. Range Manag.*, **36**, 492–494.

Chan, K.L., Teo, C.K.H., Jinadasa, S. and Yuen, K.H. (1995) Selection of high artemisinin yielding *Artemisia annua*. *Planta Med.*, **61**, 285–287.

Chandrashekar, V., Krishnamutri, M. and Seshadri, T.R. (1965) Chemical investigation of *Artemisia sacrorum*. *Curr. Sci.*, **34**, 609.

Chen, X.Y. and Xu, Z.H. (1996) Recent progress in biotechnology and genetic engineering of medicinal plants in China. *Med. Chem. Res.*, **6**, 215–224.

Chen, Y.L., Huang, H.C., Weng, Y.I., Yu, Y.J. and Lee, Y.T. (1994) Morphological evidence for the antiatherogenic effect of scoparone in hyperlipidaemic diabetic rabbits. *Cardiovascular Research*, **28**(11), 1679–1685.

Chialva, F. (1985) La coltivazione delle artemisie nella pianura piemontese: aspetti economici ed agronomici. In *Artemisie, ricerca ed applicazione*. Quaderno Agricolo, supplementó 2, Federagrario, Torino, pp. 41–52.

Clements, F.E. and Hall, H.M. (1923) *The phylogenetic method in taxonomy. The North American species of Artemisia, Chrysotamnus and Atriplex*. Publs. Carnegie Inst., pp. 355.

Corsi, G. and Nencioni, S. (1995) Secretory structures in *Artemisia nitida* Berthol. (Asteraceae). *Isr. J. Plant Sci.*, **43**, 359–365.

Croteau, R. (1987) Biosynthesis and catabolism of monoterpenoids. *Chem. Rev.*, **87**, 929–954.

Croteau, R. (1992) Biochemistry of monoterpenes and sesquiterpenes of the essential oils, In L.E. Craker and J.E. Simon (eds.), *Herbs, spices, and medicinal plants: recent advances in botany, horticulture, and pharmacology*, Food Product Press, New York, pp. 81–134.

Cubukcu, B. and Melikogu, G. (1995) Flavonoids of *Artemisia austriaca. Planta Med.*, **61**, 488.

Dalla Torre, C.G. and Harms, H. (1907) *Genera Siphonogamarum*, Lipsiae, Berlin.

De Candolle, A.P. (1837) *Prodomus Systematis Naturalis Regni Vegetabilis*, Treuttel and Wurtz, Paris, vol. VI.

Deans, S.G. and Svoboda, K.P. (1988) Antibacterial activity of French terragon (*Artemisia dracunculus* L.) essential oil and its constituents during ontogeny. *J. Hortic. Sci.*, **63**, 135–140.

Deans, S.G. and Svoboda, K.P. (1990) Essential oils profiles of several temperate and tropical aromatic plants: their antimicrobial and antioxidative properties. *Proceedings 75th Intern. Symp. of the Res. Inst. for Med. Plants, Budakalasz, Hungary*: 25–27.

Delgado, G. (1996) Bioactive constituents of some Mexican medicinal Compositae. In D.J.N. Hind and H.J. Beentje (eds.), *Compositae: Systematics. Proceedings of the International Compositae Conference*, Royal Botanic Gardens, Kew, vol. II, chapter. 39, pp. 505–515.

Dewick, P.M. (1995a) The biosynthesis of C5–C20 terpenoid compounds. *Nat. Prod. Rep.*, **12**, 507–534.

Dewick, P.M. (1995b) The biosynthesis of shikimate metabolites. *Nat. Prod. Rep.*, **12**, 101–134.

Diettert, R.A. (1961) The morphology of *Artemisia tridentata* Nutt. *Lloydia*, **1**, 3–74.

Dominguez, X.A. and Cardenas, E.A. (1975) Achillin and desacetylmatricarin from two *Artemisia* species. *Phytochemistry*, **14**, 2511–2512.

Duffey, S.S. (1986) Plant glandular trichomes: their partial role in defence against insects, In B.E. Juniper and T.R.E. Southwood (eds.), *Insects and plant surfaces*, Arnold, London, pp. 151–172.

Duke, S.O. (1994) Glandular trichomes – A focal point of chemical and structural interactions. *Int. J. Plant Sci.*, **155**, 617–620.

Duke, S.O. and Paul, R.N. (1993) Development and fine structure of the glandular trichomes of *Artemisia annua. Int. J. Plant Sci.*, **154**, 107–118.

Dukes, S.O., Vaughn, K.C., Croom, E.M.Jr. and Elsohly, H.N. (1987) Artemisinin, a constituent of annual wormwood (*Artemisia annua*), is a selective phytotoxin. *Weed Sci.*, **35**, 499–505.

Elhag, H.M., El-Domiaty, M.M., El-Feraly, F.S., Mossa, J.S. and El-Olemy, M.M. (1992) Selection and micropropagation of high artemisinin producing clones of *Artemisia annua* L. *Phytother. Res.*, **6**, 20–24.

Facknath, S. and Kawol, D. (1993) Antifeedant and insecticidal effects of some plant extracts on the cabbage webworm, *Crocidolomia binotalis. Insect Science & its Application*, **14**(5–6), 571–574.

Fahn, A. (1979) *Secretory tissues in plants*, Academic Press, London.

Fahn, A. (1988). Tansley review No. 14. Secretory tissues in vascular plants. *New Phytologist*, **108**, 229–257.

Ferreira, J.F.S. and Janick, J. (1995) Floral morphology of *Artemisia annua* with special reference to trichomes. *Int. J. Plant Sci.*, **156**, 807–815.

Ferreira, J.F.S. and Janick, J. (1996) Immunoquantitative analysis of artemisinin from *Artemisia annua* using polyclonal antibodies. *Phytochemistry*, **41**, 97–104.

Ferreira, J.F.S., Simon, J.E. and Janick, J. (1995a) Developmental studies of *Artemisia annua*. Flowering and artemisinin production under greenhouse and field conditions. *Planta Med.*, **61**, 167–170.

Ferreira, J.F.S., Simon, J.E. and Janick, J. (1995b) Relationship of artemisinin content of tissue-cultured, greenhouse-grown, and field-grown plants of *Artemisia annua*. *Planta Med.*, **61**, 351–355.

Filotava, N.S. (1986) The system of the warmwood of *Seriphidium* (Bess.) Peterm. (*Artemisia* L., Asteraceae) of Eurasia and North Africa. *Novosti Sist. Nizsh. Rast.*, **23**, 217–239.

Fraga, B.M. (1992) Natural sesquiterpenoids. *Nat. Prod. Rep.*, **9**, 217–242.

Fraga, B.M. (1993) Natural sesquiterpenoids. *Nat. Prod. Rep.*, **10**, 397–419.

Fraga, B.M. (1994) Natural sesquiterpenoids. *Nat. Prod. Rep.*, **11**, 533–554.

Fulzele, D.P., Heble, M.R., Rao, P.S. (1995) Production of terpenoid from *Artemisia annua* L. plantlet cultures in bioreactor. *J. Biotechnol.*, **40**, 139–143.

Gautheret, R., Leddet, C. and Paupardin, C. (1984) Comparison of the essential oils of some *genipi* clones isolated and grown *in vitro*. *C. R. Acad. Sci*, Ser. 3, **299**, 621–624.

Ghan, K.L., Teo, C.K.H., Jinadasa, S. and Yuen, K.H. (1995) Selection of high artemisinin yielding *Artemisia annua*. *Planta Med.*, **61**(3), 285–287.

Gilani, A.H., Janbaz, K.H., Lateef, A. and Zaman, M. (1994) Ca²⁺ channel blocking activity of *Artemisia scoparia* extract. *Phytotherapy Research*, **83**(3), 161–165.

Graham, A. (1996) A contribution to the geologic history of the Compositae. In D.J.N. Hind and H. Beentje (eds.), *Compositae: Systematics. Proceedings of the International Compositae Conference*, Royal Botanic Gardens, Kew, vol. I, chapter. 10, pp. 123–140.

Graven, E.H., Deans, S.G., Svoboda, K.P., Mavi, S. and Gundidza, M.G. (1992) Antimicrobial and antioxidant properties of the volatile (essential) oil of *Artemisia afra* Jacq. *Flav. Fragr. J.*, **7**, 121–123.

Greger, H. (1977) Anthemideae – chemical review. In V.H. Heywood, J.B. Harborne, and B.L. Turner, (eds.), *The biology and chemistry of the Compositae*, Academic Press, London, pp. 899–941.

Guardia, T, Guzman, J.A., Pestchanker, M.J., Guerriero E. and Giordano O.S. (1994) Mucus synthesis and sulphydryl groups in cytoprotection mediated by dehydroleucodine, a sesquiterpene lactone. *Journal of Nat. Prod.*, **57**, 507–509.

Gulati, A., Bharel, S., Jain, S.K., Abdin, M.Z. and Srivastava, P.S. (1996) *In vitro* micropropagation and flowering in *Artemisia annua*. *J. Plant Biochem. Biotechnol.*, **5**, 31–35.

Gunther, R.T. (1968) *The Greek Herbal of Dioscorides*. Haefner, London.

Gupta, M.M., Jain, D.C., Mathur, A.K., Singh, A.K., Verma, R.K. and Kumar, S. (1996) Isolation of high artemisinic acid containing plant of *Artemisia annua*. *Planta Med.*, **62**, 280–281.

Hahlbrock, K. (1981) Flavonoids, In E.E. Conn (ed.), *The biochemistry of plants*. Vol. 7. *Secondary plant products*, Academic Press, New York, pp. 425–456.

Halligan, J.P. (1975) Toxic terpenes from *Artemisia californica*. *Ecology*, **56**, 999–1003.

Hayakawa, Y, Hayashi, T., Niiya, K. and Sakuragawa, N. (1995) Selective activation of heparine cofactor II by a sulfated polysaccharide isolated from the leaves of *Artemisia princeps*. *Blood Coagulation & Fibrinolysis*, **6**(7), 643–649.

Hegnauer, R. (1977) The chemistry of the Compositae. In V.H. Heywood, J.B. Harborne, and B.L. Turner, (eds.), *The Biology and Chemistry of the Compositae*, Academic Press, London, pp. 283–335.

Herout, V. and Sorm, F. (1954) On terpenes. LXI. Contribution to the constitution of prochamazulenogen, the natural precursor of chamazulene in *Artemisia absinthium* L. *Coll. Czech. chem. Commun.*, **19**, 792–797.

Herout, V. and Sorm, F. (1959) Isolation and structure of Costunolide from *Artemisia balchanorum*. *Chem. Ind.*, 1067.

Herz, W. (1977) Asteraceae-chemical review. In V.H. Heywood, J.B. Harborne and B.L. Turner, (eds.), *The Biology and Chemistry of the Compositae*, Academic Press, London, vol. II, chapter 11, pp. 567–576.

Herz, W., Bhat, S.V. and Santhanam, P.S. (1970) Coumarins of *Artemisia dracunculoides* and 3′,6-dimethoxy-4′,5,7-trihydroxyflavone in *A. arctica*. *Phytochemistry*, **9**, 891–894.

Heywood, V.H. and Humphries, C.J. (1977) Anthemideae-systematic review. In V.H. Heywood, J.B. Harborne and B.L. Turner, (eds.), *The Biology and Chemistry of the Compositae*, Academic Press, London, vol. II, chapter 31, pp. 852–888.

Heywood, V.H., Harborne, J.B. and Turner, B.L. (1977) An overture to the Compositae. In V.H. Heywood, J.B. Harborne and B.L. Turner, (eds.), *The Biology and Chemistry of the Compositae*, Academic Press, London, vol. I, chapter1, pp. 1–19.

Hoffmann, O. (1894) Compositae. In A. Engler and R. Prantl (eds.), *Die naturliche Pflanzenfamilien*, W. Eugelmann, Leipzig, **4** (5), 87–391.

Holton, T.A. and Cornish, E.C. (1995) Genetics and biochemistry of anthocyanin biosynthesis. *Plant Cell*, **7**, 1071–1083.

Hooker, W.J. (1840) *Flora Boreali-Americana*, H.G. Bohn, London, vol. I.

Hopkins, D.M. (1967) In *Flora of Kamtchaka and Adjacent Islands*, Almqvist and Wiksells Boktryckery-A.-B., Stockolm, vol. IV, pp. 176–191.

Hu, J.-F., Bai, S.-P. and Jia, W.-Z. (1996a) Eudesmane sesquiterpenes from *Artemisia eriopoda*. *Phytochemistry*, **43**, 815–817.

Hu, J.-F., Zhu, Q.-X., Bai, S.-P. and Jia, W.-Z. (1996b) New eudesmane sesquiterpene and other constituents from *Artemisia mongolica*. *Planta Med.*, **62**, 477–478.

Hu, S.Y. (1965) The Compositae of China I. *Q.J. Taiwan Mus.*, **18**, 1–136.

Huang, Y.P., and Ling, Y.R. (1996) In D.J.N. Hind and H.J. Beentje (eds.), *Compositae: Systematics. Proceedings of the International Compositae Conference*, Royal Botanic Gardens, Kew, vol. II, chapter 36, pp. 431–451.

Janssen, A.M., Scheffer, J.J.C., Baerhein-Svendsen, A. and Svendsen, A.B. (1987) Antimicrobial activities of essential oils. A 1976–1986 literature review on possible applications. *Pharmaceut. Weekbl.*, **9**, 193–197.

Javadi, I. (1989) Studies on extracts of some medicinal plants as insect repellents. *37th Annual Congress in Braunschweig*. The Society for Medicinal Plant Research, Germany, pp. 265.

Jeffrey, C. (1978) Compositae. In V.H. Heywood (ed.), *Flowering Plants of the World by Families*, Elsevier International Progects, Oxford.

Kelsey, R.G. and Shafizadeh, F. (1979) Sesquiterpene lactones and systematics of the genus *Artemisia*. *Phytochemistry*, **18**, 1591–1611.

Kelsey, R.G. and Shafizadeh, F. (1980) Glandular trichomes and sesquiterpene lacones of *Artemisia nova* (Asteraceae). *Biochem. Syst. Ecol.*, **8**, 371–377.

Kitamura, S. (1939) A classification of *Artemisia*. *Acta Phytotax. Geobot.*, **8**, 62–66.

Kitamura, S. (1940) Compositae Japonicae. Pars Seconda. *Mem. Coll. Sci. Kyoto Univ.*, **25**(3) art. 9, 286–446.

Klayman, D.L., Lin, A.J., Acton, N., Scovil, J.P., Hoch, J.M., Milhous, W.K. *et al.* (1984) Isolation of artemisinin (qinghaosu) from *Artemisia annua* growing in the United States. *J. Nat. Prod.*, **47**, 715–717.

Koekemoer, M (1996) In D.J.N. Hind and H.J. Beentje (eds.), *Compositae: Systematics. Proceedings of the International Compositae Conference*, Royal Botanic Gardens, Kew, vol. I, chapter.8, pp. 95–110.

Konovalova, O.A. and Sheichenko, V.I. (1991) Chemical composition of *Artemisia frigida. Khim. Prir. Soedin.*, **1**, 143–145.

Krascheninnikov, I.M. (1948) The experience of phylogenetic analysis of some Euroasiatic groups of genus *Artemisia* L. in connection with features of paleogeography of Eurasia. *Materiali po istorii flory i rastitel'nosti SSSR*. Moskwa, Leningrad, **2**, 87–96.

Krascheninnikov, I.M. (1958) The role and meaning of Angar floristic centre in the phylogenetic development of the general Eurasiatic groups of the warmwood of *Euartemisia* subgenus. *Materiali po istorii flory i rastitel'nosti SSSR*. Moskwa, Leningrad, **23**, 62–128.

Kroes, B.H., Vanufford, H.C.Q., Tinbergendeboer, R.L., Vandenberg, A.J.J., Beukelmann, C.J., Labadie, R.P. *et al.* (1995) Modulatory effects of *Artemisia annua* extracts on human complement, neutrophil oxidative burst and proliferation of T lymphocytes. *Phytotherapy Research*, **9**(8), 551–554.

Laughlin, J.C. (1994) Agricultural production of artemisinin-A review. *Transact. Royal. Soc. Trop. Med. & Hygiene*, **88** (suppl. 1), 21–22.

Lawrence, B.M. (1992) Wormwood oil, In progress in essential oils. *Perf. Flav.*, **17**, 39–44.

Lee, K.R., Hong, S.W., Kwak, J.H., Pyo, S. and Jee, O.P. (1996) Phenolic constituents from the aerial parts of *Artemisia stolonifera. Arch. Parmac. Res.*, **19**, 231–234.

Lessing, C.F. (1832) Anthemideae. in *Synopsis Generum Compositarum*, Berlin, pp. 274–269.

Liu, Q., Yang, Z.Y., Deng, Z.B., Sa, G.H. and Wang, X.J. (1988) Preliminary analysis on chemical constituents of essential oil from inflorescence of *Artemisia annua* L. *Acta Bot. Sinica*, **30**, 223–225.

Maffei, M. (1994) Discriminant analysis of leaf wax alkanes in the Lamiaceae and four other plant families. *Biochem. Syst. Ecol.*, **22**, 711–728.

Maffei, M. (1996a) Chemotaxonomic significance of leaf wax alkanes in the Gramineae. *Biochem. Syst. Ecol.*, **24**, 53–64.

Maffei, M. (1996b) Chemotaxonomic significance of leaf wax alkanes in the Umbelliferae, Cruciferae and Leguminosae (subf. Papilionoideae). *Biochem. Syst. Ecol.*, **24**, 531–545.

Maffei, M. (1996c) Chemotaxonomic significance of leaf wax alkanes in the Compositae, In D.J.N. Hind and H.J. Beentje, (eds), *Compositae: Systematics. Proceedings of the International Compositae Conference, Kew, 1994*, Royal Botanic Gardens, Kew, vol. 1, pp. 141–158.

Maffei, M., Meregalli, M. and Scannerini, S. (1997) Chemotaxonomic significance of surface wax alkanes in the Cactaceae. *Biochem. Syst. Ecol.*, **25**, 241–253.

Maffei, M., Peracino, V. and Sacco, T. (1993a) Multivariate methods for aromatic plants. An application to mint essential oils. *Acta Horticult.*, **330**, 159–169.

Maffei, M., Mucciarelli, M. and Scannerini, S. (1993b) Environmental factors affecting the lipid metabolism in *Rosmarinus officinalis* L. *Biochem. Syst. Ecol.*, **21**, 765–784.

Marco, J.A., Sanz, J.F., Yuste, A., Carda, M. and Jakupovic, J. (1991) Sesquiterpene lactones from *Artemisia barrelieri. Phytochemistry*, **30**, 3661–3668.

Marco, J.A., Sanz-Cervera, J.F., Morante, M.D., García-Lliso, V., Vallès-Xirau, J. and Jakupovic, J. (1996) Tricyclic sesquiterpenes from *Artemisia chamaemelifolia. Phytochemistry*, **41**, 837–844.

Marco, J.A., Sanz-Cervera, J.F., Sancenon, F., Arno, M. and Vallès-Xirau, J. (1994a) Sesquiterpene lactones and acetylenes from *Artemisia reptans. Phytochemistry*, **37**, 1095–1099.

Marco, J.A., Sanz-Cervera, J.F., Pereira, J.M., Sancenon, F. and Vallès-Xirau, J. (1994b) Sesquiterpene lactones from North African *Artemisia* species. *Phytochemistry*, **37**, 477–485.

Marco, J.A., Sanz-Cervera, Manglano, E., Sanceon, F., Rustaiyan, A. and Kardar, M. (1993) Sesquiterpene lactones from Iranian *Artemisia* species. *Phytochemistry*, **34**, 1561–1564.

McArthur, E.D. (1979) In G.F. Gifford, F.E. Busby and J.K. Shaw (eds.) *The Sage-brush Ecosystem: A Symposium*, Utah State Univ. Press, Logan, pp. 22–26.

McArthur, E.D. and Plummer, A.P. (1978) In *Intermountain Biogeography Symposium*, K.T. Harper and J.L. Reveal (eds.) Great Basin Nat. Memoirs, No. 2, Brigham Young University Press, Provo, Utah, pp. 229–243.

Mericli, A.H., Jakupovic, J., Bohlmann, F., Damadyan, B., Ozhatay, N.and Cubukcu, B. (1988) Eudesmanolides from *Artemisia santonicum. Planta Med.*, **54**, 447–449.

Mitsouka, S. and Ehrendorfer, F. (1972) Cytogenetics and evolution of Matricaria and related genera (Asteraceae-Anthemideae). *Ost. Bot. Z.*, **120**, 155–200.

Moss, E.H. (1940) Interxylary cork in *Artemisia* with reference to its taxonomic significance. *Am. J. Bot.*, **27**, 726–768.

Mucciarelli, M., Caramiello, R., Chialva, F. and Maffei, M. (1995) Essential oils from some *Artemisia* species growing spontaneously in North West Italy. *Flav. Fragr. J.*, **10**, 25–32.

Murray, R.D.H. (1995) Coumarins. *Nat. Prod. Rep.*, **12**, 477–505.

Nin, S., Morosi, E., Schiff, S. and Bennici, A. (1996) Callus cultures of *Artemisia absinthium* L. Initiation, growth optimization and organogenesis. *Plant Cell Tiss. & Org. Cult.*, **45**, 67–72.

Novotny, L., Herout, V. and Sorm, F. (1960) On terpenes. CIX. A contribution to the structure of absinthin and anabsinthin. *Coll. Czech. Chem. Commun.*, **25**, 1492–1499.

Nuñez, D.R. and De Castro, C.O. (1996) Paleoethnobotany of Compositae in Europe, North Africa and the Near East. In D.J.N. Hind and H.J. Beentje (eds.), *Compositae: Systematics. Proceedings of the International Compositae Conference*, Royal Botanic Gardens, Kew, vol. II, chapter. 40, pp. 517–545.

Oliva, M. and Vallès, J. (1993) Karyological studies in some taxa of the genus *Artemisia* (Asteraceae). *Can. J. Bot.*, **72**, 1126–1135.

Ovezdurdyev, A., Zakirov, S.Kh., Yusupov, M.I., Kasymov, Sh.Z., Abdusamatov, A. and Malikov, V.M. (1987) Sesquiterpene lactones of two *Artemisia* species. *Khim. Prir. Soedin.*, **4**, 607–608.

Ozhatay, N. and Cubukcu, B. (1990) Constituents of *Artemisia marschalliana. Fitoterapia*, **61**, 145–146.

Paniego, N.B. and Giulietti, A.M. (1994) *Artemisia annua* L. – Dedifferentiated and differentiated cultures. *Plant Cell Tiss.& Org. Cult.*, **36**(2), 163–168.

Paniego, N.B. and Giulietti, A.M. (1996) Artemisinin production by *Artemisia annua* L. transformed organ cultures. *Enz. Microb. Technol.*, **18**, 526–530.

Poljakov, P.P. (1961a) Systematic studies in the genus *Artemisia* L. *Trudy Ins. Bot. Akad. Nauk. Kazakh, SSR. Alma Acta*, **11**, 134–177.

Poljakov, P.P. (1961b) In V.L. Komarov (ed.) *Flora of the U.S.S.R.*, Izdatelstvo Akad. NauK SSSR, Leningrad, vol. XXVI, pp. 425–631.

Poljakov, P.P. (1967) Origin and classification of the Compositae. *Akad. Nauk. Kazach. SSR Inst. Bot.* pp. 335

Putievsky, E., Ravid, U., Dudai, N. and Katzir, I. (1992) Variations in the essential oil of *Artemisia judaica* L. chemotypes related to phenological and environmental factors. *Flav. Fragr. J.*, **7**, 253–257.

Raven, P.H. and Axelrod, D.I. (1974) Angiosperm biogeography and past continental movements. *Ann. Mo. Bot. Gdn*, **61**, 539–673.

Ravid, U., Putievsky, E. and Ikan, R. (1992) Determination of the enantiomeric composition of terpinen-4-ol in essential oils using a permethylated b-cyclodextrin coated chiral capillary column. *Flav. Fragr. J.*, **7**, 49–52.

Rebele, F., Surma, A., Kuznik, C., Bornkamm, R and Brej, T. (1993) Heavy metal contamination of spontaneous vegetation and soil around the copper smelter Legnica. *Acta Societatis Botanicorum Poloniae*, **62**(1–2), 53–57.

Reitbrecht, F. (1974) Fruchanatomie und Systematik der Anthemideae (Asteraceae). *Dissertation*, University of Vienna.

Reucker, G., Detlef, M. and Sibylle, W. (1992) Peroxides as constituents of plants. Part 10; Homoditerpene peroxides from *Artemisia absinthium*. *Phytochemistry*, **31**, 340–342.

Rodriguez, E., Carman, N.J., Vander Velde, G., McReynolds, J.H., Mabry, T.J., Irwin, M.A. *et al.* (1972) Methoxylated flavonoids from *Artemisia*. *Phytochemistry*, **11**, 3509–3514.

Rodriguez, E., Healey, P.L. and Mehta, I. (1984) *Biology and chemistry of plant trichomes*. Plenum Press, New York.

Rojatkar, S.R., Pawar, S.S., Pujar, P.P., Sawaikar, D.D., Gurunath, S., Sathe, V.T. *et al.* (1996) A germacranolide from *Artemisia pallens*. *Phytochemistry*, **41**, 1105–1106.

Ronse, A.C. and De Pooter, H.L. (1990) Essential oil productivity by Belgian *Artemisia alba* (Turra) before and after micropopagation. *J. Essent. Oil Res.*, **2**, 237–242.

Safranova, I.N. (1996) In D.J.N. Hind and H.J. Beentje (eds.), *Compositae: Systematics. Proceedings of the International Compositae Conference*, Royal Botanic Gardens, Kew, vol. II, chapter.8, pp. 105–110.

Saleh, N.A.M. and Mosharrafa, S. (1996) Flavonoids and chemosystematics of some African Compositae-A review, In D.J.N. Hind and H.J. Beentje, (eds), *Compositae: Systematics. Proceedings of the International Compositae Conference, Kew, 1994*, Royal Botanic Gardens, Kew, vol. 1, pp. 187–205.

Sangwan, R.S., Agarwal, K., Luthra, R., Thakur, R.S. and Singh-Sangwan, N. (1993) Biotransformation of arteannuic acid into arteannuin-B and artemisinin in *Artemisia annua*. *Phytochemistry*, **34**, 1301–1302.

Sanz, J.F., Garcis-Lliso, V., Marco, J.A. and Valles-Xirau, J. (1991) A cadinane derivative from *Artemisia crithmifolia*. *Phytochemistry*, **30**, 4167–4168.

Satar, S. (1986) Chemische charakterisierung ätherischer Öle aus mongolischen Arten der Gattung *Artemisia* L. *Pharmazie*, **41**, 819–820.

Seeligmann, P. (1996) Flavonoids of the Compositae as evolutionary parameters in the tribes which synthesize them: a critical approach. In D.J.N. Hind and H.J. Beentje, (eds), *Compositae: Systematics. Proceedings of the International Compositae Conference, Kew, 1994*, Royal Botanic Gardens, Kew, vol. 1, pp. 159–167.

Shafizadeh, F. and Melnikoff, A.B. (1970) Coumarins of *Artemisia tridentata* subsp. *vaseyana*. *Phytochemistry*, **9**, 1311–1316.

Shah, N.C. (1996) In D.J.N. Hind and H.J. Beentje (eds.), *Compositae: Systematics. Proceedings of the International Compositae Conference*, Royal Botanic Gardens, Kew, vol. II, chapter 40, pp. 415–422.

Shah, N.C. and Thakur, R.S. (1992) Chemical composition of leaf/inflorescence of *Seriphidium brevifolium* (Wall) Y. Ling et Y.R. Ling folklore incense from Kumaon Himalaya. *India. J. Essen. Oil. Res.*, **4**, 25–28.

Sharma, S.K., Ali, M. and Singh, R. (1996) New 9-b-lanostane-type triterpenic and 13, 14-seco-steroidal esters from the roots of *Artemisia scoparia*. *J. Nat. Prod.*, **59**, 181–184.

Shulte, K.E., Rücker, G. and Perlick, J. (1967) Das Vorkommen von Polyacetylen-Verbindungen in *Echinacea purpurea* Moench und *Echinacea angustifolia* DC. *Arzneim. Forsch. (Drug Res.)*, **17**, 825.

Shultz, L.M. (1984) Taxonomic and geographic limits of *Artemisia* subgenus *Tridentatae* (Beetle) McArthur. *Proc. Symposium on the Biology of Artemisia and Crysothamnus*. Gen. Tech. Rpt. 200. USDA, Forest Service. Ogden, USA, pp. 20–28.

Skvarla, J.J. and Larson, D.A. (1965) An electron microscope study of pollen morphology in the Compositae with special reference to the Ambrosinae. *Grana Palynol.*, **6**(2), 211–269.

Skvarla, J.J., Turner, B.L., Patel, V.C. and Tomb, A.S. (1977) Pollen morphology in the Compositae and in morphologically related families. In V.H. Heywood, J.B. Harborne and B.L. Turner, (eds.), *The Biology and Chemistry of the Compositae*, Academic Press, London, vol. I, chapter 8, pp. 143–249.

Slone, J.H. and Kelsey, R.G. (1985) Isolation and purification of glandular secretory cells from *Artemisia tridentata* (ssp. *vaseyana*) by percoll density gradient centrifugation. *Am. J. Bot.*, **172**, 1445–1451.

Small, J. (1919) The origin and development of the Compositae. *New Phytol.*, **18**, 129–176.

Solomon, A.M. (1983) Pollen morphology and plant taxonomy of red oaks in eastern North America. *Am. J. Bot.*, **70**, 495–507.

Stebbins, G.L. (1974) *Flowering Plants*. Belknap Press, Cambridge, Massachussets.

Stebbins, G.L. (1977) Developmental and comparative anatomy of the Compositae. In V.H. Heywood, J.B. Harborne and B.L. Turner, (eds.), *The Biology and Chemistry of the Compositae*, Academic Press, London, vol. I, chapter 5, pp. 92–107.

Stix, E. (1960) Pollenmorphologische Untersuchungen and Compositae. *Grana Palynol.*, **2**, 41–114.

Tan, R.X., Wang, W.Z., Yang, L. and Wei, J.H. (1995) A new eudesmenoic acid from *Artemisia phaeolepis*. *J. Nat. Prod.*, **58**, 288–290.

Tantaoui-Elaraki, A., Ferhout, F. and Errifi, A. (1993) Inhibition of the fungal asexual reproduction stages by three Moroccan essential oils. *J. Essent. Oil Res.*, **5**, 535–545.

Teoh, K.H., Weathers, P.J., Cheetham, R.D. and Walcerz, D.B. (1996) Cryopreservation of transformed (hairy) roots of *Artemisia annua*. *Cryobiol.*, **33**, 106–117.

Thakur, R.S. and Misra, L.N. (1989) Essential oils of Indian *Artemisia*. In S.C. Bhattacharyya, N. Sen and K.L. Sethi, (eds.) *11th International Congress of Essential Oils, Fragrances and Flavours, New Delhi, India 12–16 Nov. 1989*, Oxford & IBH Publishing Co., PVT. LTD., New Delhi, pp. 127–135.

Todorova, M.N. and Krasteva, M.L. (1996) Sesquiterpene lactones from *Artemisia lerchiana*. *Phytochemistry*, **42**, 1231–1233.

Todorova, M.N., Tsankova, E.T., Trendafilova, A.B. and Gussev, C.V. (1996) Sesquiterpene lactones with the uncommon rotundane skeleton from *Artemisia pontica*. *Phytochemistry*, **41**, 553–556.

Trendafilova, A.B., Todorova, M.N. and Gussev, C.V. (1996) Eudesmanolides from *Artemisia pontica*. *Phytochemistry*, **42**, 469–471.

Turner, B.L. (1977) Fossil history and geography. In V.H. Heywood, J.B. Harborne and B.L. Turner, (eds.), *The Biology and Chemistry of the Compositae*, Academic Press, London, vol. I, chapter 2, pp. 22–38

Tutin, T.G. and Persson K. (1976) CLXIX Compositae. 88. *Artemisia* L. In T.G. Tutin, N.A. Burges, D.M. Moore, D.H. Valentine, S.M. Walters, D.A. Webb and V.H. Heywood, (eds.), *Flora Europaea*, Cambridge University Press, Cambridge, vol. IV, pp. 178–186.

Valant-Vetschera, K.M. and Wollenweber, E. (1995) Flavonoid aglycones from the leaf surfaces of some *Artemisia* spp. (Compositae-Anthemideae). *Z. Naturforsch.*, **50**, 353–357.

Vallès, J., Blanché, C., Bonet, M.A., Agelet A., Muntané, J., Raja, D. *et al.* (1996) In D.J.N. Hind and H.J. Beentje (eds.), *Compositae: Systematics. Proceedings of the International Compositae Conference*, Royal Botanic Gardens, Kew, vol. II, chapter 36, pp. 453–466.

Van den Berghe, D.R., Vergauwe, A.N., Van Montagu, M. and Van den Eckhout, E.G. (1995) Simultaneous determination of artemisinin and its bioprecursors in *Artemisia annua. J. Nat. Prod.*, **58**, 798–803.

Vergauwe, A., Cammaert, R., Van den Berghe, D., Genetello, C., Inze, D., Van Montagu, *et al.* (1996) *Agrobacterium tumefaciens*-mediated transformation of *Artemisia annua* L. and regeneration of transgenic plants. *Plant Cell Rep.*, **15**, 929–933.

Vezey, E.L., Watson, L.E., Skavarla, J.J. and Estes, J.R. (1993) Plesiomorphic and apomorphic pollen structure characteristic of Anthemideae (Asteroideae, Asteraceae). *Am. J. Bot.*, **81**(5), 648–657.

Wagenitz, G. (1976) Systematics and phylogeny of the Compositae. *Plant Syst. Evol.*, **125**, 29–46.

Wagner, H. (1977) Pharmaceutical and economic uses of the Compositae. In V.H. Heywood, J.B. Harborne and B.L. Turner, (eds.), *The Biology and Chemistry of the Compositae*, Academic Press, London, vol. II, chapter 14, pp. 412–428.

Ward, G.H. (1953) *Artemisia* section *Seriphidium* in North America. A cytotaxonomic study. *Contr. Dudley Herb.*, 4(6), 155–205.

Watson, L.E. (1996) Molecular systematics of the tribe Anthemideae. In D.J.N. Hind and H.J. Beentje (eds.), *Compositae: Systematics. Proceedings of the International Compositae Conference*, Royal Botanic Gardens, Kew, vol. I, chapter 23, pp. 341–348.

Welch, B.L., Briggs, S.F. and Young, S.A. (1994) Pine Valley Ridge source. A superior selected germplasm of black Sagebrush. *USDA Forest Service Intermountain Research Station Research Paper*, (474):U1–U9.

Weyerstahl, P., Schneider, S. and Marschall, H. (1993) The essential oil of *Artemisia sieberi* Bess. *Flav. Fragr. J.*, 8, 139–145.

Weathers, P.J., Cheetham, R.D., Follansbee, E. and Teoh, K. (1994) Artemisinin production by transformed roots of *Artemisia annua. Biotechnol. Lett.*, 16, 1281–1286.

White, N.J. (1994) Artemisinin-Current status. *Transact. Royal. Soc. Trop. Med.& Hygiene*, 88 (suppl.1), 3–4.

Wilson, E.O. (1986) *Biodiversity*. National Academic Press, Washington.

Wittmack, H. (1903) Die Pompeji Gefundenen Pflanzlichen Reste. *Bot. Jahrb.*, 33 Beibl., 73, 38–66.

Woerdenbag, H.J., Ros, R., Salomons, M.C., Hendriks, H., Pras, N. and Malingré, Th. M. (1992) Volatile constituents of *Artemisia annua. Planta Med.*, 58, 682.

Woerdenbag, H.J., Ros, R., Salomons, M.C., Hendriks, H., Pras, N. and Malingré, Th. M. (1993) Volatile constituents of *Artemisia annua* (Asteraceae). *Flav. Fragr. J.*, 8, 131–137.

Wollenweber, E. and Valant-Vetschera, K.M. (1996) New results with exudate flavonoids in Compositae, In D.J.N. Hind and H.J. Beentje, (eds), *Compositae: Systematics. Proceedings of the International Compositae Conference, Kew, 1994*, Royal Botanic Gardens, Kew, vol. 1, pp. 169–185.

Xu, Z.Q. and Jia, J.F. (1996) Callus formation from protoplasts of *Artemisia sphaerocephala* Krasch and some factors influencing protoplast division. *Plant Cell Tiss. & Org. Cult.*, 44, 129–134.

Yang, S.-L., Roberts, M.F., O'Neill, M.J., Bucar, F. and Phillipson, J.D. (1995) Flavonoids and chromenes from *Artemisia annua*. *Phytochemistry*, **38**, 255–257.

Young, J.A. and Mayeux, H. (1996) In D.J.N. Hind and H.J. Beentje (eds.), *Compositae: Systematics. Proceedings of the International Compositae Conference*, Royal Botanic Gardens, Kew, vol. II, chapter 7, pp. 93–104.

Young, J.A., Longland, W.S. and Blank, R.R. (1996) In D.J.N. Hind and H.J. Beentje (eds.), *Compositae: Systematics. Proceedings of the International Compositae Conference*, Royal Botanic Gardens, Kew, vol. II, chapter 22, pp. 277–290.

Yun, K.W., Kil, B.S. and Han, D.M. (1993) Phytotoxic and antimicrobial activity of volatile constituents of *Artemisia princeps* var. *orientalis*. *J. Chem. Ecol.*, **19**, 2757–2766.

Zheng, G.Q. (1994) Cytotoxic terpenoids and flavonoids from *Artemisia annua*. *Planta Medica*, **60**(1), 54–57.

Zheng, W.F., Tan, R.X., Yang, L. and Liu, Z.L. (1996) Two flavones from *Artemisia giraldii* and their antimicrobial activity. *Planta Med.*, **62**, 160–162.

2. ANALYSIS AND QUALITY CONTROL OF COMMERCIAL *ARTEMISIA* SPECIES

H.J. WOERDENBAG[1] AND N. PRAS[2]

[1]*Groningen Research Institute of Pharmacy (GRIP), Groningen University Institute for Drug Exploration (GUIDE), Antonius Deusinglaan 2, 9713 AW Groningen, The Netherlands.*

[2]*Department of Pharmaceutical Biology, Groningen Research Institute of Pharmacy, (GRIP), Groningen University Institute for Drug Exploration (GUIDE), Antonius Deusinglaan 1, 9713 AV Groningen, The Netherlands.*

INTRODUCTION

The genus *Artemisia*, which comprises about 400 species, belongs to the Asteraceae family and is the largest of the flowering plants. Most representatives are aromatic herbs or shrubs. They are mainly found in the northern hemisphere. Several *Artemisia* species have been or are used medicinally and hence are of more or less commercial value. In Western herbal medicine they include, among others, *Artemisia abrotanum* L., *Artemisia absinthium* L., *Artemisia cina* Berg, *Artemisia dracunculus* L., *Artemisia maritima* L., *Artemisia pontica* L. and *Artemisia vulgaris* L. (Frohne and Jensen 1992, Evans 1996). In traditional Chinese herbal medicine the following *Artemisia* species are used: *Artemisia annua* L., *Artemisia* argyi Levl. et Vant., *Artemisia scoparia* Waldst. et Kit. and *Artemisia capillaris* Thunb. (Tang and Eisenbrand 1992). In addition, *Artemisia annua* L. is a source of artemisinin, which is the mother compound of a novel class of antimalarial drugs (Woerdenbag *et al.*, 1990, 1994a).

Analytical aspects and other criteria for quality control of the above mentioned *Artemisia* species are the subject of this chapter. Procedures to assay secondary metabolites which are characteristic for a given species, as described in the literature, are presented and discussed.

Despite the rather broad medicinal use of *Artemisia* species, only few of them are official and have a monograph included in one or more pharmacopoeias. Pharmacopoeial aspects for the crude drug generally include a definition, and macroscopical and microscopical description of the crude drug, identity and purity reactions, and an assay for a quantitative determination. Demands for non official *Artemisia* species as presented in this chapter, have been adopted from several textbooks. In addition, several general (pharmacopoeial) procedures described for a given species may be used for other, non official, species as well. They include the determination of the essential oil content and its composition, determination of the ash, bitter value and storage conditions. Before the state of the art of various

Artemisia species is discussed separately, generally applicable procedures are described first.

GENERAL PROCEDURES

Essential Oil

Isolation

All medicinally used *Artemisia* species are fragrant plants which contain essential oil. The medicinal use is sometimes based upon the oil. The essential oil content as well as the composition are therefore valid criteria for the quality of the crude drug.

In several (European) pharmacopoeias an assay is included for the determination of essential oils in vegetable drugs. This is a hydrodistillation in a specially designed apparatus. The distillate is collected in a calibrated tube, whereas the aqueous phase is automatically returned to the distillation flask.

According to the German Pharmacopoeia, "Deutsches Arzneibuch", DAB 9 (1986), 50.0 g of freshly powdered *Herba Absinthii* (source plant: *A. absinthium*), is distilled in a 1000 ml flask, with 500 ml of water as distillation liquid and 0.50 ml of xylene as the collection liquid. The distillation rate is 2–3 ml per min and the distillation time 3 hours. This procedure is applicable for all *Artemisia* species discussed in this chapter.

After determination of the volume, yielding the essential oil content as percentage v/m, the samples can be stored at –20° C until analysed for their composition using gas chromatography (GC) and gas chromatography coupled with mass spectrometry (GC-MS). Below, the experimental conditions are given for GC and GC-MS analysis as applied successfully in our laboratory for essential oils (Woerdenbag *et al.*, 1993a, 1994b, Bos *et al.*, 1996). Prior to GC and GC-MS analysis, the samples are diluted 50 times with cyclohexane.

The identity of the components is assigned by comparison of their retention indices, relative to C_9-C_{19} *n*-alkanes, and mass spectra with corresponding data from reference compounds and from the literature (Adams 1989, Tucker and Maciarello 1993). The concentration of the components is calculated from the GC peak areas, using the normalization method.

Gas Chromatography (GC)

GC analysis is performed on a Hewlett Packard 5890 Series II gas chromatograph equipped with a 7673 injector and a Hewlett Packard 3365 Series II Chemstation, under the following conditions: column, WCOT fused-silica CP-Sil 5 CB, 25 m × 0.32 mm i.d.; film thickness, 0.25 μm (Chrompack, Middelburg, The Netherlands); oven temperature programme, 50–290° C at 4° C/min; injection temperature, 250° C; detector (FID) temperature, 300° C; carrier gas, nitrogen; inlet pressure, 5 psi; linear gas velocity, 26 cm/s; split ratio, 1:56; injected volume, 1.0 μl.

Gas Chromatography mass/spectrometry (GC-MS)

GCMS (EI) is performed on a Unicam 610/Automass 150 GC-MS system. The GC conditions are as described above, except: column, 25 m × 0.25 mm i.d.; carrier gas, helium; linear gas velocity, 32 cm/s; split ratio, 1:20. MS conditions: ionization energy, 70 eV; ion source temperature, 250° C; interface temperature, 280° C; scan speed, 2 scans/s; mass range, 34–500 amu; injected volume, 1.0 μl.

Sulphated Ash

Assay

According to the European Pharmacopoeia 2nd edn (1993) the sulphated ash is determined as follows. Heat a silica or platinum crucible to redness for 30 min. Allow to cool in a desiccator and weigh. Place 1.00 g of powdered drug in the crucible and add 2 ml of dilute sulphuric acid (9.8% (m/v)). Heat at first on a water-bath, than cautiously over a flame, then progressively to about 600° C. Continue the incineration until all black particles have disappeared and allow the crucible to cool. Add a few drops of dilute sulphuric acid (9.8% (m/v)), heat and incinerate as before and allow to cool. Add a few drops of a 15.8% (m/v) ammonium carbonate solution. Evaporate and incinerate carefully, allow to cool, weigh, and repeat the ignition for 15 min to constant mass.

Comment

When organic matter is incinerated as such, the residue found will depend on the temperature used. For instance, alkali chlorides and earth-alkali carbonates are volatile at certain temperatures. In the presence of sulphuric acid non-volatile sulphates are formed. Because pyrosulphates may be formed during the heating procedures, ammonium carbonate is added at the end. The sulphated ash is a measure for the total amount of inorganic matter in the plant material (Böhme and Hartke 1978). Only for *A. absinthium* demands exist with respect to sulphated ash, but the method is applicable for all species included in this chapter.

Ash Insoluble in Hydrochloric Acid

Assay

According to the European Pharmacopoeia 2nd edn (1993) the ash insoluble in hydrochloric acid is determined as follows. To the crucible containing the residue from the determination of sulphated ash, add 15 ml of water and 10 ml of hydrochloric acid (25% (m/v)). Cover with a watch glass, boil the mixture gently for 10 min and allow to cool. Filter through an ashless filter paper, wash the residue with hot water until the filtrate is neutral, dry, ignite to dull redness, allow to cool in a desiccator and weigh. Repeat until the difference between two consecutive weighings is not more than 1 mg.

Comment

The ash insoluble in hydrochloric acid is the residue obtained after extracting the sulphated ash with hydrochloric acid, calculated with reference to 100 g of crude drug. Non-volatile, inorganic impurities are determined, such as soil and sand. It is a control of the washing procedure of the crude plant material (Hartke and Mutschler 1987). No demands exist for any of the *Artemisia* species discussed in this chapter with respect to the ash insoluble in hydrochloric acid, but in fact it is a general procedure for crude plant material and is broadly applicable.

Bitter Value

A parameter to determine and compare the bitterness of a crude drug or an isolated compound is the so-called bitter value. The bitter value is defined as the reciprocal value of the lowest concentration of a drug that possesses a bitter taste. Thus, if the lowest concentration required for a bitter taste is 1 g drug in 10,000 ml of water, the bitter value is 10,000. 1 g plant material is extracted with 1000 ml of boiling water for 30 min. Subsequently, dilutions are made and evaluated for the presence or absence of a bitter taste. Quinine hydrochloride is used as a reference compound. This alkaloid has a bitter value of 200,000.

Storage Conditions

According to most pharmacopoeias, a crude drug should be stored in a well-closed container, protected from light. Powdered fragrant plant material rapidly loses its essential oil and will no longer comply to quality standards. In addition, the drying as well as the storage temperature are important. At higher temperatures (exceeding room temperature) the crude drug will lose its oil faster. Storage in plastic is best avoided because of a possible interaction of the essential oil with plastics.

ARTEMISIA ABROTANUM L.

The Plant, its Use and Principal Constituents

Artemisia abrotanum L., southernwood, is an aromatic-bitter plant, of which the dried leaves and blooming stem parts are used medicinally and culinary, in liqueurs and in the perfume industry. The drug is known as *Herba Abrotani*.

In folk medicine, southernwood is used because of its reputed emmenagogue, anthelmintic, antiseptic, antipyretic, appetite stimulant, spasmolytic and choleretic effects. In addition, it is used to alleviate bronchial disorders. For the choleretic activity the coumarins, isofraxidin, scopoletin and umbelliferone, are held responsible (Steinegger and Hänsel 1972). Flavonols, such as casticin, centaureidin and quercetin derivatives, possess weak spasmolytic activity and explain the plant's use in the treatment of bronchial diseases (Bergendorff and Sterner 1995). The herb has been shown to possess antimalarial activity *in vitro*. This activity is ascribed to isofraxidine and a bisabolol oxide derivative (Cubukcu *et al.*, 1990). The alkaloid abrotine (content: 2–3%) has has been shown to possess stimulant and antipyretic properties (Berger 1954). The herb contains about 0.45% of essential oil.

Description of the Drug

No description of the drug has been found in the literature.

Criteria for Quality

No criteria for the quality assurance of *A. abrotanum* have been recovered from the literature.

ARTEMISIA ABSINTHIUM L.

The Plant, its Use and Principal Constituents

Artemisia absinthium L., wormwood, is a very aromatic herbaceous plant, common in the Mediterranean area, in Europe, Asia and North Africa (Bruneton 1995). The crude drug, *Herba Absinthii*, consisting of the dried leaves and blooming tops of the plant, is imported from the Soviet Union, Bulgaria, the former Yugoslavia, Hungary and Poland (Wichtl 1989). It has a characteristic, penetrant aromatic odour. Its taste is aromatic and intensely bitter.

The principal medicinal use of wormwood is as a bitter tonic, diaphoretic, anthelmintic, antibacterial, antipyretic, emmenagogue and even schizonticide, but without clinical data to support this (Tyler 1993, 1994, Bruneton 1995). The plant is furthermore used to stimulate the appetite and to flavour some alcoholic beverages, including vermouth. The plant is also applied in perfume industry.

The plant's essential oil and bitter principles underlie its medicinal and commercial significance. Essential oil contents between 0.2 and 1.5% in the crude drug have been reported in the literature (Wichtl 1989). The time of harvesting is very important for the quality of the drug (Steinegger and Hänsel 1992). The maximal essential oil content is found just before blooming. More than 90 compounds have been identified in the oil (Nin *et al.*, 1995). The essential oil mainly consists of the monoterpene ketones α- and β-thujone (= (–)-thujone and (+)-thujone, respectively) and the corresponding alcohol thujol (Fig. 1). These so-called "thujones" may comprise as much as 35% of the total oil (Bruneton 1995). Furthermore, *trans*-sabinyl acetate, *cis*-epoxycymene and chrysanthenyl acetate can be abundant, depending of the origin of the plant material.

The oil is often blue-coloured due to the presence of the sesquiterpene chamazulene, which arises during the distillation through decomposition of sesquiterpenes, such as artabsin and absinthin. Oil without this artifact is dark green (Steinegger and Hänsel 1972).

A. absinthium contains 0.15–0.4% of bitter substances which belong to the class of sesquiterpene lactones. They include mainly absinthin (a dimer guaianolide

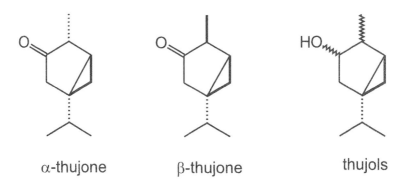

α-thujone   β-thujone   thujols

Figure 1   Thujones and thujols

structure) (Fig. 2) and artabsin (Fig. 3). Small amounts of matricin may be present. Anabsinthin is an artifact, formed after isomerisation of absinthin (Steinegger and Hänsel 1992). In full bloom, the amount of bitter substances is almost doubled (Schneider and Mielke 1979).

The plant also contains flavonoids, e.g. artemisitin, polyacetylenes, pelanolides (non-bitter sesquiterpenes) and phenylcarbolic acids (Wichtl, 1989).

Thujones are neurotoxic and cause absinthism. They affect the central nervous system, cause dizziness, convulsions, epileptiform seizures and delirium. The use of the drug is therefore considered as not safe. Ingestion of excessive doses over long periods may lead to digestive and urinary disorders. Although only relatively small amounts of thujones have been found in aqueous preparations (teas), these are best not used either (Tyler 1994, Bruneton 1995). *A. absinthium* is listed as an unsafe herb in a report of the American Food and Drug Administration (FDA) from 1975 (De Smet 1993).

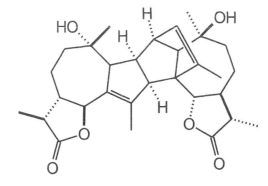

Absinthin

Figure 2   Absinthin

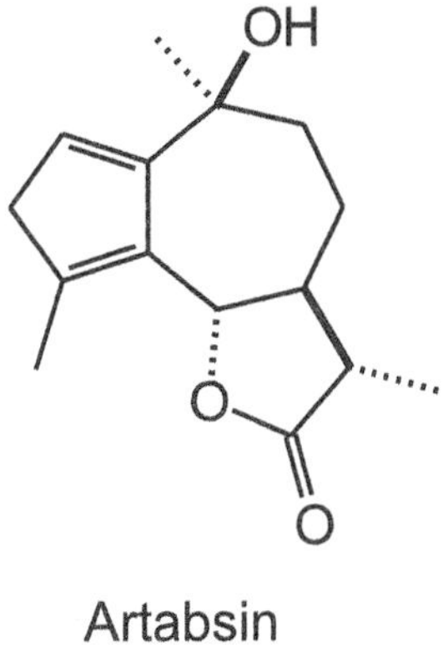

Artabsin

Figure 3   Artabsin

Formerly, *A. absinthium* was used in, among others, France, Switzerland, Belgium and Germany for the preparation of the strong alcoholic beverage absinth. The manufacture and consumption of this drink is currently banned because of the high toxicity of the thujones (Samuelsson 1992, Tyler 1994).

Description of the Drug

A. absinthium is a perennial herb, 60–120 cm tall, densely covered with hairs on the stems and leaves, giving the plant a greyish appearance (Steinegger and Hänsel 1972). The inflorescences are racemes of small globulous capitulums of yellow flowers. The leaves are highly divided and have a characteristic silvery colour (Bruneton 1995).

Microscopically, the presence of epidermal trichomes, which look like T-shaped hairs, is very characteristic. They occur on both sides of the leaves. The colour of the powdered drug is greyish-green (Stahl 1970, Wichtl 1989).

Criteria for Quality

Demands for the essential oil content are found in several (older) pharmacopoeias: the German Pharmacopoeia, "Deutsches Arzneibuch", DAB 9 and DAB 10 (>0.2%), the Swiss Pharmacopoeia, "Pharmacopoeia Helvetica", Ph. Helv. VI (>0.2%) and the Austrian Pharmacopoeia, "Österreichisches Arzneibuch", ÖAB 9 (>0.3%). According to Bruneton (1995) the official drug must contain not less than 0.3% and not more than 1.3% of oil.

Demands for the bitter value of the crude drug are given in various (older) pharmacopoeias: ÖAB 9 and the Pharmacopoeia of the former German Democratic Republic (East Germany), "Arzneibuch der Deutschen Demokratischen Republik", AB-DDR: >10,000, DAB 9 and DAB 10: >15,000, Ph. Helv. VI: >25,000. The bitter value of absinthin is 12,700,000 and of artabsin 486,000. Thus the ratio of the bitter values of absinthin and artabsin is 26:1. As the usual ratio of the contents of these sesquiterpenes in the plant material is 3:1 respectively, about 99% of bitter value comes from absinthin (Stahl 1970, Hartke and Mutschler 1987).

According to DAB 9 and DAB 10, herb of pharmaceutical quality may not contain more than 5% stems with a diameter >4 mm. Foreign matter may not exceed 2%. The sulphated ash may not be greater than 12%. Loss of weight upon drying (2 h at 100–105° C) may not exceed 10%.

Analytical Procedures

Identity of the plant material

The presence of azulenogenic bitter substances (artabsin and absinthin) is used as of proof of identity, according to DAB 9. Powdered plant material, 0.1 g, is extracted for 2 min in 10 ml of chloroform with frequent shaking at room temperature. After filtration, the volume of the extract is reduced to about 1 ml. After adding 5 ml dimethylaminobenzaldehyde reagent, the mixture is heated on a boiling water bath for 5 min. The solution turns blue-green. After treatment with acid, azulenes are formed that can react with dimethylaminobenzaldehyde yielding coloured products (Hartke and Mutschler 1987).

The identity can also be proved by thin layer chromatography (TLC). For the test solution, 0.1 g of powdered plant material is extracted for 2–3 min with 10 ml of methylene chloride and subsequently filtrated. The filtrate is concentrated to about half of its original volume. Because of the lability of the sesquiterpene lactones, especially of artabsin, the test solution should be prepared freshly and used immediately for the analysis. As a reference solution, 50 μl thujone, 2.0 mg methyl red, 2.0 mg phloroglucinol and 2.0 mg resorcinol are dissolved in 10 ml methanol.

On the silica gel plate, 30 μl of the test solution and 10 μl of the reference solution are applied as bands of 20 × 3 mm. The plate is developed over 10 cm, using a mixture of acetone-methylene chloride 1:9 as the mobile phase. After evaporation of the mobile phase, detection is achieved after spraying with anisaldehyde reagent (10 ml for a plate of 200 × 200 mm). The chromatogram is evaluated without heating after 2–3 min. The test solution shows, at about the height of methyl red of the reference solution, a blue-violet zone of artabsin. The colour gains in intensity after heating. Sometimes a blue-grey zone is seen under that of artabsin, originating from matricin. Subsequently, the plate is heated at 100–105° C for 5–10 min. A brown zone of absinthin appears, located between the yellow-orange zone of resorcinol (upper) and the zone of phloroglucinol (lower) of the reference solution. There should be a very clear red-violet zone in the chromatogram of the test solution due to hydroxypelanolide, located a little above the resorcinol zone of the reference solution. In the upper part of the chromatogram a red-violet zone of thujone is visible, which has under UV 365 nm a characteristic "brick" red colour. Higher in the chromatogram zones of esters of sabinyl and thujyl alcohol are visible. Furthermore, the chromatogram contains faint red, blue or blue-violet zones. In particular, the presence of hydroxypelenolide, one of the pelenolides, proves the identity of the plant material by TLC.

Bitter substances, sesquiterpene lactones

In addition to the indirect quantitative analysis of bitter substances, based on the determination of the bitter value, several other analytical procedures have been described in the literature. The presence of bitter substances in the crude drug can be shown histochemically. When the powder is treated with a few drops of 6 N HCl, a blue colour appears (Schneider and Mielke 1978, 1979).

The sesquiterpene lactones can be analysed spectrophotometrically, as a ferri-hydroxamate-complex. This is based on the principle that, in alkaline solution with hydroxylamine, esters and lactones yield hydroxamic acids, which form red to violet coloured complexes with ferric chloride (Schneider and Mielke 1979).

Bitter subtances can also be analysed by TLC according to DAB 8. A chloroform or light petrol extract is chromatographed on a silica gel plate with a mixture of methylene chloride-acetone 85:15 as the mobile phase. After spraying with 6 N HCl, the following spots are seen: absinthin, red, Rf = 0.31; matricin, blue, Rf = 0.51; artabsin, blue, Rf = 0.65.

Essential oil

See under "General procedures".

ARTEMISIA ANNUA L.

The Plant, its Use and Principal Constituents

Artemisia annua L., sweet wormwood, is an annual herbaceous plant with a strong fragrance. *A. annua* is of Asiatic and eastern European origin and has also become naturalized in North America. The plant is widely used in traditional herbal medicine in countries in South East Asia. *Herba Artemisiae annuae* or *Qing Hao* is contained in the Pharmacopoeia of the Peoples Republic of China (1988). The official drug consists of the dried aerial parts of *A. annua*. The drug is collected in autumn when in full blossom, divested of the older stems and dried in the shade.

The drug is used to relieve fever and to stop malarial attacks. These ethnopharmacological applications led to the isolation of the sesquiterpene lactone endoperoxide artemisinin (also named qinghaosu) (Fig. 4). Nowadays, artemisinin and semi-synthetic derivatives (dihydroartemisinin, artemether, arteether, artelinate and artesunate) are being developed as a novel class of antimalarial drugs, with activity against multidrug-resistant strains of *Plasmodium falciparum* and cerebral malaria (Klayman 1985, Woerdenbag *et al.*, 1990, 1994a, Zhu and Woerdenbag 1995).

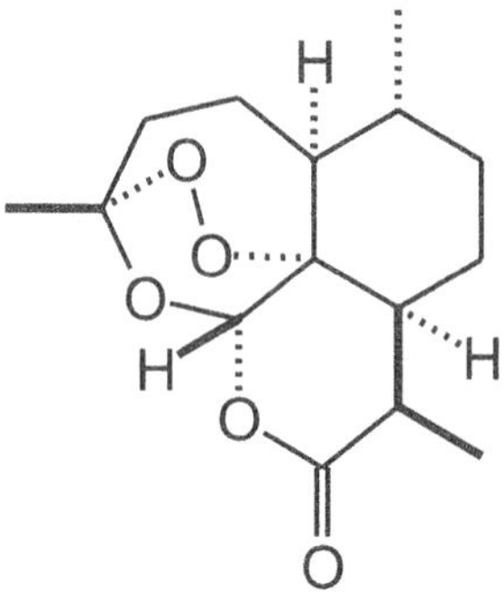

artemisinin

Figure 4   Artemisinin

The total organic synthesis of artemisinin has been achieved, but is very complicated and only very low yields are obtained that are not commercially attractive. Therefore, the plant is still the only valid source for artemisinin (Woerdenbag *et al.*, 1990, 1994a).

In the plant, artemisinin is biosynthesised *via* artemisinic acid (see chapter 12). Two new biosynthetic intermediates, dihydroartemisinic acid and its hydroperoxide derivative have been identified in *A. annua* plants of Vietnamese origin (Wallaart *et al.*, 1999). Very recently the occurrence of chemotypes within the species *A. annua* has been described; a chemotype with a high artemisinin level was found to have a low artemisinic acid level, while conversely, a chemotype with low levels of artemisinin contained a high artemisinic acid level (Wallaart *et al.*, 2000).

A. annua is also valued for its essential oil of which the characteristic sweet aroma has been described as grassy, fresh, and bitter with a camphoraceous nuance. The commercial significance of the oil is limited, but it is sometimes used in fragrances, in perfumery and cosmetic products. The oil has been reported to possess antimycotic and antimicrobial activities (Woerdenbag *et al.*, 1993a). Contents of up to 4% in the dried leaves have been reported. Depending on the variety, artemisia ketone or camphor are the principal constituents of the oil. Further important constituents are borneol, 1,8-cineole, pinene, carvacrol, thymol, caryophyllene and caryophyllene oxide (Woerdenbag *et al.*, 1990).

Description of the Drug

In the Pharmacopoeia of the People's Republic of China (1988) the drug is described as follows. The stems are cylindrical, much branched above, 30–80 cm long, 2–6 mm in diameter. Externally they are yellowish-green or brownish yellow, ribbed longitudinally. The texture is slightly hard, the fracture short. The fractured surface exhibits pith in the central part. Leaves are alternate, dark green or brownish-green, rolled and crumpled, easily broken. When spread out the entire ones are 3-pinnately parted. Segments and lobes are oblong or long-elliptical. They are pubescent on both surfaces. According to Bruneton (1995) the plant is furthermore characterized by large panicles of small globulous capitulums, 2–3 mm in diameter, with a whitish involucre. The leaves disappear after the blooming period.

Criteria for Quality

Despite its inclusion in the Pharmacopoeia of the People's Republic of China (1988), no standards are given with respect to quality assurance of *A. annua*, except that foreign matter should be eliminated. No further criteria for the quality of *A. annua* have been found in the literature. Usually, contents of about 0.1% artemisinin have been reported in the dried herb. For commercial purposes, however, contents as high as about 0.5% are desired. Via selection procedures such high producing plants may be obtained (Woerdenbag *et al.*, 1994a).

Analytical Procedures

Several analytical methods to assay artemisinin have been described in the literature. They include thin layer chromatography (TLC), high performance liquid chromatography (HPLC), gas chromatography (GC), gas chromatography coupled with mass spectrometry (GC-MS), and immunoassays. In addition, analytical procedures for biosynthetic precursors of artemisinin as well as metabolites in biological fluids have been developed.

Thin layer chromatography

Artemisinin and related sesquiterpenes in *A. annua* can be analysed by means of TLC on silica gel plates. Various eluent mixtures have yielded a successful separation of the compounds: petroleum ether-ethyl acetate 1:1, ethyl acetate-chloroform 7.5:92.5, hexane-acetone 87:13, cyclohexane-diethyl ether 1:1, petroleum ether (60–90° C)-diethyl ether 1:2, petroleum ether (40–60° C)-diethyl ether 1:1. Spots are visualized by spraying with a 2%-phosphoric acid solution, a 2%-vanillin solution, anisaldehyde reagent, *p*-dimethylaminobenzaldehyde reagent or by exposure to iodine vapour (Tu *et al.*, 1982, Klayman *et al.*, 1984, Niu *et al.*, 1985, Rücker *et al.*, 1986, Kudakasseril *et al.*, 1987, Roth and Acton 1989, Pras *et al.*, 1991). TLC – densitometry can be used for quantitative purposes (Gupta *et al.*, 1996).

In our laboratory, the following TLC method is applied for the analysis of artemisinin and can be used for semi-quantitative purposes. Plant material is extracted with toluene. With this solvent, artemisinin and related sesquiterpenes (biosynthetic precursors) are extracted rapidly and completely. Of this extract 1 μl is applied on GF254 silica gel plates (Merck, Darmstadt, Germany). The plates are developed with petroleum ether (40–60° C)-diethyl ether 1:1 as the mobile phase. The spot of artemisinin is visualized after 10 s immersion in a freshly prepared mixture of 50 ml glacial acetic acid, 1 ml concentrated sulphuric acid and 0.5 ml anisaldehyde (anisaldehyde reagent), followed by 10 min heating at 105° C. Under daylight artemisinin (Rf = 0.40–0.45) is purple-red coloured; under UV 366 nm it is fluorescent orange. Pure artemisinin as reference is commercially available (Pras *et al.*, 1991).

High performance liquid chromatography

Artemisinin has no chromophoric group and only absorbs at the low end of the UV spectrum, below 220 nm. In order to obtain a product with more specific and sensitive spectrometric characteristics, Zhao and Zeng (1985) converted artemisinin into a product named Q260. First, artemisinin is treated with alkali and converted into Q292, possessing a UV absorption at 292. Upon acidification, Q292 yields Q260, which has a strong UV absorption at 260 nm and is more stable than Q292. The reaction scheme is given in Fig. 5. Q260 can be separated by means of HPLC on a reversed phase C18 column, with phosphate buffer-methanol as the mobile phase. Artemisinin can also be analysed following a comparable postcolumn derivatization (ElSohly *et al.*, 1987).

In addition to hydrolysis in alkaline or acidic solution, the peroxide bridge of artemisinin can undergo electrochemical reduction. Based on this reaction, an HPLC detection procedure has been developed by Acton *et al.* (1985), that can be used for a sensitive, selective and rapid assay of artemisinin in crude plant material. Artemisinin is detected at –0.8 V. Because molecular oxygen is reduced as well at this high cathodic potential, special precautions are required in order to deoxygenate the HPLC system, including the samples to be injected (Theoharides *et al.*, 1989). Hence, this method of detection may be less appropriate for routine analysis.

Artemisinin also gives a high response on HPLC with polarographic detection, due to the endoperoxide moiety (Zhang and Xu 1985, Zhou *et al.*, 1988).

Charles *et al.* (1990) presented a modification to the earlier described electrochemical detection of artemisinin. Their modification consisted of the use of a glassy carbon, instead of a gold/mercury electrode, without EDTA in the mobile phase, and also gave a good sensitivity for the drug.

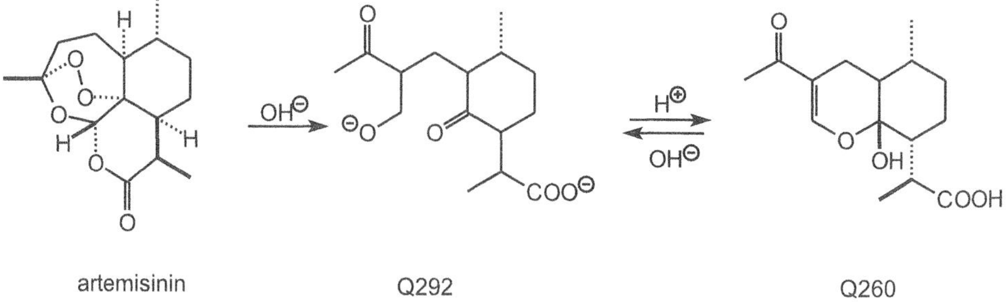

Figure 5   Derivatization of artemisinin

The closely related sesquiterpene lactone endoperoxide artemisitene, that may co-occur with artemisinin in the plant, can be assayed simply and selectively using reversed-phase HPLC with ultraviolet (UV) detection at 216 nm (El-Domiaty *et al.*, 1991). A general HPLC method with UV detection for the determination of γ-lactones, including artemisinin, in plant material has been described by Rey *et al.* (1992).

Ferreira *et al.* (1994) presented an improved method for HPLC with electrochemical detection of artemisinin. This method was compared with a gas chromatographic method, which measured artemisinin indirectly through two of its degradation products. Both methods provided a good separation of artemisinin from artemisitene, arteannuin B, and dihydroartemisinin. HPLC with electrochemical detection was about 10-fold more sensitive for artemisinin than gas chromatography, and therefore more accurate for analysing samples containing low levels of artemisinin. Compounds lacking the peroxide moiety could only be analysed by gas chromatography.

An HPLC method, with combined electrochemical and UV detection at 228 nm has been described for the simultaneous detection of artemisinin and its biosynthetic precursors, arteannuic acid, arteannuin B and artemisitene, in plant material (Vandenberghe *et al.*, 1995).

Maillard *et al.* (1993) proved that coupling liquid chromatography with thermospray mass spectrometry offers an excellent method for the qualitative analysis of natural products lacking a suitable chromophore for UV detection, including artemisinin. This technique was shown to be valuable in cases where very little material is available. Total ion current traces and mass chromatograms allowed the easy detection of these compounds in extracts.

In our laboratory the following HPLC procedure with precolumn derivatization for the analysis of artemisinin is applied routinely. It is the method described by Zhao and Zeng (1985), with some modifications and adaptations (Pras *et al.*, 1991, Woerdenbag *et al.*, 1992, 1993b, 1994b). About 250 mg of the powdered plant material, accurately weighed, is extracted with 10.0 ml of toluene in an ultrasonic bath for 30 min. After filtration, 500 μl of the toluene extract is evaporated to dryness and the residue redissolved in 200 μl methanol. Then, 800 μl of a 0.2% (w/v) sodium hydroxide solution is added. The mixture is vortexed and heated for 30 min in a water bath at 50° C. After cooling, 200 μl methanol and 800 μl 0.05 M acetic acid is added. The hydrolysis product of artemisinin, Q260, is assayed on a (reversed phase) Chromsep Microspher C18 column, 100 × 3 mm i.d., equipped with a guard column (Chrompack, Middelburg, The Netherlands). The mobile phase used is 0.01 M phosphate buffer-methanol 55:45, pH 7.0, at a flow rate of 1 ml/min. The artemisinin derivative elutes from the column after about 6 min and is detected at 260 nm.

Gas chromatography

Gas chromatography has also been applied for the analysis of artemisinin. However, artemisinin is a thermolabile compound and decomposes on the column.

Sipahimalani *et al.* (1991) presented a method for the detection and determination of artemisinin in plants and plant tissue cultures, in which a linear relationship was obtained between the concentration of artemisinin and the respective peak areas for either of its two thermally degradated products. In that particular study, no complete separation of arteannuin B from one of the decomposition products of artemisinin was afforded.

We have developed a gas chromatographic method, used to analyse artemisinin as well as its biosynthetic precursors, artemisinic acid, arteannuin B and artemisitene (Woerdenbag *et al.*, 1991a). The method was validated using mass spectroscopy coupled with the gas chromatograph. The identity of the sesquiterpenes was confirmed by comparing the mass spectra of the separated compounds eluting from the gas chromatograph with those of solid probes of reference standards. Here, artemisinin was partly measured as its pyrolysis products, but also a peak of artemisinin itself was present. Nevertheless, a reliable standard curve was obtained, since a good linearity was found between concentration and total area of the artemisinin peaks and of the peaks of the pyrolysis products. All the sesquiterpenes could be detected at nanogram levels.

Immunoassays

Radioimmunoassays have been also been applied to the detection of artemisinin (Song *et al.*, 1985).

A highly specific and sensitive ELISA method has been developed for the detection and semi-quantitative analysis of artemisinin and structurally related compounds in crude extracts of *A. annua*. The antibodies were raised in rabbits, using a 10-succinyldihydroartemisinin-BSA conjugate as the antigen. The peroxide linkage in the artemisinin molecule appeared to be critical in determining the antibody specificity. The working range was from 0.02 to 10 ng per assay (Jaziri *et al.*, 1993).

Ferreira and Janick (1996) developed an immunoquantitative analysis of artemisinin, using polyclonal antibodies. Artemisinin was derivatized to dihydoartemisinin carboxymethylether and linked to thyroglobulin or bovine serum albumin. After injection of these conjugates into rabbits, polyclonal antibodies were generated. An ELISA was developed, by which artemisinin could be detected at nanogram levels in plant material. This method was shown to be about 400-fold more sensitive than HPLC with electrochemical detection.

Other analytical methods

Caniato *et al.* (1989) presented a method for the detection of peroxides (such as artemisinin) in intact plant tissues. Upon treatment of the tissue with a methanolic solution of ferrous thiocyanate, a red brownish colour develops (modified Abraham's reaction). This technique was stated to be useful for a rapid screening of compounds of this type.

ARTEMISIA ARGYI LEVL. ET VANT.

The Plant, its Use and Principal Constituents

Artemisia argyi Levl. et Vant., argy wormwood, is used in traditional Chinese medicine. The Pharmacopoeia of the People's Republic of China (1988) includes a monograph on *Folium Artemisiae Argyi*, in China known as *Aiye*. The drug consists of the dried leaves of the plant. The leaves are collected in the summer during the late flowering period, removed from foreign matter and dried in the sun.

The leaves are carbonized and processed with vinegar before they are used medicinally. They are stir-fried until the outer part is charred, then sprayed with vinegar and stir-fried to dryness. It is recommended for use as an analgesic and hemostatic. Indications are pain in the lower abdomen, menstrual disorders, infertility, spitting of blood, epistaxis, uterine bleeding in pregnancy, excessive menstrual flow and external as an antipruritic. The drug is also used for moxibustion (Tang and Eisenbrand 1992).

Leaves of *A. argyi* contain essential oil with *trans*-carveol, α-terpineol, 4-terpineol, carvone, elemol, α-cedrene, α-phellandrene, camphene, bornyl acetate, isoborneol and carvone as principal components. Furthermore, flavonoids, sesquiterpene lactones (bitter substances) and triterpenes are present (Tang and Eisenbrand 1992).

Description of the Drug

According to the Chinese Pharmacopoeia (1988) the leaves are mostly crumpled, broken and short-petioled. The entire leaves are ovate-elliptical, pinnatipartite, with segments elliptical-lanceolate and margins irregularly dentate. The upper surface is greyish-green or deep yellow-green, sparsely pubescent and glandular-punctate. The lower surface is densely greyish-white tomentose; the texture is soft. The odour is fragrant, the taste is bitter.

The powder is greenish-brown. Microscopically, two types of non-glandular hairs are seen. One is T-shaped with an elongated and bent apical cell, unequally 2-armed and a 2–4 celled stalk. The other is uniseriate, 3–5 celled, with a very long and

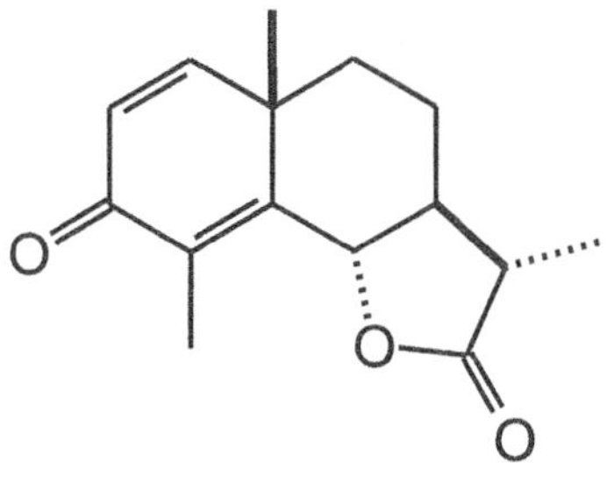

Figure 6   Santonin

twisted apical cell, generally broken. Glandular hairs are present, which consist of 4 or 6 opposite and overlapped cells. They are devoid of a stalk. Clusters of calcium oxalate, 3–73 gmm in diameter, are found in the mesophyll.

Criteria for Quality

Except that foreign matter and petioles have to be removed, no concrete criteria for quality assurance are included in the Chinese Pharmacopoeia and none have been recovered from other literature.

ARTEMISIA CINA O.C. BERG ET C.F. SCHMIDT

The Plant, its Use and Principal Constituents

Artemisia cina O.C. Berg et C.F. Schmidt, wormseed, is native to the steppe areas East of the Caspian Sea, in Afghanistan and in the Southern Ural region. The aromatic odour is characteristic. The taste is camphoraceous, unpleasant-aromatic, bitter and cooling (Evans 1989). The unexpanded flowerheads have been used medicinally because of the anhelmintic action of santonin (Fig. 6). The drug is known as *Flores Cinae*. Santonin is, however, obsolete as an anthelmintic because of its high toxicity to the host (Woerdenbag *et al.*, 1997). Furthermore, the seed has been used medicinally under the name *Semen Cinae*.

The crude drug contains 1–2% of essential oil, consisting mainly of cineole (about 80%), next to α-terpineol, terpineol, carvacrol, sesquiartemisol, and several monoterpene and sesquiterpene hydrocarbons. The plant material contains 2–3% of sesquiterpene lactones, santonin and related eudesmanolides, such as artemisin (α-hydroxy-santonin) (Bruneton 1995, Woerdenbag *et al.*, 1997).

Description of the Drug

The drug consists of oval, yellowish or brownish flowerheads, about 1.5–4 mm in length. The flowerhead consists of an involucre of about 16 bracts, which completely enclose 2–5 more or less immature tubular flowers (Evans 1989). Microscopically, the involucre is characterized by yellowish, multicellular glandular hairs. In the parenchyma of the involucre small calcium oxalate crystals are present (Hänsel *et al.*, 1992).

Criteria for Quality

The Austrian Pharmacopoeia (ÖAB9) demands not less than 2.5% santonin and at least 1% of essential oil.

Analytical Procedures

Identity of the plant material

The identity of *A. cina* can be proved by TLC. Stahl (1981) described the following procedure. To obtain the test solution, 0.1 g of powdered plant material, is extracted with 1 ml of chloroform for 10 min while shaking several times followed by filtration (cotton plug). As a reference solution 10 mg santonin and 10 μl 1,8-cineole are dissolved in 1 ml of toluene. 5 or 10 μl of the test solution is applied to a silica gel plate and 1 μl of the reference solution. The plate is developed over a path of 10 cm, using a mixture of chloroform – acetone 95:5 as the mobile phase. The chromatogram is evaluated under UV 254 nm, revealing several spots. Subsequently the plate is sprayed with phosphomolybdic acid reagent (10.0 g phosphomolybdic acid dissolved in 100 ml of ethanol) and heated for 10 min at 105° C. Santonin gives a blue spot, Rf = 0.45–0.55 and 1,8-cineole a red-blue spot, Rf = 0.70–0.75.

Picman *et al.* (1980) describe the following procedure. A chloroform extract is chromatographed on silica gel plates in an unsaturated chamber using a mixture of chloroform-acetone (6:1) as the mobile phase. After drying at room temperature, the plate is sprayed with a freshly prepared vanillin or p-dimethylaminobenzaldehyde reagent (0.5 g of vanillin or p-dimethylaminobenzaldehyde respectively, dissolved in a mixture of 9 ml ethanol (95%), 0.5 ml concentrated sulfuric acid and three drops of acetic acid) and immediately placed on a hot plate 70° C. The colours of the spots are evaluated after 10 min and after 24 h. After spraying with vanillin reagent, santonin (Rf = 0.81) yields a grey brown colour after 10 min, which turns greenish after 24 h. After spraying with p-dimethylaminobenzaldehyde reagent, santonin yields an orange-brown spot.

Quantitative determination of santonin

Several quantitative determinations for santonin exist but, probably because the drug has been abandoned nowadays, they are old-fashioned and time-consuming. In the 4th edn of *Hagers Handbuch* (List and Hörhammer 1967) the following methods are listed. After separation using paper chromatography, santonin is coloured after a reaction with a 30% sodium methylate solution. This technique has been applied semi-quantitatively. Santonin has been determined spectrophotometrically with dinitrophenylhydrazine. A titrimetric method has also been applied. The lactone ring is opened with an excess of alkali and the excess is titrated with hydrochloric acid. Finally, the content of santonin has been determined gravimetrically (procedure described in DAB 6). Of the plant material 10 g is extracted with 100 ml benzene (!) for 30 min with regular shaking. After filtration, 80 ml is evaporated to dryness and the residue taken up in 15 g of absolute ethanol and 85 g of water. The mixture is refuxed for 15 min, after which the solution is filtered. Subsequently, 0.1 g of diatomaceous earth is added and the mixture refluxed for another 15 min. After filtration, the solution is allowed to stand for 24 h at about 15° C in the dark. Santonin crystallizes and is isolated from the solution and determined gravimetrically.

Nowadays, the application of an high-performace liquid chromatographic (HPLC) method is more likely, as stated in the 5th edn of *Hagers Handbuch* (Hänsel *et al.*, 1992). In our laboratory, sesquiterpene lactones are routinely analysed using gradient HPLC on a reversed phase C18 column. A linear gradient, over 35 min, of 20–80% acetonitrile in water, containing 1% phosphoric acid, is applied as the mobile phase. Detection is done at 220 nm. (Woerdenbag *et al.*, 1991b). Reference santonin is commercially available.

ARTEMISIA DRACUNCULUS L.

The Plant, its Use and Principal Constituents

Artemisia dracunculus L., tarragon, is an herbaceous plant. Two chemical races of tarragon are distinguished. Plants originating mainly from France have yielded 0.15–3% essential oil, with methylchavicol (estragol) (Fig. 7) as the principal constituent (68–80%). Further important constituents of this oil are *cis*- and *trans*-ocimene (6–12%) and limonene (2–6%) (Zani *et al.*, 1991). This oil has an aniseed-like taste. Plants from German or Russian provenance contain 0.25–2% essential oil with sabinene, elemicine and *trans*-isoelemicin as the principal components (Albasini *et al.*, 1983, Evans 1989). This oil has a bitter chervil-like taste.

The herb is principally used as a seasoning for food and as a source of essential oil. Medicinally, aerial parts of tarragon are used to treat symptoms of various digestive ailments and as an adjunctive therapy for the painful component of spasmodic colitis (Bruneton 1995). Further medicinal properties that are claimed for the plant include antiseptic, antipyretic, emmenagogue, anthelmintic, diuretic and calming effects (Zani *et al.*, 1991).

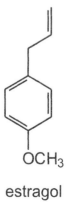

estragol

Figure 7   Estragol

Description of the Drug

No description of the drug has been found in the literature.

Criteria for Quality

No criteria for quality assurance of tarragon have been recovered from the literature.

Analytical Procedures

Methylchavicol can be determined spectrophotometrically with Millon's reagent, according to Thieme and Ngyuen Thi Tam (1972).

ARTEMISIA MARITIMA L.

The Plant, its Use and Principal Constituents

Artemisia maritima L., sea mugwort, is distributed all over the northern hemisphere of the Old World. The odour of the plant is strong and aromatic, the taste bitter and camphor-like. Dried flowering tops have been used as a source of santonin.

Sea mugwort contains 0.25–1.45% santonin, along with the sesquiterpene lactones pseudosantonin and alantolactone. The essential oil content is 0.15–0.75%, with thujone, borneol, cineole, pinene, camphor and sabinol as the main constituents (Hoppe 1975). The oil has an odour comparable to cajaput oil and camphor (Kapoor 1990).

Sea mugwort has been used extensively as anthelmintic in coastal areas of Europe. This use is very likely to be based on the presence of santonin. Further medicinal claims include antibacterial, fungistatic, antidiabetic, spasmolytic, carminative, antipyretic and abortifacient properties. It has also been used to treat jaundice and as a liver protective (Jambaz and Gilani 1995).

Description of the Drug

A. maritima is a shrub with a stout, branched rootstock. Stems are up to about 1 m high, slender, much branched from the bases and striate. Leaves are 2.5–4 cm, often quite white, 2-pinnatisect, with numerous segments. They are small, spreading, linear and obtuse. Upper leaves are simple and linear. Flowerheads are homogamous, numerous, ellipsoid, oblong or ovoid, about 2.5 mm long and 3–10 flowered in spike-shaped clusters in the axil of a small linear leaf. Receptacles are naked (Kapoor 1990).

Criteria for Quality

No criteria for quality assurance of *A. maritima* have been recovered from the literature.

ARTEMISIA PONTICA L.

The Plant, its Use and Principal Constituents

Artemisia pontica L. is a shrub-like plant, 30–45 cm high. *A. pontica* is closely related to *A. absinthium*, but in everything, smaller. *A. pontica* is more aromatic and less bitter than *A. absinthium*. The plant occurs wild in south Europe. It is used instead of *A. absinthium* to prepare vermouth. It contains bitter substances and 0.25% of essential oil. The oil contains iso-artemisiaketone, α-thujone, 1,8-cineole and *trans*-β-farnesene (Hurabielle *et al.*, 1977). It is stated by Steinegger and Hänsel (1972) that, in contrast to *A. absinthium*, *A. pontica* does not cause neurotoxicity. Because of the presence of thujone, however, this is unlikely.

Description of the Drug

No description has been found in the literature.

Criteria for Quality

No criteria for quality assurance of *A. pontica* have been recovered from the literature.

ARTEMISIA SCOPARIA WALDST. ET KIT./*ARTEMISIA CAPILLARIS* THUNB.

The Plant, its Use and Principal Constituents

In the Chinese Pharmacopoeia (1988) *Herba Artemisiae scopariae* or *Yinchen* is officially listed as a drug. It consists of the dried seedlings of *Artemisia scoparia* Waldst. et Kit. and *Artemisia capillaris* Thunb., virginate wormwood, collected in spring when they are 6–10 cm high.

The drug is used in traditional Chinese medicine to treat infectious icteric hepatitis and sores with exudation and itching. It has choleretic and anti-inflammatory properties.

A. capillaris contains essential oil with polyacetylene derivatives, such as capillene, capillone and capillin, mono- and sesquiterpenes (α- and β-pinene, ρ-cymene, α-terpineol, bornyl acetate, methyleugenol, β-elemene and β-caryophyllene. Furthermore, flavonoids and coumarins have been identified in this species.

A. scoparia contains essential oil with α- and β-pinene, myrcene, cineole, ρ-cymol, carvone, thujone, apiole, isoeugenol, cadinene, caryophyllene epoxide and vanillin. In addition, flavonoids and coumarins are present (Tang and Eisenbrand 1992).

Description of the Drug

The herb is mostly rolled into a loose mass, which is greyish-white or greyish-green, white-woolly throughout and soft like a velvet. The stems are thin, 1.5–2.5 cm long, with a diameter of 1–2 mm. After removal of the white hairs from the surface, some obvious longitudinal striations are visible. The texture is brittle, the fracture short. Leaves are petioled. Spread out, they are 1–3-pinnatisect, 1–3 cm long and 1 cm broad. Segments are ovate or slightly oblanceolate and linear, the apex is acute. The odour of the drug is fragrant, its taste bitterish.

Criteria for Quality

Older stems, roots and foreign matter must be removed according to the Chinese Pharmacopoeia (1988). No further criteria for quality assurance of have been recovered from the literature.

Analytical Procedures

According to the Chinese Pharmacopoeia (1988) the drug is identified as follows. 1 g of the powdered drug is heated under reflux with 20 ml ethanol for 30 min in a waterbath. The filtrate has a light yellowish-green colour. Under UV 365 nm a purple-red fluorescence is seen.

ARTEMISIA VULGARIS L.

The Plant, its Use and Principal Constituents

Artemisia vulgaris L., mugwort, is a very common species in Europe, Asia and North America. The drug, known as *Herba Artemisiae*, consists of the leaves and flowering tops of the plant. Mugwort has a faint bitter taste, that almost completely disappears upon drying.

A. *vulgaris* contains 0.03–0.3% essential oil, in which more than 100 components have been identified. The composition is highly variable. The main constituents depend of the origin of the plant, and are 1,8-cineole, camphor and linalool. Furthermore the sesquiterpenes vulgarin (=tauremisin), psilostachyin and psilo-stachyin C are present. Mugwort contains no sesquiterpene lactones. The plant contains flavonol glycosides, coumarins, polyacetylenes, triterpenes and carotinoids (Wichtl 1989). Thujones are only present in traces or are even absent.

The plant is traditionally used to stimulate the appetite and to relieve "painful periods". Use of extensive doses over long periods may lead to digestive and urinary disorders (Bruneton 1995). Its medicinal usefulness, however, is not documented adequately (De Smet 1993).

Description of the Drug

It is an aromatic perennial, shrub like plant, 50–150 cm high, pubescent or tormentose. Stems are branched, red-violet, ribbed, erect and rough. They contain pith. The upper leaves are smaller than the lower. They are 3-lobed or entire, lanceolate. Lower leaves are 5–10 cm long, ovate, pinnately lobed, hairy on both surfaces. The upperside of the leaves bears no hairs and is dark green. The underside of the leaves is silver-grey with a felty-like appearance, due to the presence of hairs. This is an important difference with *A. absinthii*, which has hairs on both sides of the leaves. The inflorescences of mugwort are racemes of small globulous capitulae of yellow flowers. Flowerheads are small, egg-shaped, reddish, yellow, or whitish, drooping, in branched woolly spikes, heterogamous, minute. It is a very variable plant regarding the leaves and flowerheads (Wichtl 1989, Kapoor 1990, Bruneton 1995).

Because of the presence of many T-shaped hairs, the drug occurs in small "clumps". In the lower epidermis stomata of the anomocytic-type are found (Kartnig and Brantner 1992).

Criteria for Quality

The French Pharmacopoeia requires the verification of the absence of thujone in a hexane extract of the plant (Bruneton 1995). This distinguises it from another European species, *A. verlotorum* Lamotte, with which it may be confused. *A. verlotorum* does contain thujone.

According to Bruneton (1995) the official drug should contain >0.1% of essential oil.

The bitter value of *A. vulgaris* is lower than of *A. absithium* (Steinegger and Hänsel 1972).

Analytical Procedures

Vulgarool (Fig. 8) is a unique constituent of *A. vulgaris*, and can be used to identify the plant material. Vulgarool can be determined by TLC. The essential oil of mugwort is chromatographed on silica with a mixture of chloroform-methanol 98:2

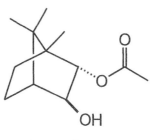

vulgarool

Figure 8   Vulgarool

as the mobile phase. Vulgarool has an Rf value of 0.45. After spraying with vanillin-sulfuric acid, a lilac colour is obtained and with Dragendorff's reagent it turns yellowish white (Nano *et al.*, 1976).

For the identification of *A. vulgaris* a TLC method has been described by Kartnig and Brantner (1992). This method yields a "fingerprint" which is compared with that of a reference sample. The test solution is obtained by extracting 1.0 g of plant material with 10.0 ml of ethanol for 10 min at 60° C under refux and stirring. After filtration, the solvent is evaporated and the residue taken up in 0.5 ml ethanol. As a reference, a methanolic solution of rutin, isoquercitrin and quercitrin is used. 10 μl of the test solution is applied as a band on silica gel plates. The plates are developed over 10 cm, with ethyl acetate-formic acid-water (88:6:6) as the mobile phase. Detection is obtained by spraying with Neu's reagent and PEG 4000. The chromatogram is evaluated under UV 366 nm. Rutin has an Rf-value of 0.1, isoquercitrin of 0.34 and quercitrin of 0.54.

REFERENCES

Acton, N., Klayman, D.L. and Rollman, I.J. (1985) Reductive electrochemical HPLC assay for artemisinin (qinghaosu). *Planta Med.*, **51**, 445–446.

Adams, R.P. (1989) *Identification of Essential Oils by Ion Trap Mass Spectroscopy*, Academic Press, Inc., San Diego, New York, Berkeley, Boston, London, Sydney, Tokyo, Toronto.

Albasini, A., Bianchi, A., Melegari, M., Vampa, G., Pecorari, P. and Rinadi, M. (1983) Indagini su pianti di *Artemisia dracunculus* L. s.l. (Estrogone) *Fitoterapia*, **54**, 229–235.

Bergendorff, O. and Sterner, O. (1995) Spasmolytic flavonols from *Artemisia abrotanum*. *Planta Med.*, **61**, 370–371.

Berger, F. (1954) *Handbuch der Drogenkunde: Erkennung, Wertbestimmung und Anwendung, Band IV: Herbae*, Maudrich, Vienna.

Böhme, H. and Hartke, K. (1978) *Europäisches Arzneibuch. Band I und Band II. Kommentar*, Wissenschaftliche Verlagsgesellschaft mbH, Stuttgart, Govi-Verlag GmbH, Frankfurt.

Bos, R., Woerdenbag, H.J., Hendriks, H., Sidik, Wikström, H.V. and Scheffer, J.J.C. (1996) The essential oil and valepotriates from roots of *Valeriana javanica* Blume grown in Indonesia. *Flavour Fragr. J.*, **11**, 321–326.

Bruneton, J. (1995) *Pharmacognosy, Phytochemistry, Medicinal Plants*, Lavoisier Publishing, Paris.

Caniato, R., Filippini, R., Cappelletti, E.M. and Appendino, G. (1989). Detection of peroxides in intact plant material. *Fitoterapia*, **60**, 49–51.

Charles, D.J., Simon, J.E., Wood, K.V. and Heinstein, P. (1990) Germplasm variation in artemisinin content of *Artemisia annua* using an alternative method of artemisinin analysis from crude plant extracts.20*J. Nat. Prod.*, **53**, 157–160.

Cubukcu, B., Bray, D.H., Warhurst, D.C., Mericli, A.H., Ozhatay, N. and Sariyar, G. (1990) *In vitro* antimalarial activity of crude extacts and compounds from *Artemisia abrotanum*20L. *Phytother. Res.*, **4**, 203–204.

De Smet, P.A.G.M. (1993) Legislatory outlook on the safety of herbal remedies. In *Adverse Effects of Herbal Drugs*, Vol. 2, P.A.G.M. De Smet, K. Keller, R. Hänsel and R.F. Chandler, (eds.), Springer-Verlag, Berlin, Heidelberg, p. 1–90.

El-Domiaty, M.M., Al-Meshal, I.A. and El-Feraly, F.S. (1991) Reversed-phase high-performance liquid chromatographic determination of artemisitene in artemisinin. *J. Liq. Chromatogr.*, 14, 2317–2330.

ElSohly, H.N., Croom, E.M. and ElSohly M.A. (1987) Analysis of the antimalarial sesquiterpene artemisinin in *Artemisia annua* by high-performance liquid chromatography (HPLC) with postcolumn derivatization and ultraviolet detection. *Pharm. Res.*, 4, 258–260.

European Pharmacopoeia, 2nd edn (1993), Council of Europe, Strasbourg.

Evans, W.C. (1989) *Trease and Evans' Pharmacognosy*, 13th edn., Baillière Tindall, London.

Evans, W.C. (1996) *Trease and Evans' Pharmacognosy*, 14th edn., W.B. Saunders Company Ltd., London.

Ferreira, J.F.S., Charles, D.J., Wood, K., Janick, J. and Simon, J.E. (1994) A comparison of chromatography and high performance liquid chromatography for artemisinin analysis. *Phytochem. Anal.*, 5, 116–120.

Ferreira, J.F.S. and Janick, J. (1996) Immunoquantitative analysis of artemisinin from *Artemisia annua* using polyclonal antibodies. *Phytochemistry*, 41, 97–104.

Frohne, D. and Jensen, U. (1992) *Systematik des Pflanzenreichs*, 4. Auflage, Gustav Fischer Verlag, Stuttgart, Jena, New York, p. 199.

German Pharmacopoeia, 'Deutsches Arzneibuch', 9. Auflage (1986) Deutscher Apotheker Verlag, Stuttgart; Govi-Verlag GmbH, Frankfurt.

German Pharmacopoeia, 'Deutsches Arzneibuch', 10. Auflage (1993) Deutscher Apotheker Verlag, Stuttgart; Govi-Verlag GmbH, Frankfurt.

Gupta, M.M., Jain, D.C., Verma, R.K. and Gupta, A.P. (1996) A rapid analytical method for the determination of artemisinin in *Artemisia annua*. *J. Med. Aromat. Plant Sci.*, 18, 5–6.

Hänsel, R., Keller, K., Rimpler, H. and Schneider, G. (eds.) (1992) *Hagers Handbuch der Pharmazeutischen Praxis*, 5. Auflage, Vierter Band, Springer-Verlag, Berlin, p. 368–370.

Hartke, K. and Mutschler, E. (1987) *DAB 9 – Kommentar*, Band 2, Wissenschaftliche Verlagsgesellschaft mbH, Stuttgart, Govi-Verlag GmbH, Frank furt, p. 917–921.

Hoppe, H.A. (1975) *Drogenkunde*, Band 1. Walter de Gruyter, Berlin, New York.

Hurabielle, M., Tillequin, F. and Paris, M. (1977) Étude chimique de l'huile essentielle d'*Artemisia pontica*. *Planta Med.*, 31, 97–102.

Jambaz, K.H. and Gilani, A.H. (1995) Evaluation of the protective potential of *Artemisia maritima* extract on acetaminophen- and CCl$_4$-induced liver damage. *J. Ethnopharmacol.*, 47, 43–47.

Jaziri, M., Diallo, B., Vanhaelen, M., Homès, J., Yoshimatsu, K. and Shimomura, K. (1993) Immunodetection of artemisinin in *Artemisia annua* cultivated in hydroponic conditions. *Phytochemistry*, 33, 821–826.

Kapoor, L.D. (1990) *CRC Handbook of Ayurvedic Medicinal Plants*, CRC Press, Boca Raton, p. 52–53.

Kartnig, Th. and Brantner, A. (1992) Zur Identitätsprüfung von Drogen, deren Aufnahme in das Österreichische Arzneibuch (ÖAB) erwogen wird. 2. Mitteilung. *Sci. Pharm.*, 60, 129–136.

Klayman, D.L. (1985) Qinghaosu (artemisinin): an antimalarial drug from China. *Science*, 228, 1049–1055.

Klayman, D.L., Lin, A.J., Acton, N., Scovill, J.P., Hoch, J.M., Milhous, W.K. *et al.* (1984) Isolation of artemisinin (qinghaosu) from *Artemisia annua* growing in the United States. *J. Nat. Prod.*, 47, 715–717.

Kudakasseril, G.J., Lam, L. and Staba, E.J. (1987) Effect of sterol inhibitors on the incorporation of 14C-isopentenyl pyrophosphate into artemisinin by a cell-free system from *Artemisia annua* tissue cultures and plants. *Planta Med.*, 53, 280–284.

List, P.H. and Hörhammer, L. (eds.) (1967) *Hagers Handbuch der Pharmazeutischen Praxis*, 4. Auflage, Springer-Verlag, Berlin, p. 927–932.

Maillard, M.P., Wolfender, J.-L. and Hostettmann, K. (1993) Use of liquid chromatography-thermospray mass spectrometry in phytochemical analysis of crude plant extracts. *J. Chromatogr.*, **647**, 147–154.

Nano, G.M., Bicchi, C., Frattini, C. and Gallino, M. (1976) On the composition of some oils from *Artemisia vulgaris*. *Planta Med.*, **30**, 211–215.

Nin, S., Arfaioli, P. and Bosetto, M. (1995) Quantitative determination of some essential oil components of selected *Artemisia absinthium* plants. *J. Essent. Oil Res.*, **7**, 271–277.

Niu, X., Ho, L. Ren, Z. and Song, Z. (1985) Metabolic fate of qinghaosu in rats; a new TLC densitometric method for its determination in biological material. *Eur. J. Drug. Metab. Pharmacokinet.*, **10**, 55–59.

Picman, A.K., Ranieri, R.L., Towers, G.H.N. and Lam, J. (1980) Visualization reagents for sesquiterpene lactones and polyacetylenes on thin-layer chromatograms. *J. Chromatogr.*, **189**, 187–198.

Pharmacopoeia of the People's Republic of China (English Edition) (1988), People's Medical Publishing House, Beijing.

Pras, N., Visser, J.F., Batterman, S., Woerdenbag, H.J., Malingré, Th.M. and Lugt, Ch.B. (1991) Laboratory selection of *Artemisia annua* L. for high artemisinin yielding types. *Phytochem. Anal.*, **2**, 80–83.

Rey, J.-P., Levesque, J. and Pousset, J.L. (1992) Extraction and high-performance liquid chromatographic methods for the γ-lactones parthenolide (*Chrysanthemum parthenium* Bernh.), marrubiin (*Marrubium vulgare* L.) and artemisinin (*Artemisia annua* L.). *J. Chromatogr.*, **605**, 124–128.

Roth, R.J. and Acton, N. (1989) The isolation of sesquiterpenes from *Artemisia annua*. *J. Chem. Educ.*, **66**, 349–350.

Rücker, G., Mayer, R. and Manns, D. (1986) Isolierung von Quinghaosu aus *Artemisia annua* europäischer Herkunft. *Planta Med.*, **52**, 245.

Samuelsson, G. (1992) *Drugs of Natural Origin*, Swedish Pharmaceutical Press, Stockholm, p. 160.

Schneider, G. and Mielke, B. (1978) Absinth, Artabsin und matricin aus *Artemisia absinthium* L. I. *Dtsch. Apoth. Ztg.*, **118**, 469–472.

Schneider, G. and Mielke, B. (1979) Absinth, Artabsin und matricin aus *Artemisia absinthium* L. II. *Dtsch. Apoth. Ztg.*, **119**, 977–982.

Sipahimalani, A.T., Fulzele, D.P. and Heble, M.R. (1991) Rapid method for the detection and determination of artemisinin by gas chromatography. *J. Chromatogr.*, **538**, 452–455.

Song, Z.Y., Zhao, K.C., Liang, X.T., Liu, C.X. and Yi, M.G. (1985) Radio-immunoassay of qinghaosu and artesunate. *Yaoxue Xuebao*, **20**, 610–614.

Stahl, E. (1970) *Chromatographische und mikroskopische Analyse von Drogen*, Fischer-Verlag, Stuttgart, p. 161.

Stahl, E. (1981) *Pharmazeutische Biologie 4. Drogenanalyse II: Inhaltstoffe und Isolierungen.* Gustav Fischer Verlag, Stuttgart, New York.

Steinegger, E. and Hänsel, R. (1972) *Lehrbuch der Pharmakognosie auf phytochemischer Grundlage*, 3. Auflage, Springer-Verlag, Berlin.

Steinegger, E. and Hänsel, R. (1992) *Pharmakognosie*, 5. Auflage, Springer-Verlag, Berlin.

Tang, W and Eisenbrand, G. (1992) *Chinese Drugs of Plant Origin*, Springer-Verlag, Berlin, Heidelberg.

Theoharides, A.D., Peggins, J.O., Brewer, T.G., Melendez, V. and Boyd, J.M. (1989) Automated sample deoxygenation and injection using a programmable autoinjector and a valve-switching apparatus for HPLC with reductive electrochemical detection. *LC-GC*, 7, 925–928.

Thieme, H. and Nguyen Thi Tam (1972) Untersuchungen über die Akkumulation und die Zusammensetzung der ätherischen Öle von *Satureja hortensis* L., *Satureja montana* L. und *Artemisia dracunculus* L. im Verlauf der Ontogenese (2). *Pharmazie* 27, 324–331.

Tu, Y.Y., Ni, M.Y., Zhong, Y.R., Li. L.N., Cui, S.L., Zhang, M.Q., *et al.* (1982) Studies on the constituents of *Artemisia annua*. Part II. *Planta Med.*, 44, 143–145.

Tucker, A.O. and Maciarello, M. (eds.) (1993) *Mass Spectral Library of Flavor & Fragrance Compounds, Volume I–XXVII*, Department of Agriculture & Natural Resources, Delaware State University, Dover, Delaware.

Tyler, V.E. (1993) *The Honest Herbal*, 3rd edn, Pharmaceutical Products Press, New York, p. 315–317, p. 350–351.

Tyler, V.E. (1994) *Herbs of Choice*, Pharmaceutical Products Press, New York, p. 45–46.

Vandenberghe, D.R., Vergauwe, A.N., Van Montagu, M. and Van den Eeckhout, E.G. (1995) Simultaneous determination of artemisinin and its bioprecursors in *Artemisia annua*. *J. Nat. Prod.*, 58, 798–803.

Wallaart, T.E., Van Uden, W., Lubberdink, H.G.M., Woerdenbag, H.J., Pras, N. and Quax, W.J. (1999) Isolation and identification of dihydroartemisinic acid from *Artemisia annua* and its possible role in the biosynthesis of artemisinin. *J. Nat. Prod.* 62, 430–433.

Wallaart, T.E., Pras, N., Beekman, A.C. and Quax, W.J. (2000) Seasonal variation of artemisinin and its biosynthetic precursors in plants of *Artemisia annua* of different geographical origin: proof of the existence of chemotypes. *Planta Med.* 66, 57–62.

Wichtl, M. (1989) *Teedrogen*, 2. Auflage, Wissenschaftliche Verlagsgesellschaft mbH, Stuttgart.

Woerdenbag, H.J., Lugt, Ch.B. and Pras, N. (1990) *Artemisia annua* L.: a source of novel antimalarial drugs. *Pharm. Weekbl. Sci.*, 12, 169–181.

Woerdenbag, H.J., Pras, N., Bos, R., Visser, J.F., Hendriks, H. and Malingré, Th.M. (1991a) Analysis of artemisinin and related sesquiterpenoids from *Artemisia annua* L. by combined gas chromatography/mass spectrometry. *Phytochem. Anal.*, 2, 215–219.

Woerdenbag, H.J., Hendriks, H. and Bos, R. (1991b) *Eupatorium cannabinum* L. – Der Wasserdost oder Wasserhanf. *Z. Phytother.*, 12, 28–34.

Woerdenbag, H.J. Pras, N., Van Uden, W., De Boer, A., Batterman, S., Visser, J.F. *et al.* (1992) High peroxidase activity in cell cultures of *Artemisia annua* with minute artemisinin contents. *Nat. Prod. Lett.*, 1, 121–128.

Woerdenbag, H.J., Bos, R., Salomons, M.C., Hendriks, H., Pras, N. and Malingré, Th.M. (1993a) Volatile constituents of *Artemisia annua* L. (Asteraceae). *Flavour Fragr. J.*, 8, 131–137.

Woerdenbag, H.J., Lüers, F.J.F., Van Uden, W., Pras, N., Malingré, Th.M. and Alfermann, A.W. (1993b) Production of the new antimalarial drug artemisinin in shoot cultures of *Artemisia annua* L. *Plant Cell Tiss. Org. Cult.*, 32, 247–257.

Woerdenbag H.J., Pras, N., Van Uden, W., Wallaart, T.E., Beekman, A.C. and Lugt, Ch.B. (1994a) Progress in the research of artemisinin related antimalarials: an update. *Pharm. World Sci.*, 16, 169–180.

Woerdenbag, H.J., Pras, N., Nguyen Gia Chan, Bui Thi Bang, Bos, R., Van Uden, W. *et al.* (1994b) Artemisinin, related sesquiterpenes, and essential oil in *Artemisia annua* during a vegetation period in Vietnam. *Planta Med.*, 60, 272–275.

Woerdenbag, H.J., Van Uden, W and Pras, N. (1997) *Artemisia cina*. In *Adverse Effects of Herbal Drugs*, Vol. 3, P.A.G.M. de Smet, K. Keller, R. Hänsel and R.F. Chandler, (eds.), Springer-Verlag, Berlin, Heidelberg, p. 15–22.

Zani, F., Massimo, S., Bianchi, A., Albasini, A., Melegari, M., Vampa, G. *et al.* (1991) Studies on the genotoxic properties of essential oils with *Bacillus subtilis rec*-assay and *Salmonella*/microsome reversion assay. *Planta Med.*, 57, 237–241.

Zhang, X.Q. and Xu, L.X. (1985) Determination of qinghaosu (arteannuin) in *Artemisia annua* L. by pulse polarography. *Yaoxue Xuebao*, 20, 383–386.

Zhao, S.S. and Zheng, M.Y. (1985) Spektrometrische Hochdruck-Flüssigkeits chromatographische (HPLC) Untersuchungen zur Analytik von Qinghaosu. *Planta Med.*, 51, 233–237.

Zhou, Z., Huang, Y., Xie, G., Sun, X., Wang, Y., Fu, L. *et al.* (1988) HPLC with polarographic detection of artemisinin and its derivatives and application of the method to the pharmacokinetic study of artemether. *J. Liq. Chromatogr.*, 11, 1117–1137.

Zhu, Y.-P. and Woerdenbag, H.J. (1995) Traditional Chinese herbal medicine. *Pharm. World Sci.*, 17, 103–112.

3. *ARTEMISIA ABSINTHIUM*

STANLEY G. DEANS AND ALAN I. KENNEDY

Aromatic & Medicinal Plant Group, Scottish Agricultural College Auchincruive, Ayr KA6 5HW, Scotland, UK.

INTRODUCTION

Artemisia absinthium L. is a member of the family Compositae (Asteraceae) and is known by the common names wormwood (UK), absinthe (France) and wermut (Germany). The name *Artemisia* is derived from the Goddess Artemis, the Greek name for Diana, who is said to have discovered the plant's virtues (Simon *et al.*, 1984), while *absinthium* comes from the Greek word *apinthion* meaning "undrinkable", reflecting the very bitter nature of the plant. The plant is also known by a number of synonyms which include: Absinthium, Wermutkraut, Absinthii Herba, Assenzio, Losna, Pelin, Armoise, Ajenjo and Alsem. The herb is native to warm Mediterranean countries, usually found growing in dry waste places such as roadsides, preferring a nitrogen-rich stoney and hence loose soil. It is also native to the British Isles and is fairly widespread (Rodway, 1979). Wormwood has been naturalised in northeastern North America, North and West Asia and Africa.

BRIEF BOTANICAL DESCRIPTION

The stem of this shrubby perennial herb is multibranched and firm, almost woody at the base, and grows up to 130 cm in height (Kybal, 1980). The root stock produces many shoots which are covered in fine silky hairs, as are the leaves. The leaves themselves are silvery grey, 8 cm long by 3 cm broad, abundantly pinnate with linear, blunt segments (Gabriel, 1979). Flowers, produced from July to October, are small and globular, arranged as loose clusters of small yellow umbels on erect panicles (Simon *et al.*, 1984). The fruit produced is a cylindrical, slightly flattened acheme, with no pappus (Fig. 1).

Wormwood can be propagated from cuttings, by root division in autumn or by seeds sown in autumn and grows well in even relatively poor, dry soils. Plantations of wormwood last from seven to ten years, peaking in production during the second or third year. The plant can be harvested twice a year, during the late spring and during full bloom (Simon *et al.*, 1984).

Figure 1   *Artemisia absinthium.*

TRADITIONAL USES

The ancient Egyptians were acquainted with wormwood and used it as a medicinal plant, as well as in their religious ceremonies (Rodway, 1979). It is a very old herbal remedy, and is frequently grown in herb gardens. The dried leaves, flowering tops and essential oil of wormwood have traditionally been used as an anthelmintic, antiseptic, antispasmodic, carminative, sedative, stimulant, stomachic and tonic (Simon *et al.*, 1984). It was once used for many disorders, but nowadays is usually used on its own, or in tea mixtures, for various digestive disorders. The very bitter essential oil has also found use as a vermifuge. Folk remedies have also mentioned worm-wood extracts as being useful against colds, rheumatism, fevers, jaundice, diabetes and arthritis. *Artemisia absinthium* was used to flavour beer before the common use of hops.

MODERN USES

Wormwood is sometimes grown as an ornamental (Gabriel, 1979) and when dried, also used in flower arrangements. Culinary uses are limited because of the very bitter flavour of the plant. Tender foliage and non-woody top parts are used either fresh or dried as seasoning usually with boiled or roasted fatty meats, improving flavour as well as making them more digestible (Kybal, 1980). The essential oil extracted from wormwood has very limited uses in the field of fragrance and cosmetics (Heath, 1977) and in some external analgesics (Lawrence, 1977). Extracts are nowadays taken rarely for medicinal purposes, and then only for digestive complaints. More recently, interest has focused on potential medicinal benefits from individual and groups of compounds (Perez-Souto, 1992), rather than crude extracts from the plant (Hernandez *et al.*, 1990). Wormwood is still used in small quantitites as a flavouring agent in alcoholic beverages (Martindale, 1978), such as absinthe, bitters, tonics, liqueurs and vermouth, the latter being a blend of wines containing traces of *Artemisia absinthium* and other flavours.

Absinthe is an interesting example. This liquorice-tasting liqueur was invented in Switzerland but became the French national drink in the 1890s. The essential ingredient of absinthe is wormwood, which is mixed with hyssop, fennel, anise, badiane, angelica and other herbs, producing a poisonous combination. In heavy drinkers, it induced stupor, interrupted spasmodically by epileptic fits, and often proved fatal. It was also highly addictive, eventually assuming cult status in France, where it is said to have inspired an entire culture. There is good evidence that the post-impressionist artist Vincent van Gogh was addicted to absinthe (Arnold, 1989) and there are flamelike images of thuja trees in some of his Auvers paintings. Van Gogh was also believed to have suffered intermittent porphyria through malnutrition and absinthe abuse (Bonkovsky *et al.*, 1992). It is thought that his psychotic depressions were exacerbated by his intake of thujone and that his fits with hallucinations contributed to his suicide in 1890.

Like many "drugs", absinthe came to be viewed as a major social problem. By 1910, 20 million litres were being consumed annually, while in Switzerland, absinthe-related crime resulted in its ban in 1907. In the USA, it was banned in 1912, and was finally outlawed in France due to pressure exerted by army generals who were desperate to place blame elsewhere for their lack of success in the First World War. In addition to the problems that can be caused by alcoholic beverages containing wormwood extracts as a flavouring agent, toxic effects can also be seen if wormwood is used for certain medical purposes. If used over a long period, or in large doses, it can become habit forming (Simon *et al.*, 1984), causing restlessness, vomiting, convulsions and even brain damage, all classic signs of narcotic poisoning.

Extracts of *A. absinthium* have been shown to possess a range of biological activities, including insecticidal action of an alcoholic extract against the stored crop pest *Sitophilus granarius* (Ignatowicz and Wesolowska, 1994) and nematocidal action against *Meloidogyne incognata* (Walker, 1995) and *Helicotylenchus dihystera* (Korayem *et al.*, 1993). The antimalarial activity, against *Plasmodium falciparum*, of two diastereomeric homoditerpene peroxides from the aerial parts of *A. absinthium* was demonstrated (Rucker *et al.*, 1991; 1992) while Zafar *et al.* (1990) screened aqueous and alcoholic extracts against a strain of *Plasmodium berghei* in mice, demonstrating their pronounced schizontocidal properties. Other interesting properties attributable to wormwood extracts include the hepatoprotective effects of an aqueous/methanolic preparation wherein the mode of action was suggested to involve partly the inhibition of microsomal drug metabolising serum transaminase enzymes (Gilani and Jambaz, 1995). In the field of oncology, extracts from the aerial parts of wormwood failed to show direct antitumour activities against sarcoma 180, Erlichs carcinoma, melanoma B-16, Louis' lung carcinoma and Pliss' lymphosarcoma, but showed a definite antimetastatic effect which could be exploited as a corrective against homeostasis disturbance (Gribel and Pashinskii, 1991).

PHYTOCHEMISTRY AND COMMERCIAL IMPORTANCE

Various secondary metabolites and other products have been isolated from *Artemisia absinthium*, perhaps the most important being the bitter essential oil obtained from glands on the leaves and flowering tops (Plates 1, 2). Other compounds include:

- 13 tetrahydrofurofuran lignans extracted from the root system of *A. absinthium* (Greger and Hofers, 1980);
- Several flavonoglucosides in the leaves including quercetin 3-rhamnoglucoside, quercetin 3-glucoside, isorhamnetin 3-glucoside and patuletin 3-glucoside (Hoffman and Herrmann, 1982);
- Oligosaccharides (non-reducing oligofructosides) identified in the root system (Kennedy *et al.*, 1988);
- Merck Index (1983) names the chief bitter principle of wormwood as absinthin, a $C_{30}H_{46}O_6$ (Fig. 2) compound with a molecular weight of 496.62, isolated by chromatography (Bissett, 1994).

A few studies have investigated the dark green to brown essential oil obtained by steam distillation from the dried leaves and flowering tops of *A. absinthium*, sometimes referred to as absinthol. The oil is only of minor importance and is used in limited quantities in fragrance compounding and in some external analgesics (Lawrence, 1977). The major constituent of wormwood oil is thujone, sometimes present as thujyl alcohol and thujyl acetate, present at levels of approximately 40–70% of the oil (Heath, 1977; Tucker *et al.*, 1993). Other constituents present at significant levels include myrcene (<35%)(Fig. 2), α-pinene (6%) (Fig. 2) and nerol

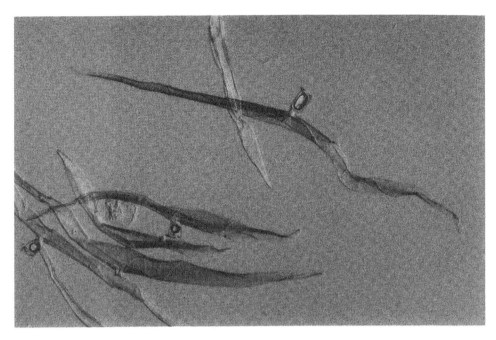

Plate 1   *Artemisia absinthium* non-secretory trichomes [interference contrast microscopy].

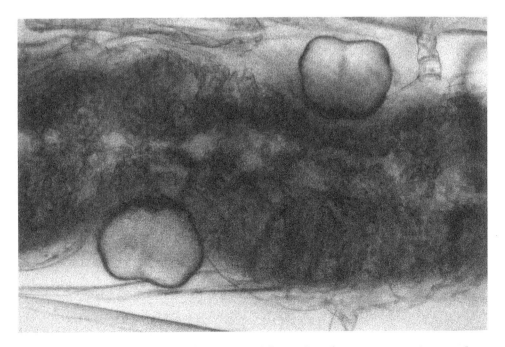

Plate 2   *Artemisia absinthium* sessile secretory trichomes [interference contrast microscopy].

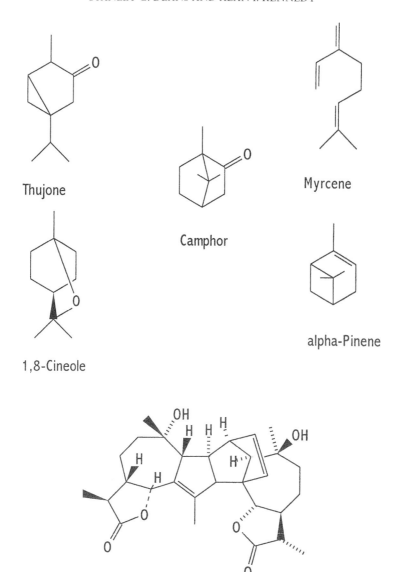

Figure 2 Chemical structures of major components of *Artemisia absinthium*.

(3%) in wormwood oil of Russian origin (Goraev *et al.*, 1962); camphor (6%) (Fig. 2), *p*-cymene (4%), limonene (4%) and α-pinene (4%) in Spanish worm-wood oil (Mugica and Ochoa, 1974); β-phellandrene (10%), α-humachalene (7%) and β-caryophyllene (5%) (Kaul *et al.*, 1979). Carnat *et al.* (1992) found cis-chrysanthenol as the major component in the oil of plants grown in central France while thujane and camphane, along with monoterpenes, were found to be the major constituents in Indian *Artemisia* species (Thakur *et al.*, 1990).

The essential oil, having a very strong odour and acrid taste, is described as neurotoxic due to the high thujone content (Miller *et al.*, 1981) and its use is proscribed in most countries (Tyler *et al.*, 1981). The structure of thujone (Fig. 2) has been compared to that of tetrahydrocannabinol (del Castillo *et al.*, 1975) and it has been postulated that thujone and tetrahydrocannabinol interact with a common receptor in the central nervous system (Martindale, 1982). In addition, Tosi *et al.* (1991) have reported that *A. absinthium* may contain potentially toxic levels of photoactive di- and ter-thiophenes. Ravid *et al.* (1992) undertook chiral gas chromatography analysis of enantiomerically pure (–)-fenchone from *A. absinthium* oil, while Mucciarella *et al.* (1995) suggested a taxonomic scheme utilising oil composition data to partition *Artemisia* species into two sub-groups: one group, containing *A. absinthium*, was characterised by the presence of camphor and 1,8-cineole (Fig. 2); the second smaller group all contained α-thujone. In addition to these constituents, Salnikova *et al.* (1993) isolated, for the first time from aerial parts of several *Artemisia* species, the following phenolic acids: vanillic, *p*-hydroxybenzoic, neochlorogenic, ferulic and *p*-coumaric acids. Earlier, Salnikova *et al.* (1992) showed the presence of several novel coumarins, scopoletin, umbelliferone, esculetin and esculin from flowering *A. absinthium*.

Although there is currently a limited commercial application for the volatile oil from this plant, it was decided that experimental investigations be undertaken into the use of *Artemisia absinthium* transformed root cultures as a model system for complex plant secondary metabolites. Subsequently, this species has been successfully grown as a transformed ("hairy") root culture (Kennedy, 1992; Kennedy *et al.*, 1993). This was initiated by infecting surface-sterilised (using dilute sodium hypochlorite solution) explants of stem with *Agrobacterium rhizogenes*, a Gram negative soil bacterium. Stem explants have proved most successful and are subsequently wounded with a sterile hypodermic syringe which is also used to deliver a small volume of *A. rhizogenes* culture, at a concentration of approximately 4×10^8 viable cells ml^{-1}. Hairy roots develop within 21 days and can subsequently be aseptically excised and placed in liquid growth medium such as Gamborg B5 0 supplemented with ampicillin to eliminate residual bacteria. Growth in liquid culture on an orbital incubator can be rapid as an initial inoculum of less than 0.1 g fresh weight can reach over 5 g in 25 days. It is a feature of many transformed root cultures that they begin to grow very rapidly soon after subculture into fresh medium (Hamill *et al.*, 1987) and produce a high number of lateral branches and hence new root tips. Culture scale-up for hairy root growth has proved problematical much beyond the 1–5 L volume (Wilson *et al.*, 1987), but a 500 L pilot plant of novel design has been built wherein the developing root cultures are immobilised on a matrix of stainless steel barbs, sprayed with sterile growth medium (Wilson *et al.*, 1990).

Table 1 shows typical yields of plant material as well as oil yield in transformed and conventionally grown *A. absinthium*. The majority of studies on hairy root cultures have reported that the level and spectrum of secondary metabolites

Table 1 Yield of plant material and essential oil from field grown and transformed roots of *Artemisia absinthium* L.

A. absinthium	Fresh weight (g)	Dry weight (g)	Oil yield (ml)	Oil yield (%v/w)
Leaves	294.4	42.9	0.4	0.9
Root	70.8	14.1	0.2	1.4
Transformed Roots	155.1	13.8	0.1	0.7

produced by these cultures mirrors closely those produced by the "parent" plant root system *in vivo* (Hamill *et al.*, 1986; Kamada *et al.*, 1986). The profiles of the oils obtained from the normal and transformed root of *A. absinthium* appear to be completely different (Table 2). The neryl isovalerate ester making up the majority of the hairy root oil lies on the same metabolic pathway as α-fenchene, the major component of the normal root oil, and is a distant precursor thereof. It appears that, in the hairy root of wormwood, the pathway has been interupted or stopped at this point, and the root cannot continue its metabolism through cyclisation, towards the production of fenchene.

CONCLUSIONS

Natural products from plants will continue to be regarded as important sources of biologically active compounds, flavourings, colourings and agrichemicals. Many of the relevant plants have yet to be fully exploited and it is reasonable to expect that even more novel and valuable compounds await discovery. With this in mind, many governments are now undertaking a more detailed evaluation of their endemic species. Advances being made in analytical techniques, sophisticated bioassays and biotechnological exploitation should provide the means by which these important plants continue to play a key role to the benefit of Man and his environment (Deans and Kennedy, 1991; Deans and Svoboda, 1993).

Table 2 Analysis of the essential oils in field grown and transformed roots of *Artemisia absinthium* L.

Field grown root		Transformed root	
Constituent	%v/v	Constituent	%v/v
α-Fenchene	53	Neryl Isovalerate	47
β-Myrcene	6	Neryl Butyrate	6
Endo Bornyl Acetate	2		
β-Pinene	1		

REFERENCES

Arnold, W.N. (1989) Absinthe. *Scientific American*, **260**, 86–91.

Bissett, N.G. (Ed.) (1994) *Herbal Drugs and Phytopharmaceuticals*. CRC Press.

Bonkovsky, H.L., Cable, E.E., Cable, J.W., Donohue, S.E., White, E.C., Greene, Y.J. *et al.* (1992) Porphyrogenic properties of the terpenes camphor, pinene and thujone. *Biochemical Pharmacology*, **43**, 2359–2368.

Carnat, A.P., Madesclaire, M., Chavignon, O. and Lamaison, J.L. (1992) Cis-Chrysanthenol, a main component in the essential oil of *Artemisia absinthium* L. growing in Auvergne (Massiv Central), France. *Journal of Essential Oil Research*, **4**, 487–490.

Deans, S.G. and Kennedy, A.I. (1991) Biologically active plant secondary metabolites from transformed roots. *Agro-Industry Hi-Tech*, **6**, 73–76.

Deans, S.G. and Svoboda, K.P. (1993) Biotechnology of aromatic and medicinal plants. In: *Volatile Oil Crops: Their Biology, Biochemistry and Production* (Eds. R.K.M. Hay and P.G. Waterman), pp. 113–136. Longman Group UK, Harlow, Essex, UK.

Del Castillo, R. (1975) Letter. *Nature (London)*, **253**, 365.

Gabriel, I. (1979) *Herb Identifier Handbook*. Sterling Publishing Company, London.

Gilani, A.H. and Janbaz, K.H. (1995) Preventative and curative effects of *Artemisia absinthium* on acetaminophen and CCl_4 induced hepatotoxicity. *General Pharmacology*, **26**:309–315.

Goraev, M.I., Bazalitskaya, U.S. and Lishtvanova, L.N. (1962) The terpene portion of the essential oil of *Artemisia absinthium*. *Zeitschrift fur Naturforschung*, **35**, 2799–2802.

Greger, H. and Hofer, O. (1980) New unsymmetrical substituted tetrahydrofurofuran lignans from *Artemisia absinthium*. *Tetrahedron*, **36**, 3551–3558.

Gribel, N.V. and Pashinskii, V.G. (1991) New data on the antitumour activity of *Artemisia absinthium* L. Extract. *Rastitel'nye-Resursy*, **27**, 65–69.

Hamill, J.D., Parr, A.J., Robins, R.J., and Rhodes, M.J.C. (1986) Secondary product formation by cultures of *Beta vulgaris* and *Nicotiana rustica* transformed with *Agrobacterium rhizogenes*. *Plant Cell Reports*, **5**, 111–114.

Hamill, J.D., Parr, A.J., Rhodes, M.J.C., Robins, R.J. and Walton, N.J. (1987) New routes to plant secondary products. *Biotechnology*, **5**, 800–804.

Heath, H.B. (1977) Herbs: Their use in cosmetics and toiletries. *Cosmetics and Toiletries*, **92**, 19–24.

Hernandez, H., Mendiola, J., Torres, D., Garrido, N. and Perez, N. (1990) Effect of aqueous extracts of *Artemisia* on the *in vitro* culture of *Plasmodium falciparum*. *Fitoterapia*, **41**, 540–541.

Hoffmann, B. and Herrman, K. (1982) Flavonol glucosides of mugwort (*Artemisia vulgaris* L.), tarragon (*Artemisia dracunculus* L.) and absinth (*Artemisia absinthium* L.). *Zeitschrift fur Lebensmittel Untersuchung und Forschung*, **174**, 211–215.

Ignatowicz, S. and Wesolowska, B. (1994) Insecticidal and deterrent properties of extracts from herbaceous plants. *Ochrona Roslin*, **38**, 14–15.

Kamada, H., Okamura, N., Satake, M., Harada, H. and Shimomura, K. (1986) Alkaloid production by hairy root cultures of *Scopolia japonica*. *Agricultural Biology and Chemistry*, **50**, 2715–2722.

Kaul, V.K., Nigam, S.S. and Banerjee, A.K. (1979) Thin-layer and gas chromatographic studies of the essential oil of *Artemisia absinthium* L. *Indian Perfumer*, **23**, 1–7.

Kennedy, A.I. (1992) *Transformed Root Culture of Aromatic and Medicinal Plants: Production of Biologically Active Secondary Metabolites*. PhD Thesis, University of Strathclyde, Glasgow, Scotland, UK.

Kennedy, A.I., Deans, S.G., Svoboda, K.P., Gray, A.I. and Waterman, P.G. (1993) Volatile oils from normal and transformed root of *Artemisia absinthium* L. *Phytochemistry*, **32**, 1449–1451.

Kennedy, J.F., Stevenson, D.L., White, C.A., Lombard, A. and Buffa, M. (1988) Analysis of the oligosaccharides from the roots of *Arnica montana* L., *Artemisia absinthium* L. And *Artemisia dracuncula* L. *Carbohydrate Polymers*, **9**, 277–285.

Korayem, A.M., Hasabo, S.A. and Ameen, H.H. (1993) Effects and mode of action of some plant extracts on certain plant parasitic nematodes. *Anzeiger fur Schadlingskunde Pflanzenschutz Umweltschtz*, **66**, 32–36.

Kybal, J. (1980) *Herbs and Spices*. Hamlyn Publishing Group Limited, London.

Lawrence, B.M. (1977) Recent progress in essential oils. *Perfumer and Flavorist*, **2**, 29–32.

Martindale, (1982) *The Extra Pharmacopoeia*, 28TH Edition. The Pharmaceutical Press, London.

Merck Index (1976) 9TH Edition. Merck & Co. Inc., New Jersey, USA.

Miller, Y., Jouglard, J., Steinmetz, M.D., Tognetti, P., Joanny, P. and Ardetti, J. (1981) Toxicity of some plant essential oils: Clinical and experimental study. *Clinical Toxicology*, **18**, 1485–1497.

Mucciarelli, M., Caramiello, R. and Maffei, M. (1995) Essential oils from some *Artemisia* species growing spontaneously in North-West Italy. *Flavour and Fragrance Journal*, **10**, 25–32.

Mugica, M.G. and Ochoa, J.T. (1974) Aceite esencial de *Artemisia absinthium* L. In: *Contribucion al estudio de los aceites esenciales Espanoles II. Aceite esenciales de la provincia de Guadalajara*. Instituto Nacionale Investigaciones Agrarias, Ministry of Agriculture, Madrid.

Perez-Souto, N., Lynch, R.J., Measures, G. and Hann, J.T. (1992) Use of high-performance liquid chromatographic peak deconvolution and peak labelling to identify antiparasitic components in plant extracts. *Journal of Chromatography*, **593**, 209–215.

Ravid, U., Putievsky, E., Katzir, I. and Ikan, R. (1992) Chiral GC analysis of enantiomerically pure fenchone in essential oils. *Flavour and Fragrance Journal*, **7**, 169–172.

Rodway, A. (1979) *Herbs and Spices*. Macdonald Education Limited, London.

Rucker, G., Walter, R.D., Manns, D. and Mayer, R. (1991) Antimalarial activity of some natural peroxides. *Planta Medica*, **57**, 295–296.

Rucker, G., Manns, D. and Wilbert, S. (1992) Homoditerpene peroxides from *Artemisia absinthium*. *Phytochemistry*, **31**, 340–342.

Salnikova, E.N., Komissarenko, N.F., Dmitruk, S.E. and Kalinkina, G.I. (1992) Coumarins of wormwoods of the order Frigidae. *Chemistry of Natural Compounds*, **28**, 115.

Salnikova, E.N., Komissarenko, N.F., Derkach, A.I., Dmitruk, S.E. and Kalinkina, G.I. (1992) Phenolic acids of wormwoods of the order Frigidae. *Chemistry of Natural Compounds*, **29**, 678.

Simon, J.E., Chadwick, A.F. and Craker, L.E. (1984) *Herbs: An Indexed Bibliography* 1971–1980, pp. 7–9, 59–61, 91–92, 99–100. Elsevier Science Publishers, Amsterdam.

Thakur, R.S., Misra, L.N., Bhattacharya, S.C., Sen, N. and Sethi, K.L. (1990) Essential oils of Indian *Artemisia*. Proceedings of the 11TH International Congress on Essential Oils, Fragrances and Flavours, New Dehli, India, vol 4, pp. 127–135. Aspect Publishing, London.

Tosi, B., Bonora, A., Dall'Olio, G. and Bruni, A. (1991) Screening for toxic thiophene compounds from crude drugs of the family *Compositae* used in Northern Italy. *Phytotherapy Research*, **5**, 59–62.

Tucker, A.O., Maciarello, M.J. and Sturtz, G. (1993) The essential oils of *Artemisia* "Powis Castle" and its putative parents *A. absinthium* and *A. arborescens*. *Journal of Essential Oil Research*, 5, 239–242.

Tyler, V.E., Brady, L.R. and Robbers, J.E. (1981) *Pharmacognosy* 8TH Edition. Lea & Febiger, Philadelphia, USA.

Walker, J.T. (1995) Garden herbs as hosts for southern rootknot nematode (*Meloidogyne incognita* (Kofoid & White) Chitwood race 3). *HortScience*, 30, 292–293.

Wilson, P.D.G., Hilton, M.G., Robins, R.J. and Rhodes, M.J.C. (1987) Fermentation studies of transformed root cultures. In: *Bioreactors and Biotransformations* (Eds G.W. Moody and P.B. Baker) pp. 38–51. Elsevier Applied Science Publishers, Amsterdam.

Wilson, P.D.G., Hilton, M.G., Meehan, P.T.H., Waspe, C.R. and Rhodes, M.J.C. (1990) The cultivation of transformed roots from laboratory to pilot. Proceedings of the VIITH International Conference on Plant Cell and Tissue Culture, Amsterdam, The Netherlands.

Zafar, M.M., Hamdard, M.E. and Hameed, A. (1990) Screening of *Artemisia absinthium* for antimalarial effects on *Plasmodium berghei* in mice: A preliminary report. *Journal of Ethnopharmacology*, 30, 223–226.

4. *ARTEMISIA DRACUNCULUS*

STANLEY G. DEANS AND ELISABETH J.M. SIMPSON

Aromatic & Medicinal Plant Group, Scottish Agricultural College, Auchincruive, South Ayrshire KA6 5HW, Scotland, UK.

INTRODUCTION

Artemisia dracunculus L., French Tarragon, is a perennial herb, native to Europe, Russia, Siberia, China and western and central North America where it grows wild, especially along river banks. It was introduced to Britain in the mid-fifteenth century. This aromatic plant has an extensive fibrous root system which spreads by runners and stems which reach a maximum heigh of around 1 metre. The generic name is derived from the Greek Goddess Artemis who was believed to have given this group of plants to Chiron the centaur, while the specific name is derived from the Latin *dracunculus* meaning small dragon or snake, probably in reference to the long tongue-shaped leaves (Rodway, 1979). Its common name of tarragon is thought to be a corruption of the Arabic *tarkhun* also meaning a little dragon. French tarragon is used mainly as a culinary plant, although its value and popularity in cooking doubtless stems from it medicinal use as an aid to digestion whereby it can be taken as an infusion, or *digestif*, for poor digestion, intestinal distension, nausea, flatulence and hiccups, not to mention its claimed abilities to improve rheumatism, gout and arthritis as well as acting as a vermifuge and an agent to soothe toothache (Philips and Fox, 1990).

BOTANY AND PHYTOCHEMISTRY

French tarragon is a member of the Compositae (Asteraceae) that grows best in sunny sheltered sites, growing easily on rich, light and well drained soil (Bremners, 1997). Attempts have even been made to grow French tarragon on waste ground with only 10–15 cm layers of soil (Stepanovic *et al.*, 1989). Propagation is by cuttings or division of roots in spring. French tarragon does not set seed, so all propagation is by these techniques (Stickland, 1986). The grey-green leaves are long, smooth and shiny and mostly entire although the lower leaves are 3-toothed at their tips while the unobtrusive yellow globose flowers are clustered in a spike, drooping on downcurved stalks, blooming from June to August. In greenhouse and phytotron studies, French tarragon was grown under various temperature and daylength regimes, with the highest yield of herbage and volatile oil being realised under long daylength and constant temperature. It was found that under these conditions, there was an elevated level of elemicin content in the volatile oil, with a corresponding reduction in levels of methyl chavicol, ocimene, β-pinene and sabinene (Fig. 1) (Suchorska *et al.*, 1992). In a second study, comparing French, Russian and Polish forms of tarragon, Olszewska-Kaczynska and Suchorska (1996a) recorded that the elevated levels of methyl chavicol found in *A. dracunculus* resulted in a strong anise aroma which they considered unacceptable. In other reports on Russian tarragon, (Kostrzewa and Karkowska, 1994; Olszewska-Kaczynska and Suchorska, 1996b; Pino *et al.*, 1996), elemicin was the major component in the volatile oil. Werker and co-workers (1994) reported that the leaves of French tarragon had both glandular and non-glandular hairs, the glandular hairs secreting volatile oil when the leaves were young, where the major component in the oil was methyl chavicol. Chalvia *et al.* (1992) reported the main oil constituent in leaves of *Persea gratissima* to be methyl chavicol at levels up to 78%, making this a better commercial source of this compound for the flavour and fragrance industry compared with *A. dracunculus*.

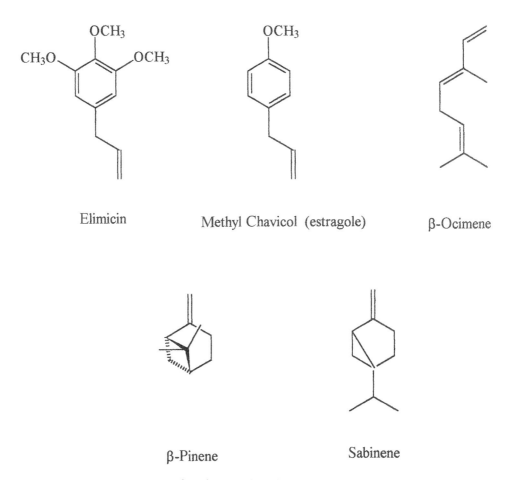

Figure 1   Major constituents of *A. dracunculus* oil.

Harvesting can be undertaken up to three times per year whereby 20–30 cm long sections from the tops of the shoots are removed before the flower buds open. It is at this stage of development that the plant leaves contain the greatest amount of volatile oil, giving the herb its pleasant aroma and slightly bitter peppery taste with a subtle undertone of anise (Bremners, 1997; Thieme and Nguyen-Thi-Tam, 1972a, b), although the comments by Olszewska-Kaczynska and Suchorska (1996a) should be noted wherein they state that high levels of methyl chavicol resulted in an unacceptably strong flavour. A number of other compounds present in the oil make for caution in their use as crude drugs. Tosi *et al.* (1991) cite the significant levels of photoactive thiophenes which may be toxic. Equally important, tarragon oil with an estragole (methyl chavicol) content of ~60% exhibited genotoxicity as evaluated in the *Salmonella typhimurium* bioassay (Bianchi-Santamaria *et al.*, 1993). In contrast, Zani *et al.* (1991) reported that of a number of volatile oils tested by the *Bacillus subtilis* rec-assay for genotoxicity, only the oil from *A. dracunculus* cv *Piemontese* was active in this bioassay, but was not found to be active in the *Salmonella* / microsome reversion assay. The DNA-damaging activity was shown to be due to the estragole component of the oil. The authors note the advantages of the combined use of both microbial bioassays in such genotoxicity studies. Aqueous extracts of a number of plants have also been shown to possess antimutagenic properties, with activity being evident in members of the Labiatae, Cruciferae, Umbelliferae and Compositae, including *A. dracunculus* (Ueda *et al.*, 1991). Antioxidant activity was observed in tarragon extracts using an iron/ascorbate-induced lipid peroxidation model system with microsomes enriched with specific cytochrome P450 isoenzymes (Plumb *et al.*, 1997).

Cotton *et al.* (1991a, b) describe cell culture of *A. dracunculus* in which the accumulation of volatile oil at different stages of differentiation is described. The phenylpropene allylanisole represented a large proportion of the total oil, some 50%, this being significantly higher than that realised in Russian tarragon. In contrast, 16-week old callus cultures showed no significant differences in allylanisole content, this being the first report of allylanisole production in disorganised culture. This compound along with methyl eugenol were accumulated at low levels with autotoxicity as well as the absence of suitable inert accumulation sites being responsible for their poor production.

TRADITIONAL USES

French tarragon, unlike the closely related *Artemisia annua* and *A. absinthium* is used solely as a culinary herb. As such, it is widely used in sauces, *fines herbes* mixtures, marinades, vinegars and preserves, particularly in Chinese and French cuisine. It is best used fresh or with the flavour preserved in oil or vinegar. Where drying is the only option for long-term storage, it should be undertaken as quickly as possible at a temperature of less than 35° C, otherwise the volatile oil is lost and the leaves turn brown

(Deans and Svoboda, 1992). Drying also substantially reduces the microflora associated with fresh plant material (Deans *et al.*, 1988). Freezing is another option which allows for French tarragon to be available over the winter months. Although not strictly a medicinal plant, its beneficial properties have long been recorded. The 13th century Spanish physician and botanist Ibn Baithar stated that fresh shoots of French tarragon were cooked with vegetables and the juice of the tarragon used to flavour beverages. He further stated that French tarragon sweetens the breath, dulls the taste of medicines while promoting sleep (Kybal, 1980).

MODERN USES

French tarragon is an important culinary herb generally used in mild-flavoured egg and poultry dishes, mayonnaise, salad dressings, light soups, herb butter for vegetables, steak and grilled fish (Chialva, 1982).

BIOLOGICAL PROPERTIES

A number of studies have been undertaken into the bioactive properties of the volatile oil from *A. dracunculus* (Tan *et al.*, 1998). In two studies into insect responses to the volatile oil and its constituents, *Papilio* spp. evoked different reactions to the oil (Baur and Feeny, 1995; Thompson *et al.*, 1990). Using GC coupled electroantennograms, they were able to demonstrate which components from the oil were active in determining oviposition preference and larval performance, areas where plant extracts could be used in insect control. In a further study, the attractiveness or repulsiveness of *A. dracunculus* volatile oil towards insects infected with tick-borne encephalitis virus allowed not only those infected to be distinguished from those uninfected, but to identify individuals with varying degrees of virus replication (Alekseev *et al.*, 1991).

The antimicrobial properties of various species of *Artemisia* are well recorded (Mehrotra *et al.*, 1993) where a number of Gram positive and Gram negative bacteria were inhibited in their growth. The test bacteria are all capable of infection, and include *Escherichia coli*, *Klebsiella pneumoniae*, *Pseudomonas aeruginosa*, *Staphylococcus aureus* and *Streptococcus* (*Enterococcus*) *faecalis*. In a similar study of *A. dracunculus* oil, Deans and Svoboda (1988) included the anaerobic bacterium *Clostridium sporogenes* as well as *Salmonella pullorum* and *Yersinia enterocolitica*, the latter organism having the ability to produce an enterotoxin under conditions of refrigeration. A larger group of bacteria was tested against a number of Italian species, including *A. dracunculus*, and found to be very active at preventing the growth of human pathogens, food spoilage/poisoning types as well as animal pathogens (Piccaglia *et al.*, 1993).

Mehrotra *et al.* (1993) reported the marked antifungal activity of *A. dracunculus* volatile oil against *Candida albicans* and *Sporotrichum schenkii*, while Margina and Zheljazkov (1996) have highlighted the susceptibility of *A. dracunculus* to the pathogenic rust fungus *Puccinia dracunculina*.

CONCLUSIONS

French Tarragon has a long history of use as a culinary and medicinal plant. It grows in a wide variety of habitats in various types of soil, preferring light and well aired sites. The composition of the volatile oil varies widely according to geographical location, climate, day length, soil type and cultivar. Its traditional ethnopharmacological applications have been enhanced by more recent studies into the phytochemical nature of the individual oil components. These investigations have revealed the extensive antibacterial, antimycotic and insect-interactive properties of the volatile oil.

REFERENCES

Alekseev, A.N., Burenkova, L.A. and Chunikhin, S.P. (1991) The use of scents of vegetable origin as indicators of the prevalence and intensity of infection of ixodid ticks with the tick-borne encephalitis virus. *Meditsinskaya Parazitologiya Parazitarnye Boleznii*, 3, 10–14.

Baur, R. and Feeny, P. (1995) Comparative electrophysiological analysis of plant odour perception in females of three *Papilio* species. *Chemoecology*, 5–6, 26–36.

Bianchi-Santamaria, A., Tateo, F. and Santamaria, L. (1993) In *Food and Cancer Prevention: Chemical and Biological Aspects* (Eds. K.W. Waldron and I.T. Johnson), pp. 75–81. Royal Society of Chemistry, Cambridge, UK.

Bremners, L. (1997) *Garden Herbs*. Dorling Kindersley Ltd, London.

Chialva, F. (1982) *The Production of Aromatic Plants in the Pancalieri Area* In *Aromatic Plants: Basic and Applied Aspects*. (Eds N. Margaris, A. Koedan and D. Vokou). Martinus Nijhoff, The Hague.

Chialva, F., Monguzzi, F., Manitto, P., Speranza, G. and Akgul, A. (1992) Volatile constituents of the leaves of *Persea gratissima* Gaertner. A source of methyl chavicol. *Journal of Essential Oil Research*, 4, 631–633.

Cotton, C.M., Evans, L.V. and Gramshaw, J.W. (1991a) The accumulation of volatile oils in whole plants and cell cultures of tarragon (*Artemisia dracunculus*). *Journal of Experimental Botany*, 42, 365–375.

Cotton, C.M., Gramshaw, J.W. and Evans, L.V. (1991b) The effect of α-naphthalene acetic acid (NAA) and benzylaminopurine (BAP) on the accumulation of volatile oil components in cell cultures of tarragon (*Artemisia dracunculus*). *Journal of Experimental Botany*, 42, 377–386.

Deans, S.G. and Svoboda, K.P. (1988) Antibacterial activity of French tarragon (*Artemisia dracunculus* L.) essential oil and its constituents during ontogeny. *Journal of Horticultural Science*, 63, 503–508.

Deans, S.G., Svoboda, K.P. and Ritchie, G.A. (1988) Changes in microflora during oven-drying of culinary herbs. *Journal of Horticultural Science*, **63**, 137–140.

Deans, S.G. and Svoboda, K.P. (1992) Effect of drying regime on volatile oil and microflora of aromatic plants. *Acta Horticulturae*, **306**, 450–452.

Kostrzewa, E. and Karwowska, K. (1994) Studies on the taste and aroma components of tarragon cultivated in Poland. *Prace Instytuow Laboratoriow Badawczych Przemyslu Spozywczego*, **48**, 43–57.

Kybal, J. (1980) *Herbs and Spices*. Hamlyn Publishing Group, London.

Margina, A. and Zheljazkov, V. (1996) Fungal pathogens from Uredinales on some medicinal and aromatic plants in Bulgaria and their control. *Acta Horticulturae*, **426**, 333–343.

Mehrotra, S., Rawat, A.K.S. and Shome, U. (1993) Antimicrobial activity of the essential oils of some Indian *Artemisia* species. *Fitoterapia*, **64**, 65–68.

Olszewska-Kaczynska, I. and Suchorska, K. (1996a) Evaluation of three cultivated forms of tarragon (*Artemisia dracunculus* L.). *Folia Horticulturae*, **8**, 29–37.

Olszewska-Kaczynska, I. and Suchorska, K. (1996b) Characterization of tarragon (*Artemisia dracunculus* L.) cultivated in Poland. *Herba Polonica*, **42**, 5–10.

Phillips, R. and Fox, N. (1990) *Herbs*. Pan Books Ltd, London.

Piccaglia, R., Marotti, M., Giovanelli, E., Deans, S.G. and Eaglesham, E. (1993) Antibacterial and antioxidant properties of Mediterranean aromatic plants. *Industrial Crops and Products*, **2**, 47–50.

Pino, J.A., Rosado, A., Correa, M.T. and Fuentes, V. (1996) Chemical composition of the essential oil of *Artemisia dracunculus* L. from Cuba. *Journal of Essential Oil Research*, **8**, 563–564.

Plumb, G.W., Chambers, S.J., Lambert, N., Wanigatunga, S. and Williamson, G. (1997) Influence of fruit and vegetable extracts on lipid peroxidation in microsomes containing specific cytochrome P450s. *Food Chemistry*, **60**, 161–164.

Rodway, A. (1979) *Herbs and Spices*. MacDonald Education Limited, London.

Stepanovic, B., Jovanovic, M. and Knezevic, D. (1989) Studies on the possibilities of growing medicinal and aromatic plants on waste land of the Pljevlja coal mine. *Zemljiste Biljka*, **38**, 57–62.

Stickland, S. (1986) *Planning the Organic Garden*. Thorsons Publishing Group, New York.

Suchorska, K., Jedraszko, B. and Olszewska-Kaczynska, I. (1992) Influence of daylength on the content and composition of the essential oil from tarragon (*Artemisia dracunculus*). *Annals of Warsaw Agricultural University*, **16**, 79–82.

Tan, R.X., Zheng, W.F. and Tang, H.Q. (1998) Biologically active substances from the genus *Artemisia*. *Planta Medica*, **64**, 295–302.

Thieme, H. and Nguyen-Thi-Tam (1972a) Studies on the accumulation and composition of volatile oils in *Satureja hortensis* L., *Satureja montana* L. and *Artemisia dracunculus* L. during ontogenesis. 1. Review of literature, thin layer and gas chromatographic studies. *Pharmazie*, **27**, 255–265.

Thieme, H. and Nguyen-Thi-Tam (1972b) Studies on the accumulation and composition of volatile oils in *Satureja hortensis* L., *Satureja montana* L. and *Artemisia dracunculus* L. during ontogenesis. 2. Changes in the content and composition of the volatile oil. *Pharmazie*, **27**, 324–331.

Thompson, J.N., Wehling, W. and Podolsky, R. (1990) Evolutionary genetics of host use in swallowtail butterflies. *Nature* (London), **344**, 148–150.

Tosi, B., Bonora, A., Dall'Olio, G. and Bruni, A. (1991) Screening for toxic thiophene compounds from crude drugs of the family Compositae used in northern Italy. *Phytotherapy Research*, **5**, 59–62.

Ueda, S., Kuwabara, Y., Hirai, N., Sasaki, H and Sugahara, T. (1991) Antimutagenic capacities of different kinds of vegetables and mushrooms. *Journal of the Japanese Society for Food Science and Technology*, **38**, 507–514.

Werker, E., Putievsky, E., Ravid, U. Dudai, N. and Katzir, I. (1994) Glandular hairs, secretory cavities, and the essential oil in leaves of tarragon (*Artemisia dracunculus* L.). *Journal of Herbs, Spices and Medicinal Plants*, **2**, 19–32.

Zani, F., Massimo, G., Benvenuti, S., Bianchi, A., Albasini, A., Melegari, M. *et al.* (1991) Studies on the genotoxic properties of essential oils with *Bacillus subtilis* rec-assay and *Salmonella* / microsome reversion assay. *Planta Medica*, **57**, 237–241.

5.   ARTEMISIA HERBA-ALBA

PETER PROKSCH

Institut für Pharmazeutische Biologie, Universitatsstr. 1., Gebaude 26.23, 40225 Dusseldorf, Germany.

INTRODUCTION

The genus *Artemisia* is a member of the large and evolutionary advanced plant family Asteraceae (Compositae). More than 300 different species comprise this diverse genus which is mainly found in arid and semi-arid areas of Europe, America, North Africa as well as in Asia (Heywood and Humphries, 1977). *Artemisia* species are widely used as medicinal plants in folk medicine. Some species such as *A. absinthium, A. annua* or *A. vulgaris* have even been incorporated into the pharmacopoeias of several European and Asian countries (Proksch, 1992).

Sesquiterpene lactones are among the most prominent natural products found in *Artemisia* species and are largely responsible for the importance of these plants in medicine and pharmacy. For example, the antimalarial effect of the long known Chinese medicinal plant Qing Hao (*A. annua*) is due to the sesquiterpene lactone artemisinin which is active against *Plasmodium falciparum* (Proksch, 1992). Another sesquiterpene lactone, absinthin, is the bitter tasting principle found in *A. absinthium* formerly used to produce an alcolohic beverage called "absinth" (Proksch, 1992). In addition to sesquiterpene lactones volatile terpenoids that constitute the so called essential oils are also characteristic metabolites of *Artemisia* species. Essential oils are primarily responsible for the use of *Artemisia* species as spices (Proksch, 1992).

In common with many other members of this genus, *Artemisia herba-alba* which is characteristic of the steppes of the Middle East and North Africa (Feinbrun-Dothan, 1978) is used in folk medicine in order to treat various ailments that include toothache, respiratory diseases, enteritis, intestinal disturbances and diabetes mellitus (Al-Shamaony *et al.*, 1994; Marrif *et al.*, 1995; Twaij and Al-Badr, 1988; Yashphe *et al.*, 1987; Ziyyat *et al.*, 1997).

BRIEF BOTANICAL DESCRIPTION

A. herba-alba is a perenial herb grows 20–40 cm in height; it is a chamaeophyte (i.e. the buds giving rise to new growth each year are borne close to the ground). The stems are rigid and erect. The grey leaves of sterile shoots are petiolate, ovate to orbicular in outline whereas leaves of flowering stems are much smaller. The flowering heads are sessile, oblong and tapering at base. The plants flower from September to December. Plants are oblong and tapering at base. Plants are found on the steppes of the Middle East and of North Africa where they are common and sometimes stand-forming (Feinbrun-Dothan, 1978). It has been suggested that North African *A. herba-alba* is the same as *A. inculta* but a study of the sesquiterpenes from *A. inculta* does not support this (Marco *et al.*, 1997).

PHYTOCHEMISTRY

Various secondary metabolites have been isolated from *A. herba-alba*, perhaps the most important being the sesquiterpene lactones that occur with great structural diversity within the genus *Artemisia*. Additional studies have focussed on flavonoids and essential oils.

Several structural types of sesquiterpene lactones were found in the aerial parts of *A. herba-alba*. Eudesmanolides followed by germacranolides seem to be the most abundant types of lactones found in this species (Ahmed *et al.*, 1990; Boriky *et al.*, 1996; Gordon *et al.*, 1981; Marco, 1989; Marco *et al.*, 1990; Marco *et al.*, 1994; Sanz *et al.*, 1990; Sanz and Marco, 1991; Segal *et al.*, 1977; Segal *et al.*, 1983).

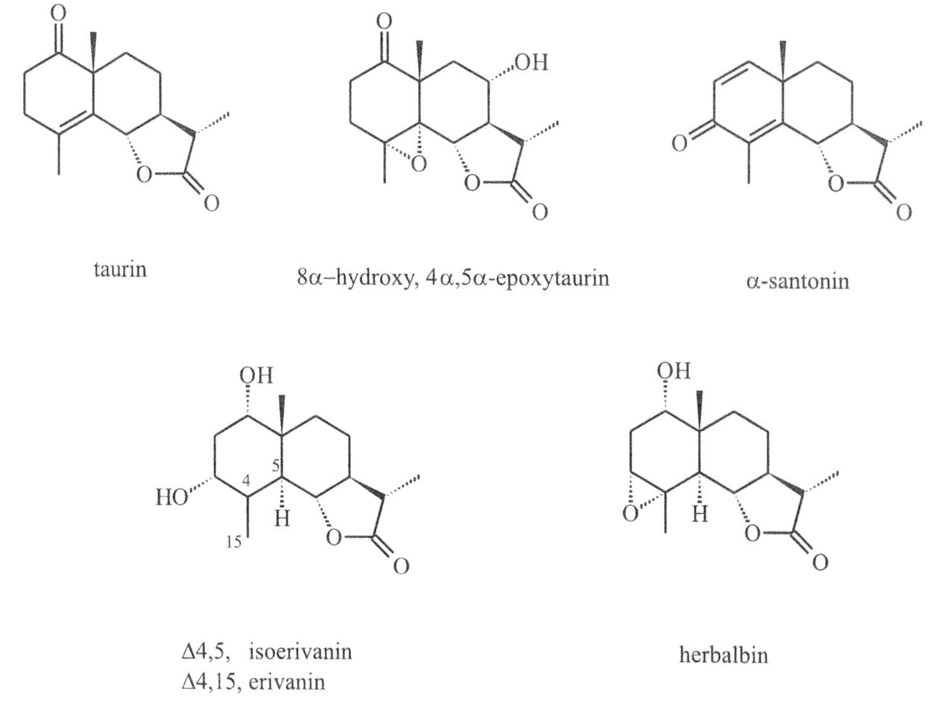

taurin   8α–hydroxy, 4α,5α-epoxytaurin   α-santonin

Δ4,5, isoerivanin   herbalbin
Δ4,15, erivanin

Figure 1   Some eudesmanolides isolated from *A. herba-alba*

Fig. 1 shows the structures of a number of eudesmanolides isolated from Moroccan *A. herba-alba* (Boriky *et al.*, 1996). Chemical variability with regard to the pattern of sesquiterpene lactones is considerable as found also for other members of this genus. Based on their sesquiterpene lactones at least 6 different chemotypes of *A. herba-alba* growing in arid zones of Egypt and Israel could be discerned (Gordon *et al.*, 1981; Segal *et al.*, 1977; Segal *et al.*, 1983).

The flavonoids detected in *A. herba-alba* also show considerable structural diversity ranging from common flavone and flavonol glycosides to more unusual highly methylated flavonoids (Saleh *et al.*, 1985; Saleh *et al.*, 1987). The flavonoid glycosides include common O-glycosides such as quercetin 3-glucoside but also flavone C-glycosides that are of a rarer occurrence within the genus *Artemisisa* as well as within the whole of the Asteraceae.

Besides sesquiterpene lactones and flavonoids phytochemical analysis has focussed on the composition of essential oils from *A. herba-alba*. High variability of volatile constituents was observed when different populations of *A. herba-alba* collected at various sites in Israel were compared (Yashphe *et al.*, 1987). Whereas samples of *A. herba-alba* collected at Elat contained chrysanthenyl acetate as major component (31%) followed by chrysanthenol (6.5%) and the acetophenone xanthocyclin, the essential oil of *A. herba-alba* from the Judean desert exhibited 1,8-cineole as the major compound (50%) followed by appreciable amounts of alpha- and beta-thujone (27%) and other oxygenated monoterpenes such as terpinen-4-ol (3.3%), camphor (3%) and borneol (3%), (Fig. 2), (Yashphe *et al.*, 1987). Essential oils of *A. herba-alba* collected at other localities in Israel contained some more unusual volatile terpenes including the so called non-head-to-tail monoterpene alcohols artemisia alcohol, santolina alcohol and lyratol; these are illustrated in Fig. 2 (Segal *et al.*, 1980). As different essential oil components differ also with regard to their antibiotic activity (monoterpene alcohols like terpinen-4-ol being clearly more active against several strains of bacteria than for example 1,8-cineole or the thujone isomers) (Yashphe *et al.*, 1987) the precise origin of plant material will certainly have an influence on any curative effect of *A.herba-alba* either in folk medicine or in modern phytotherapy.

REPORTED USES OF *ARTEMISIA HERBA-ALBA* IN FOLK MEDICINE

Artemisia herba-alba is used in the Middle East against a variety of ailments including enteritis and intestinal disturbance (Yashphe *et al.*, 1987). In a study aimed at revealing the rationale for the described use of these plants the essential oil of *A. herba-alba* was tested against various bacteria that reportedly cause intestinal problems as well as for antispasmodic activity on the rabbit jejunum. The essential oil of *A. herba-alba* collected in Israel showed antibacterial activity for example towards *Escherichia coli, Shigella sonnei* or *Salmonella typhosa* at concentrations of 1–2 mg/ml (Yashphe *et al.*, 1987). The highest activity was found for linalool, pinocarveol and especially terpinen-4-ol. The antispasmodic effects of the essential oil of *A. herba-alba* were approximately 100–1000 times higher than the observed

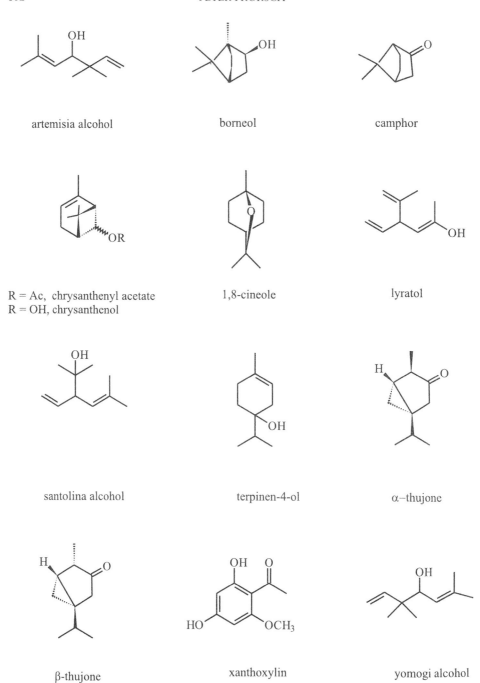

artemisia alcohol

borneol

camphor

R = Ac, chrysanthenyl acetate
R = OH, chrysanthenol

1,8-cineole

lyratol

santolina alcohol

terpinen-4-ol

α–thujone

β-thujone

xanthoxylin

yomogi alcohol

Figure 2 Some monoterpenes present in the essential oil of *A. herba-alba*

antibacterial effects (Yashpe *et al.*, 1987). The authors concluded that the extensive use of *A. herba-alba* in folk medicine especially against intestinal disturbances may be attributed to the combination of an antibacterial and antispasmodic effect (Yashphe *et al.*, 1987).

By far the most frequently cited use of *A. herba-elba* is its employment in the therapy of diabetes mellitus (Al-Shamaony *et al.*, 1994; Marrif *et al.*, 1995; Twaij *et al.*, 1988). Several authors report on the hypoglycaemic effects of aqueous extracts of *A. herba-alba* on rabbits and mice that had been made diabetic by injection of alloxan monohydrate (Marrif *et al.*, 1995; Twaij and Al-Badr, 1988). Alloxan is widely used in animal experiments focussing on diabetes mellitus since it destroys the beta-cells of the islets of Langerhans. Oral administration of an aqueous extract (0.39 g plant material/kg body weight) of *A. herba-alba* as used in folk medicine for 2–4 weeks given to treated rabbits produced significant hypoglycaemic activity (Al-Shamaony *et al.*, 1994). The extract caused a pronounced fall in plasma glucose level. Maximum effect on plasma glucose concentration (22% decrease compared to untreated diabetic animals) was observed 6 h after treatment. Oral administration of the aqueous *A. herba-alba* extract to rabbits also protected the animals from body weight loss compared with control diabetic animals that had received no treatment (Al-Shamaony *et al.*, 1994). Administration of the *A. herba-alba* extract also caused a reduction of serum lipids (Al-Shamaony *et al.*, 1994); an increase in serum lipids accompanies diabetes mellitus and is a major risk factor for coronary heart disease (Davidson, 1981). The frequently reported uses of *A. herba-alba* for the treatment of diabetes mellitus and also for hypertension (the latter may be a consequence of diabetes) may thus have a rationale. The compounds responsible for the effects observed in the animal experiments, however, remain to be elucidated.

A. herba-alba is one of a number of plants used traditionally in Jordan as an antidote to snake and scorpion venoms. An aqueous extract of *A. herba-alba* was found to inhibit the haemolytic activities of the desert viper (*Cerastes cerastes*) and the scorpion (*Leiurus quinquesteiartus*) (Sallal *et al.*, 1996). Recently, aqueous extracts of *A. herba-alba* have been reported to protect against ethanol-induced gastric lesions in rats (Gharzouli *et al.*, 1999). Protection was found to be greatest when the extract was given at the same time as the ethanol; however, the acid content of the stomach was significantly increased by the extract. *A. herba-alba* extract did not precipitate protein (ovine haemoglobin) and it is suggested that phenolic compounds present in the extract may strengthen the gastric mucosal barrier.

REFERENCES

Ahmed, A.A., Abou-El-Ela, M., Jakupovic, J., Seif El-Din, A.A. and Sabri, N. (1990) Eudesmanolides and other constituents from *Artemisia herba-alba*. *Phytochemistry*, **29**, 3661–3663.

Al-Shamaony, L., Al-Khazraji, S.M. and Twaij, H.A.A. (1994) Hypoglycaemic effect of *Artemisia herba-alba*. II. Effect of a valuable extract on some blood parameters in diabetic animals. *J. Ethnopharmacol.*, **43**, 167–171.

Boriky, D., Berrada, M., Talbi, M., Keravis, G. and Rouessac, F. (1996) Eudesmanolides form *Artemisia herba-alba*. *Phytochemistry*, **43**, 309–311.

Davidson, M.B. (1981) *Diabetes Mellitus, Diagnosis and Treatment*, pp. 27, 48, 109, 157. Wiley, New York.

Feinbrun-Dothan, N. (1978) *Flora Palaestina*, Part 3, pp. 351–353. The Israel Academy of Sciences and Humanities, Jerusalem.

Gharzouli, K., Khennouf, S., Amira, S. and Gharzouli, A. (1999) Effect of aquous extracts from q*uercus ilex* L. root bark, *Punica granatum L.* fruit peel and *Artemisia herba alba* Asso leaves on ethanol-induced gastric damage in rats. *Phytother. Res.* **13**, 42–45.

Gordon, M.M., van Derveer, D. and Zalkow, L.H. (1981) New germacranolides from *Artemisia herba alba*. *J. Nat. Prod.*, **44**, 432–440.

Heywood, V.H. and Humphries, C.J. (1977) Anthemideae – sytematic review. In: Heywood, V.H., Harborne, J.B., Turner, B.L., eds.) *The Biology and Chemistry of the Compositae*, Vol. II, pp. 851–898. Academic Press, London, New York, San Francisco.

Marco, J.A. (1989) Sesquiterpene lactones from *Artemisia herba-alba* subsp. *herba-alba*. *Phytochemistry*, **28**, 3121–3126.

Marco, J.A., Sanz, J.F., Falco, E., Jakupovic, J. and Lex, J. (1990) New oxygenated eudesmanolides from *Artemisia herba-alba*. *Tetrahedron*, **46**, 7941–7950.

Marco, J.A., Sanz-Cervera, J.F., Ocete, G., Carda, M., Rodriguez, S. and Valles-Xirau, J. (1994) New germacranolides and eudesmanolides from North African *Artemisia herba-alba*. *J. Nat. Prod.*, **57**, 939–946.

Marco, J.A., Sanz-Cervera, J.F., Garcia-Lliso, V., Guara, M. and Velles-Xirau, J. (1997) Sesquiterpene lactones from *Artemisia inculta*. *Phytochem.* **45**, 751–754.

Marrif, H.I., Ali, B.H. and Hassan, K.M. (1995) Some pharmacological studies on *Artemisia herba-alba* (Asso.) in rabbits and mice. *J. Ethnopharmacol.*, **49**, 51–55.

Proksch, P. (1992) Artemisia. In: *Hagers Handbuch der Pharmazeutischen Praxis* (Hänsel, R., Keller, K., Rimpler, H., Schneider, G., Hrsg.), Bd. 4, pp. 357–377. Springer-Verlag, Berlin.

Saleh, N.A.M., El-Negoumy, S.I., Abd-Alla, M.F., Abou-Zaid, M.M., Dellamonica, G. and Chopin, J. (1985) Flavonoid glycosides of *Artemisia monosperma* nad *A. herba-alba*. *Phytochemistry*, **24**, 201–203.

Saleh, N.A.M., El-Negoumy, S.I. and Abou-Zaid, M.M. (1987) Flavonoids of *Artemisia judaica* and *A. herba-alba*. *Phytochemistry*, **26**, 3059–3064.

Sallal, A.K.J. and Alkofahi, A. 1996. Inhibition of the haemolytic activities of snake and scorpion venoms *in vitro* with plant extracts. *Biomed. Lett.* **53**, 211–215.

Sanz, J.F., Castellano, G. and Marco, J.A. (1990) Sesquiterpene lactones from *Artemisia herba-alba*. *Phytochemistry*, **29**, 541–545.

Sanz, J.F. and Marco, J.A. (1991) New eudesmanolides related to torrentin from *Artemisia herba-alba* subsp. *valentina*. *Planta Med.*, **57**, 74–76.

Segal, R., Sokoloff, S., Haran, B., Zaitschek, D.V. and Lichtenberg, D. (1977) New sesquiterpene lactones from *Artemisia herba-alba*. *Phytochemistry*, **16**, 1237–1241.

Segal, R., Breuer, A. and Feuerstein, I. (1980) Irregular monoterpene alcohols from *Artemisia herba-alba*. *Phytochemistry*, **19**, 2761–2762.

Segal, R., Feuerstein, I., Duddeck, H., Kaiser, M. and Danin, A. (1983) The sesquiterpene lactones from two populations of *Artemisia herba alba*. *Phytochemistry*, **22**, 129–131.

Segal, R., Eden, L., Danin, A., Kaiser, M. and Duddeck, H. (1985) Sesquiterpene lactones from *Artemisia herba alba*. *Phytochemistry*, **24**, 1381–1382.

Twaij, H.A.A. and Al-Badr, A.A. (1988) Hypoglycaemic activity of *Artemisia herba-alba*. *J. Ethnopharmacol.*, **24**, 123–126.

Yashphe, J., Feuerstein, I., Barel, S. and Segal, R. (1987) The antibacterial and antispasmodic activity of *Artemisia herba-alba* Asso. II. Examination of essential oils from various chemotypes. *Int. J. Crude Drug. Res.*, **25**, 89–96.

Ziyyat, A., Legssyer, A., Mekhfi, H., Dassouli, A., Serhrouchni, M. and Benjelloun, W. (1997) Phytotherapy of hypertension and diabetes in oriental Morocco. J. *Ethnopharmacol.*, **58**, 45–54.

6.   ETHNOBOTANY, PHYTOCHEMISTRY AND BIOLOGICAL/PHARMACOLOGICAL ACTIVITIES OF *ARTEMISIA LUDOVICIANA* SSP. *MEXICANA* (ESTAFIATE)

MICHAEL HEINRICH

Centre for Pharmacognosy and Phytotherapy, The School of Pharmacy, University of London, 29/39 Brunswick Square, London, WC1N 1AX, UK.

ABSTRACT

Estafiate or iztauyatl (*A. ludoviciana* ssp. *mexicana*) is one of the most popular medicinal plants in Mexican phytotherapy and is nowadays used especially for gastrointestinal pain, as a vermifuge and as a bitter stimulant. The historical and modern uses of this species are reviewed. The first report of its medicinal use dates back to the 16th century, but at that time it was used for completely different illnesses. Only very limited pharmacological studies to evaluate these claims are available; anti-inflammatory, antibacterial and antihelmintic effects have been reported. The aerial parts contain a large number of sesquiterpene lactones, flavonoids as well as essential oil which has not yet been studied in detail.

INTRODUCTION

Estafiate or iztauyatl (*A. ludoviciana* ssp. *mexicana*) is one of the most popular remedies in Mexican phytotherapy. It is frequently sold in markets in the cities and also grown in many house gardens (Linares *et al.*, 1990; Bye, 1986; Heinrich, 1989; 1996). It is thus a locally important economic product and a phytotherapeutic resource which requires documentation of its regional or national importance as well as evaluation and monitoring for efficacy and safety. Plants generally are an important medicinal resource to many people in Mexico and other South and Central American countries (Heinrich, 1992) and some have a history that has been documented as early as the 16th century. The uses and importance of a plant may vary considerably between different regions and times; some of these may be of little or no therapeutic value and it is therefore a legitimate scientific aim to select and promulgate appropriate therapeutic uses.

BOTANY

Artemisia ludoviciana Nutt. ssp. *mexicana* Nutt. (Syn.: *A. mexicana* Willd., *Artemisia vulgaris* var. *mexicana* Torr & A. Gray; *A. vulgaris* ssp. *mexicana* H.M. Hall & Clem.) The species *A. ludoviciana* is a highly complex taxon of the tribe Anthemideae, which is regarded as including several ecogeographically significant, but wholly confluent varieties (Cronquist *et al.*, 1994) or – more frequently – as including approximately half a dozen subspecies (Kartesz and Kartesz, 1980). The material sold in Mexican markets is generally considered to include only the subspecies "*mexicana*". This subspecies is distinguished from most of the others by having a relatively loose and open inflorescence (at least in well developed plants) with more or less elongated branches and from the subspecies "*albula*" by relatively large (5–10+ cm long) and slender leaves. The leaves may be entire but are usually deeply divided into 3 or 5 narrow, elongated lobes. All subspecies of *A. ludoviciana* are aromatic, rhizomatous and perennial herbs, 30 to 100 cm high, more or less white and tomentose above ground, with a 2.5–5 mm high involucrum and inconspicuous flower head. It is native in pine and oak forests in the highlands of Mexico (1,700–2,800 m above sea level), but today it is encountered most frequently in house yards and gardens (Lozoya and Lozoya 1982). It ranges as far North as Texas and the Southern Great Plains (New Mexico and occasionally Arizona).

ETHNOBOTANY

The species is most popular in Mexico as a remedy for various gastrointestinal disorders. The modern name estafiate is derived from the Nahua term "iztauhyátl" "bitter/salty is its water" and refers to the bitter taste of the leaf extracts. In the following section the uses of this plant in historical and modern times is discussed.

Uses Before the 19th Century

For the history of the uses of this plant in the period before and after the conquest we have to rely on some colonial codices. The best known ones are the Codex Cruz Badiano and the Codex Florentino. The first is a herbal written in Nahuatl by the Aztec healer Martin de la Cruz from Tezcoco, who was at the Colegio de Santa Cruz in Tlatelolco. It was translated into Latin by Juan Badiano and given to the King of Spain Carlos I in 1552. It was written rather hastily and has numerous colour illustrations of medicinal plants. There have been several attempts to identify plants from this herbal (Viesca Treviño, 1992; Valdés *et al.*, 1992; Pineda, 1992) and most of the identifications seem to be botanically sound. The major problem with this source is that by this time the European conquest of México-Tenochtitlan 30 years previously had already had an impact. In addition the Nahuatl author attempted to show "European sophistication" in his work (Ortiz de Montellano 1990).

Another important source is the work of Fray Bernadino de Sahagún; he was a Franciscan missionary who arrived in Mexico in 1529 and worked there until his death in 1590. It is certainly the best source available for the early historical period. De Sahagún left several codices (among them the Codex Florentino, compiled ca. 1570) and on the basis of these documents he wrote the "Historia General de las Cosas de Nueva España" (publ. 1793). From an ethnobotanical point of view this source is somewhat more difficult to use than the Codex Cruz Badiano since there are fewer botanical identifications and these are less secure. The strength lies more than anything else in its description and analysis of medicinal concepts (Ortiz de Montellano, 1990; Heinrich, 1996). The third important early source is the "Historia Natural de Nueva España" by Francisco Hernandez, the personal physician to Philip II of Spain. He was sent to Mexico and between 1571–1577 gathered information on plants, animals and minerals of the New World. The complete work was never published and the original manuscript was destroyed during a fire at the Escorial palace, but several abridged and amended versions were published in later centuries.

A. ludoviciana ssp. *mexicana* is not included as a botanically identifiable drawing in the Codex Cruz Badiano, but iztauyatl (Miranda and Valdés, 1991) is mentioned several times in this source (Folios 26r, 35r, 37r, 37v, 50r, 55v) as well as in the Codex Florentino. Uses in the Codex Cruz Badiano mentioned are: for debility of the hands (together with two other plants), for rectal problems ("ani uitium emendatur herbis" and "vitium sedis"; with 5 other plants), for aching piles [with 12 other plants, 3 types of "stones" (minerals) and one type of earth], as a remedy of second choice in order to recover from tiredness (together with 4 plants and 3 types of stones), for those injured by lightning and against lice (prepared in grease together with one plant and the ashes of the head of a mouse) and for treating "excessive heat". In the Codex Florentino the plant is considered to be useful to get rid of phlegm, headache (?), "inner heat" as well as for cleaning the urine and together with another plant (cuauhyayahuae) to help with heart problems. Applied externally it is listed for abscesses of the neck and in cases of dandruff (Codex Florentino ca. 1570: X,165,140; XI149,165). Gregorio Lopez (1542–1596) advised that the plant should be taken if one feels sick or dizzy or if there is retention of urine. It can also be used as an infusion against rheumatism. Juan de Esteyneffer (1664–1716) used the plant for numerous illnesses: as an anthelmintic and stomachic, against paralysis, vomiting, constipation, liver obstruction, dropsy, for "mal de loanda" etc. (Anzures y Bolanos, 1978). In the 18th century it was said to give strength and boldness, to strengthen the spirit, to alleviate tiredness, to "correct" the menstrual cycle and for fever (Ricardo Ossado cited in Argueta, 1994). Fray Juan de Navarro, whose manuscript bears the date 1801 and who based his studies on the earlier works lists Yztauyatl as useful to treat gastrointestinal pain caused by colics and for pain of the flanks. Some combined preparations with other plants are also listed (Navarro, 1992)

Modern Internal Uses for Gastrointestinal Complaints in Mexico

In modern Mexico this species is widely used and frequently sold in markets or little shops as a medicinal plant. Mexicans generally use the infusion of the leaves as a bitter stimulant, against gastrointestinal colics and the powdered flowers as a vermifuge, stimulant and emmenagogue (Argueta V. *et al.*, 1994: 626, Aguilar *et al.*, 1994, Heinrich *et al.*, 1998). Ruiz Salazar (1989), for example, mentions the following uses: lack of appetite, stomach-ache and parasites, stomach and liver infections. On the market of Sonora (México D.F.) the plant is sold for the treatment of dysentery and vomiting. The flowers are traded as useful remedies for colic, dysentery, diarrhoea, indigestion, pain and stomach-aches (Bye 1986; Linares *et al.*, 1990). A tea, prepared from the leaves, is sold in the markets for treating various gastrointestinal disorders (Heinrich *et al.*, 1992b, Nicholson and Arzeni, 1993). The flowers and other aerial parts of the plant are used by the lowland Mixe as tea for stomach-ache and vomiting (Heinrich, 1989). The flowers are used for the treatment of intestinal parasites "lombrices" in Morelos and in numerous other parts of Mexico (Baytelman, 1979). It is also used as a laxative, but should be avoided during pregnancy because of its emmenagogic effects (Cabrera, 1984). In "Izucar de Matamoros" it was used as an infusion to cure "bilis", and stomach-ache (Rivera, 1943). The Huastec treat vomiting and gastrointestinal pain with the root (Alcorn, 1984), the Huichol the same syndrome with the leaves (Casillas Romo, 1990). Also, its uses as an anti-emetic, anthelmintic and for dysentery and dysmenorrhoea are reported from Veracruz (Amo, 1979).

Modern Internal Uses for Other Illnesses in Mexico

There are only a few scattered reports on other uses available in the literature: for colds (Rivera 1943; Morton, 1981), bronchitis (Amo, 1979), chest congestion (Bye, 1986), heart diseases (Morton, 1981), sudden fright and some other "culture-bound" syndromes (Aguilar *et al.*, 1994) as well as menstrual complaints (e.g. Ruiz Salazar, 1989).

Modern External Uses in Mexico

The leaves are frequently used to treat earache. In central Mexico it seems to be an important remedy for the folk illness "aire" which is said to be associated with headache, dizziness and vomiting (Argueta, 1994). Other uses include as an analgesic, for externally cleansing swellings, infections, and inflammations, for headaches, sudden fright, for ritual cleansing ceremonies and in the treatment of swollen feet (Alcorn, 1984; Amo, 1979; Linares *et al.*, 1990; Nicholson and Arzeni, 1993; Pimentel Tort, 1988).

PHYTOCHEMISTRY

Sesquiterpene Lactones

Detailed phytochemical studies were first undertaken in the early sixties (Sanchez-Viesca and Romo 1963) and yielded estafiatin (3). In the late sixties and early seventies several further sesquiterpene lactones (Geissman and Saitoh 1972, Lee and

Geissman 1970, Romo *et al.*, 1970, Romo and Tello, 1972, Romo de Vivar *et al.*, 1977) were isolated: arglanin (*5*), armexefolin (*9*), armexine (*7*), artemixifolin (*8*), artemorin (*1*), chrysartemin-A (*4*), douglanine (*6*), ludalbin (*13*), ludovicin-A (*10*), ludovicin-B (*11*), ludovicin-C(*12*), tulipinolide (*2*) and – a report which has been substantiated by other reseachers – α-santonin (Kelsey and Shafizadeh, 1979). Mata *et al.* (1984) additionally reported three new eudesmanolides [8 α-acetoxyarmexifolin, α-epoxyludalbin and armefolin (*14*)], the known eudesmanolides arglanin, artemixifolin, ludalbin as well as santamarin (*15*), known from several other members of the Compositae. Ruiz-Cancino, *et al.* (1993), reported the new 3 α-hydroxyreynosin as well as 1 α-,3 α-dihydroxyarbusculin B, santamarin, arglanin, artemorin, chysartemin B, armefolin and ridentin. There is some variation with regard to the compounds isolated from the different collections, but the data available are insufficient to decide whether this is due to seasonal, chemosystematic, climatic or other factors.

Flavonoids

Ruiz-Cancino, *et al.* (1993), isolated the flavonoids eupatilin (*16*) and jaceosidin (*17*). An unspecified subspecies of *A. ludoviciana* yielded naringin (Jakupovic, *et al.*, 1991).

Essential Oil

The essential oil of this taxon has not been studied in detail. According to Guillermo Delgado (Dept. de Quimica, UNAM, México, D.F.; personal communication), the essential oil of the fresh plant contains at least 70 different compounds, but most have not yet been identified. Jakupovic, *et al.* (1991), reported the presence of camphor, borneol, vanillyl alcohol, several monoterpenes, and two jonone derivatives from an unspecified subspecies of *A. ludoviciana* collected in Texas.

BIOLOGICAL AND PHARMACOLOGICAL ACTIVITIES

No detailed pharmacological study on the species is available. The plant is regarded to possess spasmolytic and antihelmintic properties, but the experimental basis for this is insufficient (Heinrich, *et al.*, 1997). The ethanolic crude extract was shown to possess antimicrobial activity against gram positive, (*Bacillus subtilis* DSM 347 and *Micrococcus luteus* DSM 348) and gram negative bacteria (*E. coli* DSM 1077) and against non-pathogenic fungi (*Cladosporium cucumerinum, Penicillium oxalicum*). This extract was inactive against *Entamoeba histoytica in vitro* (Heinrich, *et al.*, 1992a). Studies from the late 19th century indicate anthelmintic effects of the plant extract (Lozoya and Lozoya, 1982). Martinez, (1969) remarked that *Artemisia ludoviciana* ssp. *mexicana* is regarded as less toxic than, for example, *Artemisia*

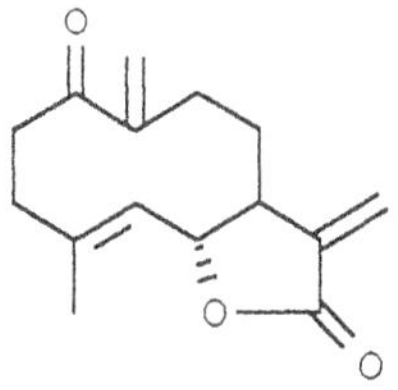

1 Artemorin

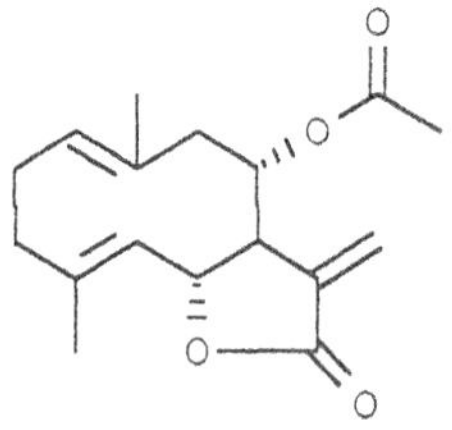

2 Tulipinolide

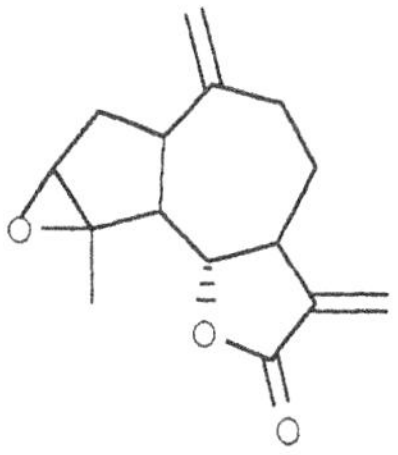

3 Estafiatin

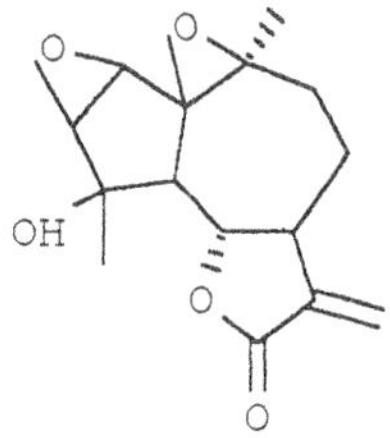

4 Chrysartemin-A

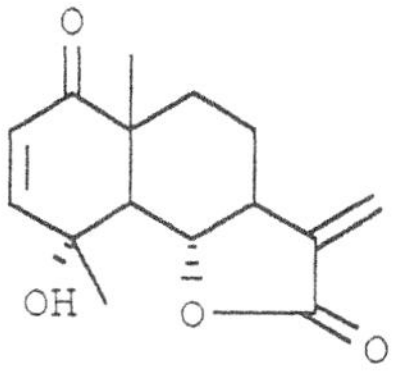

5 Arglanin

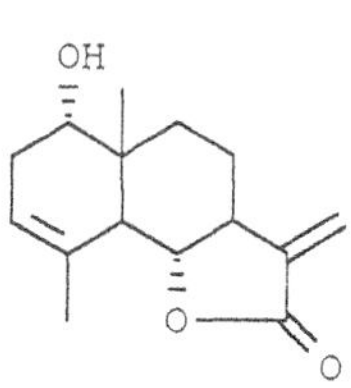

6 Douglanine

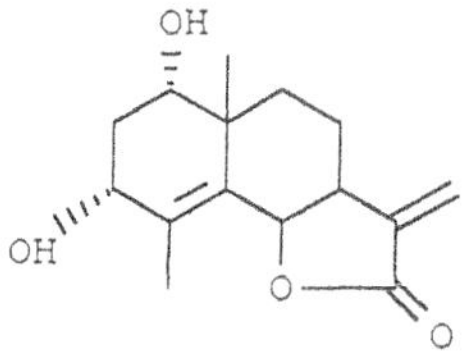

7 Armexine

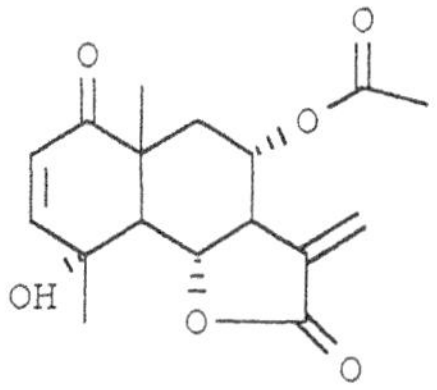

8 Artemexifolin

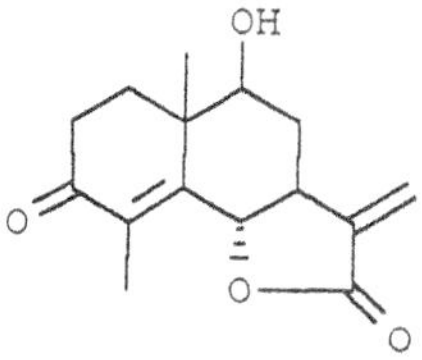

9 Armexifolin

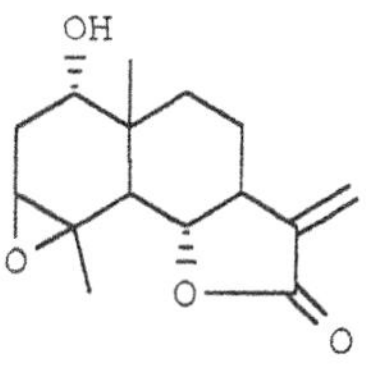

10 Ludovicin-A

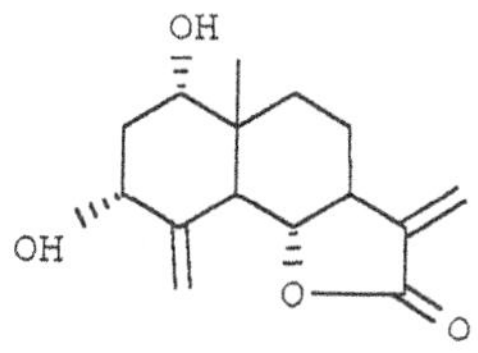

11 Ludovicin-B

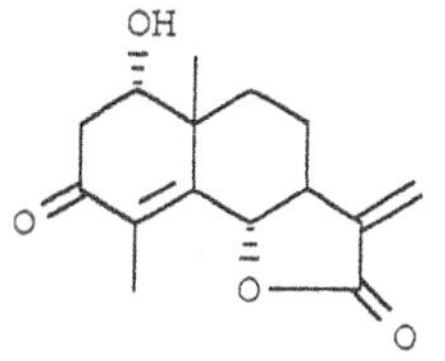

12 Ludovicin-C

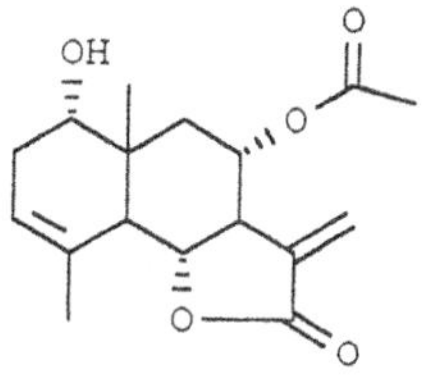

13 Ludalbin

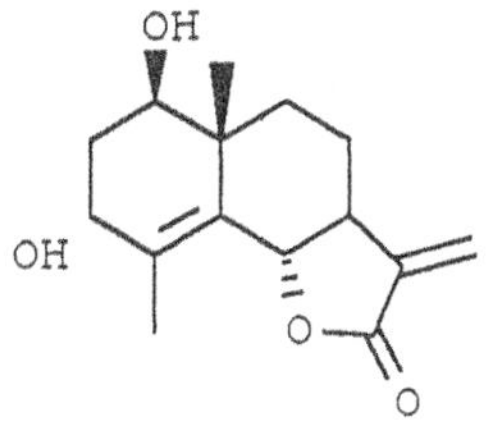

14 Armefolin

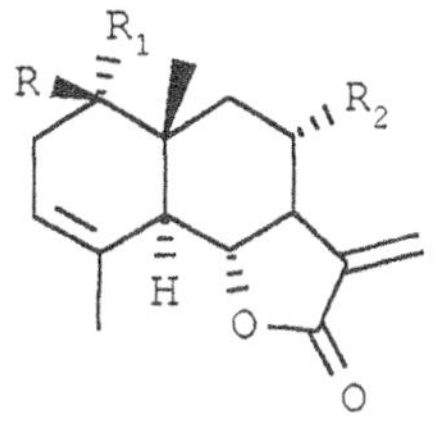

15 Santamarine

R= OH; R_1= R_2= H

Flavonoids from Artemisia ludoviciana ssp. mexicana:

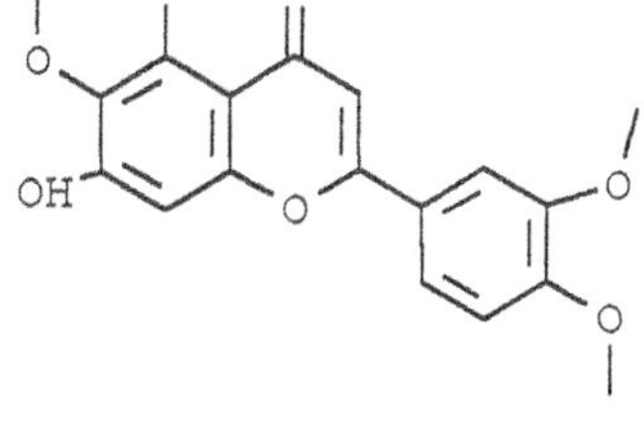

16 Eupatilin

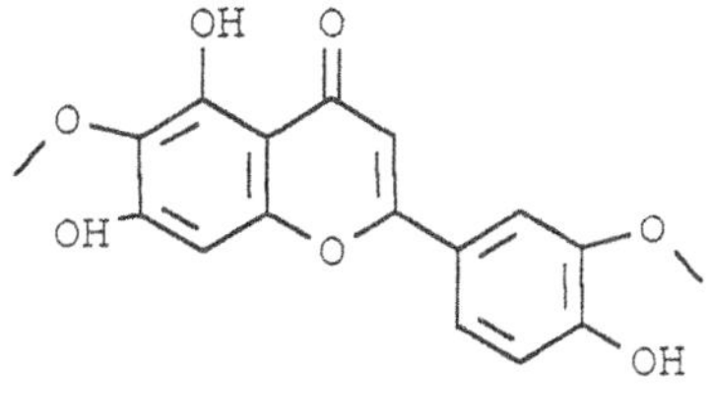

17 Jaceosidin

absinthium (cf. Villada, 1889). The plant is reported to have spasmolytic properties and camphor as an active ingredient is a mild irritant, stimulant and reliever of colics. It may thus be an effective vermicide and appetite stimulant (Messer, 1991). Recently it was shown that the ethanolic extract of the aerial parts is a potent and specific inhibitor of the transcription factor NF-κB (Bork *et al.*, 1997). Sesquiterpene lactones are considered to be compounds responsible for this activity (Bork *et al.*, 1997; Hehner *et al.*, 1998).

CONCLUSION

The taxon under discussion is a widely used and very important medicinal plant in Mexico with a well defined principal use: the treatment of various forms of gastro-intestinal cramps and pain. The plant has been used for at least 500 years, but its uses seem to have undergone considerable changes. It is likely that the sesquiter-pene lactones are pharmacologically relevant to its modern uses, not just as bitter stimulants, but also because they have several well documented pharmacologcal effects i.e. cytotoxic, antibacterial, antiinflammatory, and anthelmintic (Heinrich, *et al.*, 1997). However, at present there are insufficient pharmacological and parasitological/biological data available for this species, and therefore detailed pharmacological and/or clinical studies of standardized extracts from estafiate are required. This species is an economically and botanically important member of the genus *Artemisia* for which further evaluation would be justified.

One fascinating question is whether the use of this species as a gastrointestinal remedy was introduced into Mexico by Europeans based on the similarity of this taxon with *A. absinthium* or whether the uses developed independently in parallel to the uses of the European absinthe. The uses reported in the early sources are distinct from the modern ones. This may be explained by an introduction of the use as a gastrointestinal remedy during the Spanish colonial rule. While this is the most likely explanation, alternatively, one may speculate that the use of "iztauyatl" as a gastro-intestinal remedy was so widely distributed in the Nahua population, that it was not considered to be worth mentioning in texts such as the Codex Cruz-Badiano or the Codex Florentino.

REFERENCES

Aguilar, A., Camacho, J.R., Chino, S., Jácquez, P., and López, M.E. (1994) *Herbario Medicinal del Instituto Mexicano del Seguro Social*. México D.F. Instituto Mexicano del Seguro Social

Alcorn, J.B. (1984). *Huastec Mayan Ethnobotany*. Univ. Texas Pr., Austin.

Amo R.S. del (1979) *Plantas medicinales del Estado de Veracruz*. Xalapa (Veracruz), Instituto Nacional de Investigaciones sobre Recursos Bioticos.

Anzures y Bolanos, N. del Carmen, (eda.) (1978). *Juan de Esteyneffer: Florilegio Medicinal*. Academia nacional de Medicina, México D.F.

Argueta Villamar, A. (editor) (1994) *Atlas de las Plantas de la medicina tradicional Mexicana*. 3 vols. México D.F., Instituto Nacional Indigenista.

Baytelman, B. (1979) *Etnobotanica en el estado de Morelos*. Cuernavaca (México). Instituto Nacional de Antropología e Historia, Centro Regional de Morelos.

Bork, P.M., Schmitz, M.L., Kuhnt, M., Escher, C. and Heinrich, M. 1997. Sesquiterpene Lactone Containing Mexican Indian Medicinal Plants and Pure Sesquiterpene Lactones as Potent Inhibitors of Transcription Factor κB (NF-κB). *FEBS-Letters*, 402, 85–90.

Bye, R., (1986) Medicinal Plants of the Sierra Madre: Comparative Study of Tarahumara and Mexican Market Plants. *Econ. Bot.*, 40, 103–124.

Cabrera, L. (1984) *Plantas curativas de México*. México D.F. Libro-Mex Ed.

Casillas Romo, A. (1990) *Nosolgía Mítica de un Pueblo. Medicina Tradicional Huichola*. Guadalajara. Ed. Universidad de Guadalajara.

Codex Florentino (ca. 1570); compiled by Fray Bernardino de Sahagun, facsimilar ed. (1979). Archivo General de la Nación, México, D.F.

Cronquist, A., Holmgren, A., Holmgren, N., Reveal, J.L., and Holmgren, P.K., (1994) *Intermountain Flora. Vascular Plants of the Intermountain West, USA*. Bronx, N.Y., New York Botanical Garden

Cruz, M. de la and Badiano, J. (1991). [orig. 1552] Libellus Medicinalibus Indorum Herbis. *Fondo de la Cultura Economica/Instituto Mexicano del Seguro Social. México D.F.* 2 vols. 64 folios.

Geissman, T.A., and Saitoh, T. (1972) Ludalbin, a New Lactone from Artemisia ludoviciana. *Phytochemistry*, 11, 1157–1160.

Hehner, S.P., Heinrich, M., Bork, P.M., Vogt, M., Ratter, F., Lehmann, V., Schulze-Osthoff, K., Droge, W. and Schmitz, M.L. Sesquiterpene Lactones Specifically Inhibit Activation of NF-κB-β by Preventing the Degradation of IκB-α and IκB-β. *Journal of Biological Chemistry*, 273, 1288–1297.

Heinrich, M. (1989) *Ethnobotanik der Tieflandmixe (Oaxaca, Mexico) und phytochemische Untersuchung von Capraria biflora L. (Scrophulariaceae)*. Dissertationes Botanicae No. 144. Berlin und Stuttgart: J. Cramer in Gebr. Borntraeger Verlagsbuchhdlg.

Heinrich, M. (1992) Economic Botany of American Labiatae. IN: R.M. Harley and T. Reynolds *Advances in Labiatae Science*. Richmond: Kew Botanical Gardens. pp. 475–488

Heinrich, M. (1996) Ethnobotany of Mexican Compositae: An Analysis of Historical and Modern Sources. IN: D.J.N. Hind (Editor-in Chief). *Proceedings of the International Compositae Conference, Kew*, 1994. Vol. 2. Biology and Utilization (vol. eds. P.D.S. Caligaria and D.J.N. Hind), Royal Botanic Gardens, Kew, UK. pp. 475–503.

Heinrich, M., Kuhnt, M., Wright, C.W., Rimpler, H., Phillipson, D.P., Schandelmaier, A. and Warhurst, D.C. (1992a) Parasitological and Microbiological Evaluation of Mixe Indian Medicinal Plants (Mexico). *Journal of Ethnopharmacology*, 36, 81–85.

Heinrich, M., Rimpler, H. and Antonio N.B. (1992b) Indigenous Phytotherapy of Gastrointestinal Disorders in a Mixe Lowland Community. *Journal of Ethnopharmacology*, 36, 63–80.

Heinrich, M., Robles, M., West, J.E., Ortiz de Montellano, B. and Rodriguez, E. (1997). Ethnopharmacology of Mexican Daisies. *Annual Review of Toxicology and Pharmacology*, 38, 539–545.

Heinrich, M. A. Ankli, B. Frei, C. Weimann and O. Sticher (1998). Medicinal Plants in Mexico: Healers' Consensus and Cultural Importance. *Social Science and Medicine*, 47 1863–1875.

Jakupovic, J., Tan, R.X., Bohlmann, F., Boldt, P.E. and Jia, Z.J. (1991) Sesquiterpene Lactones from *Artemisia ludoviciana*. *Phytochemistry*, 30, 1573–157.

Kartesz, J.T. and Kartesz, R. (1980) *A Synonymized Checklist of the Vascular Flora of the United States, Canada and Greenland. Chapell Hill.* Univ. of North Carolina Pr.

Kelsey, R.G. and Shafizadeh, F. (1979) Sesquiterpene Lactones and Systematics of the Genus *Artemisa. Phytochemistry*, **18**, 1591–1611.

Lee, K.H. and Geissman, T.A. (1970) Sesquiterpene Lactones of *Artemisia*: Constituents of *A. ludoviciana* ssp. *mexicana. Phytochemistry*, **9**, 403–408.

Linares, M., Edelmira, R., Bye y, B. and Flores P. (1990). *Tes curativos de México.* UNAM, México, D.F. Cuadernos del Instituto de Biología No. 7. pp. 140

Lozoya, X. and Lozoya, M. (1982) *Flora Medicinal de Mexico.* Premiera Parte: Plantas indigenas. México, D.F. Instituto Mexicano del Seguro Social (IMSS)

Martinez, M. (1969). *Las plantas medicinales de México.* 5ª edición. Ed. Botas, México, D.F.

Mata, R., Delgado, G. and Romo de Vivvar, A. (1984) Sesquiterpene Lactones of *Artemisia mexicana* var. *angustifolia. Phytochemistry*, **23**, 1665–1668

Messer, E. (1991) Systematic and Medicinal Reasoning in Mitla Folk Botany. *J. Ethnopharmac.*, **33**, 107–128.

Miranda, F. and Valdés, J. (1991) Comentarios Botanicos. IN: Cruz, Martin de la and Juan Badiano (1991). [orig. 1552] *Libellus Medicinalibus Indorum Herbis.* Version Española con Estudios y Comentarios por diversa autores. Fondo de la Cultura Economica/Instituto Mexicano del Seguro Social. México, D.F. pp. 107–148.

Morton, J. (1981). *Atlas of Medicinal Plants of Middle America, Bahamas to Yucatan.* C. Thomas, Springfield (IL).

Navarro, F.J. (1992) *Historia Natural o Jardín Americana* [Manuscrito de 1801]. México, D.F. Universidad Nacional Autónoma de México, Instituto Mexicano del Seguro Social y Instituto de Seguridad y Servicios Sociales de los Trabajadores del Estado.

Nicholson, M.S. and Charles B.A. (1993) The Market Medicinal Plants of Monterrey, Nuevo León, México. Economic *Botany*, **47**, 184–192.

Ortiz de Montellano, B. (1990) *Aztec Medicine, Health and Nutrition.* New Brunswick. Rutgers Univ. Pr.

Pimentel Tort, J.A. (1988) *Plantas de uso medicinal entre los zoques de Tecpatán. Instituto Chiapaneco de cultura.* Tuxtla Gutiérrez, Chiapas. pp. 51

Pineda, M.E. (1992) Una nueva versión en español del *Libellus de* medicinalibus Indorum herbis. IN: Kumate, Jesús *et al. Estudios actuales sobre el libellus de medicinalibus Indorum herbis.* Secretaría de Salud. Mexico D.F., pp. 17–47.

Rivera, I.M. (1943) Algunas plantas medicinales de Izucar De Matamoros y pueblos anexos. Anales Instituto de Biología, Universidad Nacional de México, **14(1)**, 37–67.

Romo, J. and Tello, H. (1972) Estudio de la Artemisia mexicana: Armexina un nuevo santanolido cuya lactona posee fusion *cis. Rev. Latinoam. Quim.*, **3**, 112–126.

Romo, J., Romo de Vivar, A., Treviño, R., Joseph-Nathan, P. and Díaz, E. (1970) Constituents of *Artemisa* and *Chrysanthemum* species – the Structures of Chrysartemins A and B. *Phytochemistry*, **9**, 1615–1621.

Romo de Vivar, A., Vázquez, F. and Zetina, C. (1977) Lactones Sesquiterpenicas de *Artemisa mexicana* var. *angustifolia. Rev. Latinoam. Quim.*, **8**, 127–130.

Ruiz-Cancino, A., Cano, A. E and Delgado, G. (1993) Sesquiterpene Lactones and Flavonoids from *Artemisia ludoviciana* ssp. *mexicana. Phytochemistry*, **33**, 1113–1115.

Ruiz Salazar, C. (1989) Contribución al estudio de las plantas medicinales de las Delegacón Xochimilco. Mexico D.F. MEXU, Herbario Nacional (UNAM). tesis inedita, pp. 15f.

Sanchez-Viesca, F. and Romo, J. (1963) Estafiatin, a New Sesquiterpene Lactone isolated from *Artemisia mexicana. Tetrahedron*, **19**, 1285–1291.

Valdés Gutiérrez, Javier, H., Olvera, F. and Ochoterena-Booth. H. (1992) La botanica en el códice de la Cruz; IN: Kumate, Jesús *et al. Estudios actuales sobre el libellus de medicinalibus Indorum herbis*. Secretaría de Salud. Mexico D.F., pp. 129–180.

Viesca Treviño, C. (1992) El *libellus* y su contexto histórico. IN: Kumate, Jesús *et al. Estudios actuales sobre el Libellus de medicinalibus Indorum herbis*. Secretaría de Salud. Mexico D.F., pp. 49–84.

Villada, M.M. (1889) Apuntos acerca de plantas de la familia de las compuestas empleadas en la medicina. *Gaceta Médica de México*, **24**, 241–249.

7. *ARTEMISIA PALLENS*

R.N. KULKARNI

Central Institute of Medicinal and Aromatic Plants, Field Station, Allalasandra, Bangalore – 560 065, India.

INTRODUCTION

Artemisia pallens Wall. ex DC. is a small, herbaceous, aromatic plant with an exquisite aroma. It is grown in southern parts of India particularly in the states of Karnataka, Tamil Nadu and Andhra Pradesh. It is apparently native to this area in south India. However, it is not found growing wild in this area probably because its seeds are very small and the seedlings are very delicate and need to be nurtured with extreme care until they are at least one month old. *A. pallens* is locally known as davana and its essential oil is known as davana oil all over the world. *A. pallens* is, therefore, referred to as davana in this chapter.

Davana has been traditionally cultivated sporadically in gardens in south India for its delicately fragrant leaves and flower heads which are used in garlands, chaplets and religious offerings; it has been cultivated only recently for its essential oil. The fragrance of the herb and its oil is described as exquisite, deep, mellow, persistent and characteristically fruity. The cultivation of this plant on a commercial scale began only in the late 1960's after its oil caught the fancy of perfumers in the USA and Europe. Commercial quantities of davana oil on a large scale were, however, not produced until 1970 (Sugunakar, 1987). For last 10 years, davana oil has been regularly exported from India, although not in very large quantities, to the USA, France, Germany, Japan, the Netherlands and many other countries (Fig. 1). The latest figures show that the amount exported in 1998–1999 was much greater (4,060 kg) with a value of Rs 89 million. Following the recognition of the economic value of its essential oil, interest and research into various aspects of davana has increased.

BOTANY

The botanical description of davana, as given by Narayana *et al.* (1978), is as follows: Davana is an annual, erect, branched, aromatic herb, 40–60 cm high and covered with greyish-white tomentose. Leaves are bluish-green in colour, alternate, exstipulate, petiolate, lobed to pinnatisect and covered with greyish-white tomentose. Capitula are peduncled to sessile, axillary or forming lax racemes, simple, heterogamous with yellow florets. Involucre, two or more seriate, with ovate to elliptic-linear, alternating entire bracts, grey-tomentose outside, glabrous and green inside. Outer florets glabrous except for a few cottony hairs, tubular, generally three-lobed, female without pappus; stigma generally two-lobed, rarely three-lobed. Inner

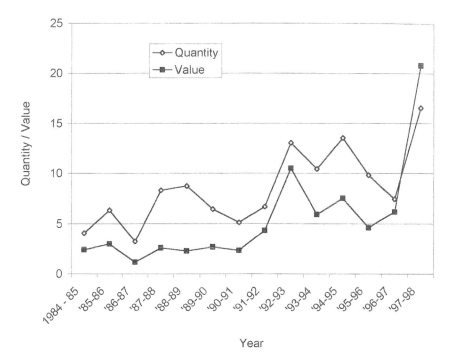

Figure 1 Quantity (× 100 kg) and value (million Rs.) of davana oil exported from India
during 1984–85 to 1997–98.

florets also glabrous except for a few cottony hairs, tubular, five-lobbed, bisexual;
stamens five with free epipetalous filaments and dithecous, introse, syngenesious
anthers; pollen sacs prolonged, tapering; style bifid.

AGRONOMY (INCLUDING DISTILLATION FOR ESSENTIAL OIL)

Davana oil is reported to have been first distilled on an experimental scale from the
herb grown near Mysore, as early as in 1921 (Sastry, 1946; Anon., 1947, 1950), at
the Government Soap Factory, Bangalore. The commercial distillation and export of
davana oil was first taken up at about the same time by Mr. M. Sundara Rao, at
M/s. Essenfleur Products Ltd., Mysore, from the davana herb grown in his estate at
Hemmanahalli, near Mysore (Sastry, 1946; Menon, 1960). However, according to
another report (Anon., 1967), davana oil was first distilled and exported to Europe
by Mr. Mavalankar, an Indian perfumer. The report also mentioned that the oil was
distilled and exported by The East Indian Sandal Oil Distilleries (Private) Ltd.,
Kuppam (S. India).

Although Sastry, (1946) Guenther, (1952) and Ranganathan, (1963) briefly
described the cultivation and distillation of davana, a detailed account was first
given by Gowda and Ramaswamy (1965). Subsequently, Gulati *et al.* (1967) gave a

similar account of cultivation and distillation of davana in the Tarai region in north India. However, the quality of davana oil from this area was not comparable with the quality of oil produced in south India. A farm bulletin describing the cultivation of davana and distillation of its oil was published in 1978 and, subsequently, revised in 1990 (Narayana *et al.*, 1978; Narayana and Dimri, 1990). The above reports appear to be based largely on empirical observations and experience rather than based on data obtained from specifically designed and systematically conducted experiments. A sudden increase in the interest in medicinal and aromatic plants and a preference for plant based rather than synthetic products all over the world during the last decade has resulted in increased scientific activity in this area and davana has also received increased attention. Different agronomic parameters including fertilizer requirement, optimum planting density, planting season etc. for maximizing productivity of davana have been investigated (Table 1). The effect of plant growth regulators like gibberellic acid, cycocel and TIBA (Tri-iodo-benzoic acid), which have been successfully used to increase the yields of several horticultural plants were also studied for their effects on the herb and oil yield of davana; however they were found to be ineffective (Farooqi *et al.*, 1993; Shenoy *et al.*, 1993).

The agronomic practices for davana are summarized as follows after taking into consideration all the published information cited above.

Soil and Season

Davana is cultivated in a very small restricted geographic area in south India. It does not withstand heavy rains and water logging. It grows well in red loamy soils with good drainage and organic matter. The crop can be grown almost throughout the year for ornamental purposes. However, for cultivation of this crop for its essential oil, the season is very important and late October to early November sown crops have been found to give maximum oil yield. Light showers, mild winters with heavy morning dew and no frost in the beginning of the season, and warm and dry weather towards harvest time are believed to contribute to higher oil yields and better quality oil.

Nursery Preparation

Fresh seed obtained from the previous crop should preferably be used as davana seeds are known to lose their viability very quickly. About 1.5 kg of seeds and an area of 500 m² are required for raising seedlings sufficient for planting an area of one hectare. As the seeds are very small, they are mixed with about 10 kg of fine sand before sowing and are sown by broadcasting the mixture in well prepared nursery beds of generally 3 m² size. Seeds germinate in three to four days and during this period nursery beds are hand-watered lightly, twice a day. Subsequently, nursery beds are lightly surface irrigated every day. The seedlings reach the transplanting stage after about five weeks.

Table 1 Optimum planting season, plant spacing and nitrogen dose for the cultivation of davana (*Artemisia pallens*) for its essential oil.

Planting season	Plant spacing (cm²)	Nitrogen (kg/ha)	Reference
—	—	82.5 (Main crop) 27.5 (Ratoon crop)	Narayana et al. (1982)
—	30 × 15	150 (Main crop) 75 (Ratoon crop)	Prakasa Rao et al. (1983)
—	—	80 (Main crop)	Rajeswara Rao et al. (1989)
—	30 × 7.5	100 (Main crop)	Rajeswara Rao et al. (1989a)
November	15 × 7.5	—	Farooqi et al. (1990)
Late October	30 × 15	—	Janardhana Rao et al. (1990)
November	15 × 7.5	160 (Main crop) 80 (Ratoon crop)	Narayana and Dimri (1990)
—	—	120 (Main crop)	Farooqi et al. (1991)

Fertilizer Requirement

It is difficult to assess the fertilizer requirement from the available literature. The maximum response has been obtained with 100 – 150 kg N per hectare for the main crop and 75 kg for the ratoon crop applied at monthly intervals, in three and two equal splits for the main and ratoon crops, respectively; (the ratoon crop is that grown from the stalks remaining after harvest of the main crop). Six tonnes of farm-yard manure, and 40 kg each of P and K per hectare are incorporated into the soil at the time of land preparation.

Transplanting and Irrigation

Five week old seedlings are transplanted in the field at a spacing of 15×7.5 cm^2; spacing wider than 30×15 cm^2 have been found to give lower oil yields. The crop is irrigated lightly, every day for about 10 days after transplanting and twice a week subsequently.

Harvesting

Harvesting is done when the crop is in full bloom. This stage is generally reached by the end of February or by first week of March i.e., about 10 weeks after transplanting. The crop is harvested at a height of about five to six cm above the ground level; the ratoon crop comes to harvest about two months later. The main and ratoon crops together yield about 10–15 tonnes of herbage and about 7.5–10 kg of oil per hectare.

Distillation

Davana herbage is either dried in shade for two to three days or for not more than 12 hours in bright sunshine before distillation. The herb is then steam distilled at a pressure of about 1.0–2.0 kg/cm^2 for about 10–15 hours; the bulk of the oil is, however, distilled in about 8 hours. The recovery of the oil from the semi-dried herbage is generally about 0.2%. The oil is then filtered, dried with anhydrous sodium sulphate and stored in aluminium containers.

Quality Aspects

Davana oil is produced in three states in south India. The oil produced in states of Karnataka and Andhra Pradesh is, however, usually preferred to the oil produced in southern parts of Tamil Nadu, as the latter is considered to be inferior in odour value. As davana oil is an expensive oil, there is a possibility of adulteration of the oil with cheaper and undesirable materials like vegetable or fixed oils. This will drastically affect the quality of the oil as well as its trade. It is therefore necessary to formulate standards for davana oil with respect to its physico-chemical characteristics and chemical constituents in order to exercise quality control.

GENETICS, BREEDING AND TISSUE CULTURE

Attention has been given to breeding aspects of davana only since 1990. Its karyo-type was studied only recently (Rekha and Kak, 1993) and classified as symmetrical, with a chromosome number of 2n = 16. The blossom biology and optimum planting season for seed set, the prerequisites for any breeding programme, were studied by Rai and Farooqi, (1990) and Farooqi *et al.* (1990a). It was found that flower heads took 19 days from visible initiation to the full bloom stage. Anthesis was observed between 7.30 a.m. to 12.30 p.m. with a peak at 10.30 a.m. Anther dehiscence also occurred during the same period with a peak at 9.30 a.m. Stigma receptivity was maximum at 7 a.m., one day after anthesis. November sown crops showed maximum seed set.

Assessment of genetic variability is basic to any plant breeding programme. Farooqi *et al.* (1990b) assessed genetic variability for 10 metric traits in five acces-sions, collected from davana growing regions. Significant genotypic variation was found for only two traits, plant height and oil content. Although, broad-sense heri-tability for oil content was 91%, the range (0.09 to 0.11%) was too narrow for taking up any breeding programme. Lack of genetic variability could have been due to the fact that farmers may have procured seed from the same source as there are not many seed suppliers. Also, farmers generally do not produce their own seed, as they harvest davana before seed set, for both ornamental and essential oil purposes, and therefore depend on seed suppliers for their requirement of seed.

To overcome the problem of lack of genetic variability for genetic improvement, gamma rays were used with a view to induce variability (Farooqi *et al.*, 1990c). However, only M_1 generation parameters were studied and no information was obtained on gamma ray induced genetic variability.

A breeding programme for improving herb yield of davana was started in 1986–87, long before the publication of the above reports on floral biology and lack of genetic variability in davana. The programme was taken up on the basic assumption that there should be sufficient genetic variability in davana for genetic improvement by selection, as davana is a cross-pollinating species and has not been subjected to any artificial selection (Kulkarni, 1991). Herb yield of davana could be increased by about 38% over the unselected population through three cycles of simple mass selection using wide plant spacing and honeycomb design advocated by Fasoulas, (1981). Selection for herb yield did not have any adverse effect on either oil content or davanone content in the oil. The study thus suggested that there was considerable intrapopulation additive genetic variation in the unselected population.

The oil content as well as herb yields of individual plants in davana are very low and, therefore, at present it is not possible to evaluate individual plants for oil con-tent and utilize the reported (Farooqi *et al.*, 1990b) high heritability of oil content in breeding programmes. The availability of a suitable distillation apparatus for the determination of oil contents in small plant samples of species with low oil content would be immensely useful in breeding for high oil content and oil yield in davana. The development of one such apparatus has been reported, recently (Whish, 1996).

A beginning has been made in the area of tissue culture of this plant by studies on micropropagation and organogenesis (Benjamin *et al.*, 1990; Sharief and Jagadishchandra, 1991).

DISEASES AND PESTS

No serious plant pathological or entomological problems have been reported in davana. However, there are reports on the occasional occurrence of damping-off of seedlings in the nursery, caused by *Rhizoctonia* sp. (Sarwar and Khan, 1971), collar rot of plants in the field, caused by *Fusarium oxysporum* (Sarwar and Khan, 1975) and root knot nematode, *Meloidogyne incognita* (Pandey, 1994) in davana.

PHYTOCHEMISTRY

The chemistry of the essential oil of davana has been more extensively studied than any other aspect of this plant. Further, most of these studies have been carried out only after 1970 and not in India, where davana oil is exclusively produced, but in other countries. The reasons for these facts are not difficult to guess. Although the cultivation of davana is restricted to a small geographical area in south India, studies on the chemistry of its essential oil are not restricted by any such agro-climatic requirements and hence could be carried out anywhere. The availability of commercial quantities of davana oil only after 1970 (Sugunakar, 1987), and relatively easy accessibility to modern analytical instrumental facilities for evaluating the quality of essential oils in developed countries, may also be a reason for more studies on this oil having been carried out after 1970 and outside India; further, the oil is hardly used in India.

Apart from being the most extensively studied aspect, the chemistry of the essential oil of davana has also been periodically reviewed (Lawrence, 1978, 1988, 1995; Verghese and Jacob, 1984; Akhila and Tiwari, 1986; Mallavarapu, 1995). Studies on davana oil prior to 1967 were mainly restricted to its physico-chemical properties (Table 2).

The first work on the chemistry of davana oil was done by Sipma and Van der Wal (1968), who isolated davanone, the major constituent of the oil. This was later shown to be cis-davanone and is odourless (Thomas and Pitton, 1971; Thomas and Dubini, 1974). Several compounds, including linalool, linalool oxides, lilac aldehydes, lilac alcohols, cinnamic acid, geraniol, nerol etc., as well as sesquiterpenes, oxygenated sesquiterpenes and dihydrofurano sesquiterpenoids have been identified from davana oil. These are compiled, mainly from review articles cited above and presented in Table 3, in chronological order. Significant findings among these appear to be identification, isolation and structure determination of compounds responsible for the characteristic odour of davana oil viz., davana ether (Thomas and Pitton, 1971), davanafurans (Thomas and Dubini, 1974a), dihydrofurans (Chandra *et al.*, 1987), α- and β-dihydrorosefurans, trans-hydroxy davanone, hydroxy

Table 2   Physico-chemical properties of davana (*Artemisia pallens*) oil.

Reference	Colour	Odour	Specific gravity	Refractive index	Optical rotation	Acid value	Ester value	Solubility	Total ketone content
Rao et al. (1937)	Dark brown viscous liquid	Pleasant balsamic odour	0.9833 (30° C)	1.4898 (30° C)	−25.8° (30° C)	2.6	19.1	–	–
Sastry (1946)	–	–	0.9605 (15.5° C)	1.4880 (20° C)	35°	2.4	52.9	Not clearly soluble in 10 volumes of 70% alcohol	–
Guenther (1952)	Too dark to determine optical rotation	–	0.961 to to 0.990 (15° C)	1.4874 to 1.5003 (20° C)	42°0' to 54°28'	1.0 to 5.6	–	Not always clearly soluble in up to 10 volumes of 80% alcohol	–
Gowda and Ramaswamy (1965)	Brown viscous liquid	Deep mellow persistent rich fruity odour	0.9722 (24° C)	1.5085 (25° C)	Too dark to determine	4.35	56.9	Opalescent in 10 volumes of 80% alcohol	–
Anonymous (1967)	Brownish viscous liquid	Pleasant balsamic persistent odour	0.960 to 0.990 (20° C)	1.4874 to 1.5003 (20° C)	35°0' to 54°28'	5	30.0 to 53.0	Not clearly soluble in 10 volumes of 70% alcohol	–
Lewis and Nambudiri (1967)	Brownish viscous liquid	Fruity, sweet, persistent odour	0.9585 to 0.9722 (25° C)	1.4880 to 1.5083 (25° C)	41° to 54°	1.0 to 5.6	49.0 to 61.0	Not always clearly soluble in up to 10 volumes of 80% alcohol	–

Table 2 Physico-chemical properties of davana (*Artemisia pallens*) oil. (*Continued*)

Reference	Colour	Odour	Specific gravity	Refractive index	Optical rotation	Acid value	Ester value	Solubility	Total ketone content
Baslas (1971)	–	–	0.9902 (15° C)	1.501 (15° C)	–	8.696	65.35	–	–
Gulati (1980) South Indian Oil	Brown	–	0.9585 to 0.9833 (20 to 25° C)	1.4874 to 1.5033 (20° C)	–	1.95 to 5.03	52.9 to 58.9	Not always soluble in up to 10 volumes of 80% alcohol	–
North Indian Oil	Reddish brown	–	0.9700 to 1.0165 (32° C)	1.4811 to 1.499 (44° C)	17.80° to 28.00°	3.16 to 8.70	61.6 to 90.03	Soluble in 50 volumes of 80% alcohol	–
Suganakar (1987)	Clear brownish yellow liquid	Very rich lingering fruity odour	0.9394 to 0.9560 (25° C)	1.4794 to 1.4917 (25° C)	34° to 41°	Less than 3.5	31.5 to 46.5	Clearly soluble in less than 1.5 volumes of 80% ethyl alcohol	36 to 56.0%

Table 3   Compounds identified from *Artemisia pallens*.

Reference	Compounds identified
1. Lewis and Nambudiri (1967)	Hydrocarbons (20%); esters (65%); alcohols and other oxygenated compounds (15%)
2. Sipma and Van der Wal (1968)	Davanone
3. Naegeli *et al.* (1970)	Artemone
4. Baslas (1971)	Fenchyl alcohol; cinnamyl cinnamate; caryophyllene; cadinene
5. Thomas *et al.* (1974)	Davanone as *cis*-davanone
6. Thomas and Pitton (1971); Thomas and Dubini (1974a)	Davana ether
7. Thomas and Ozainne (1974)	Linalool; dehydro-α-linalool; terpinen-4-ol; nordavanone
8. Thomas and Dubini (1974)	*trans*-(erythro)-davanafuran; *trans*-(threo)-davanafuran; *cis*-(erythro)-davanafuran and *cis*-(threo)-davanafuran
9. Thomas (1974)	Isodavanone
10. Gulati and Khan (1980)	Camphene; p-cymene/sabinene/γ-terpinene; 1,8-cineole; linalool; isoborneol; geraniol; borneol; eugenol; methyl isoeugenol; methyl eugenol; acetoeugenol; γ-cadinene; farnesol
11. Takahashi *et al.* (1981)	Germacrene D; mint sulphide
12. Lamparsky and Klimes (1985); Klimes and Lamparsky (1986)	Davanone; isodavanone; allo-davanone; artemone; davanol; davana ether; davana – as quoted by Lawrence (1988) furan; davanic acid; hydroxy davanone; hydroxy davana ketone; nordavanone; nordavana ether; davanyl acetate; ethyl davanate; bicyclogermacrene; bicycloelemene; aromadendrene; allo-aromadendrene; γ,γ-dimethylbutenolide; linalool; linalool oxides; lilac aldehydes; lilac alcohols; α-santalol; β-santalol; 8-oxo-nerolidol; *cis*- and *trans*-cinnamic acids; α-gurjunene; ledene; β-cubebene; β-maaliene; spathulenol; τ-cadinol; β-eudesmol; selina-6-en-11-ol; 2-acetyl-5-methyl-5vinyltetrahydrofuran; 2,5-divinyl-5-methyltetrahydrofuran; 2-ethyl-5-methyl-5-vinyltetrahydrofuran; 2,(2′-hydroxyethyl)-5-methyl-5-vinyltetrahydrofurans; 2-acetyl-5-methylfuran; 4(5′-methyl-5′-vinyltetrahydrofur-2-yl)-pentan-2-one; 4(5′-methyl-5′-vinyltetrahydrofur-2-yl)-3-oxo-pentanal; 2,5-oxido-2,6, 10-trimethyldodeca-3,5,11-trien-10-ol(2 isomers); 3-hydroxy-3,7, 11-trimethyldodeca-1,6,10-trien-8-one; lavender lactone; lavender lactol; arredouglasia oxide

Table 3 Compounds identified from *Artemisia pallens*. (*Continued*)

Reference	Compounds identified
13. Akhila *et al.* (1986)	Biosynthesis of artemone
14. Chandra *et al.* (1987)	2-(3-methylbut-2-enyl)-3-methyl 2,5-dihydrofuran; 2-(3-methylbut-2-enyl)-5-(5-cinnamoloxy-2-oxo-1,5-dimethylhex-3-enyl)-3-methyl-2,5 dihydrofuran
15. Catalan *et al.* (1990)	Davanone, 2-hydroxy-isodavanone; *cis*-3-hydroxy-allo-davanone; *trans*-3-hydroxy-allo-davanone; 3,4-epoxy-2-hydroxy-isodavanone; *cis*-3,4-epoxy-allo-davanone; 2-hydroperoxy-isodavanone; *trans*-3,4-epoxy-allo-davanone
16. Misra *et al.* (1991)	α- and β-dihydrorosefurans; furano-nor-diterpenoid; hydroxy dihydrorosefuran; *trans*-hydroxy davanone; geraniol; nerol; 11-hydroxy-8-oxo-9, 10-dehydro-10, 11-dihydronerolidol
17. Rojatkar *et al.* (1996)	Germacranolide (4,5β-epoxy-10α-hydroxy-1-en-3-one-*trans*-germancran-6α, 12-olide)

dihydrorosefuran and furano-nor-diterpenoid (Misra *et al.*, 1991). The delicate and characteristic odour of davana oil has been attributed to these compounds blended with cinnamyl cinnamates. Recently, (Z)- and (E)-ethyl cinnamates, (Z)- and (E)-methyl cinnamates and geranyl acetate were reported as new constituents of davana oil (Mallavarapu *et al.*, 1999).

Many samples of davana oil are analyzed by capillary gas chromatography (GC), every year, at CIMAP, Field station, Bangalore. The GC profile and the composition of a typical davana oil sample is given in Figure 2 and Table 4, respectively. The structures of the important constituents are shown in Figure 3. It is understood from producers/ dealers of davana oil that oils with ≥ 50% content of davanone are preferred by the users of the oil. Although davanone itself is odourless, it may be acting as a fixative and synergist lending much tenacity and smoothness to the fragrance of davana oil like the odourless constituents of jasmine absolute, phytol, isophytol, phytyl acetate and geranyl linalool, which play a decisive role in the fragrance of jasmine (Demole, 1982).

The oil content and its composition have been found to vary with the physiological stage of the plant (Narayana *et al.*, 1978; Farooqi *et al.*, 1990). Mallavarapu *et al.* (1999), investigated the effect of three plant growth stages; the essential oil content was found to be higher at the full emergence of flower heads than at anthesis and initiation of seed set stages. The contents of davanone and linalool decreased

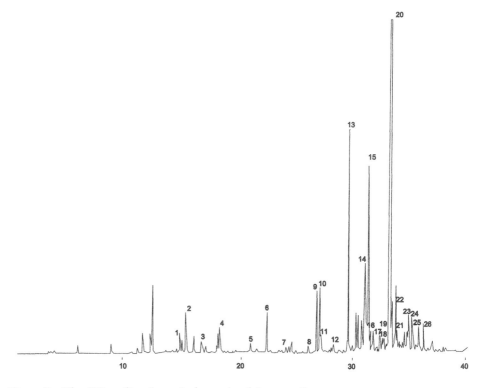

Figure 2   The GC profile of a typical sample of davana oil.

Table 4 The composition of a typical sample of davana oil.

Compound No*	Compound	RI**	Content***
1.	myrcene	984	0.1
2.	p-cymene	1012	1.1
3.	γ-terpinene	1050	0.5
4.	linalool	1086	0.7
5.	terpinen-4-ol	1166	0.4
6.	nordavanone	1207	1.1
7.	geraniol	1238	0.1
8.	(Z)-methyl cinnamate	1321	0.3
9.	(Z)-ethyl cinnamate	1346	1.8
10.	(E)-methyl cinnamate	1354	1.6
11.	geranyl acetate	1359	0.5
12.	davana furan	1389	0.1
13.	(E)-ethyl cinnamate	1437	7.4
14.	davana ether	1487	3.0
15.	bicyclogermacrene	1497	4.7
16.	δ-cadinene	1514	0.5
17.	artemone	1536	0.7
18.	artedouglasia oxide	1542	0.2
19.	(E)-nerolidol	1548	0.2
20.	davanone	1567	53.7
21.	davanol isomer	1591	0.4
22.	T-cadinol	1615	0.4
23.	β-eudesmol	1631	0.8
24.	2-hydroxyisodavanone	1644	0.9
25.	farnesol	1664	1.4
26.	β-santalol	1678	1.1

* as given in Fig. 2; **Retention index on BP-1 column; *** expressed as relative area percentage. Data from Mallavarapu et al. (1999).

while those of (Z)- and (E)-methylcinnamate, (E)-ethyl cinnamate, bicyclogerma-crene, davana ether, 2-hydroxyisodavanone and farnesol increased from the flower head emergence stage to the initiation of seed set stage. These results support the general practice of harvesting the crop at full bloom stage.

MEDICINAL PROPERTIES AND USES

Antipyretic and anthelmintic properties have been attributed to davana (Husain et al., 1992). Recently, Subramoniam et al. (1996) found that the methanol extract of davana significantly reduced blood glucose levels in glucose-fed hyperglycaemic and alloxan-induced diabetic rats. The water extract was, however, inactive.

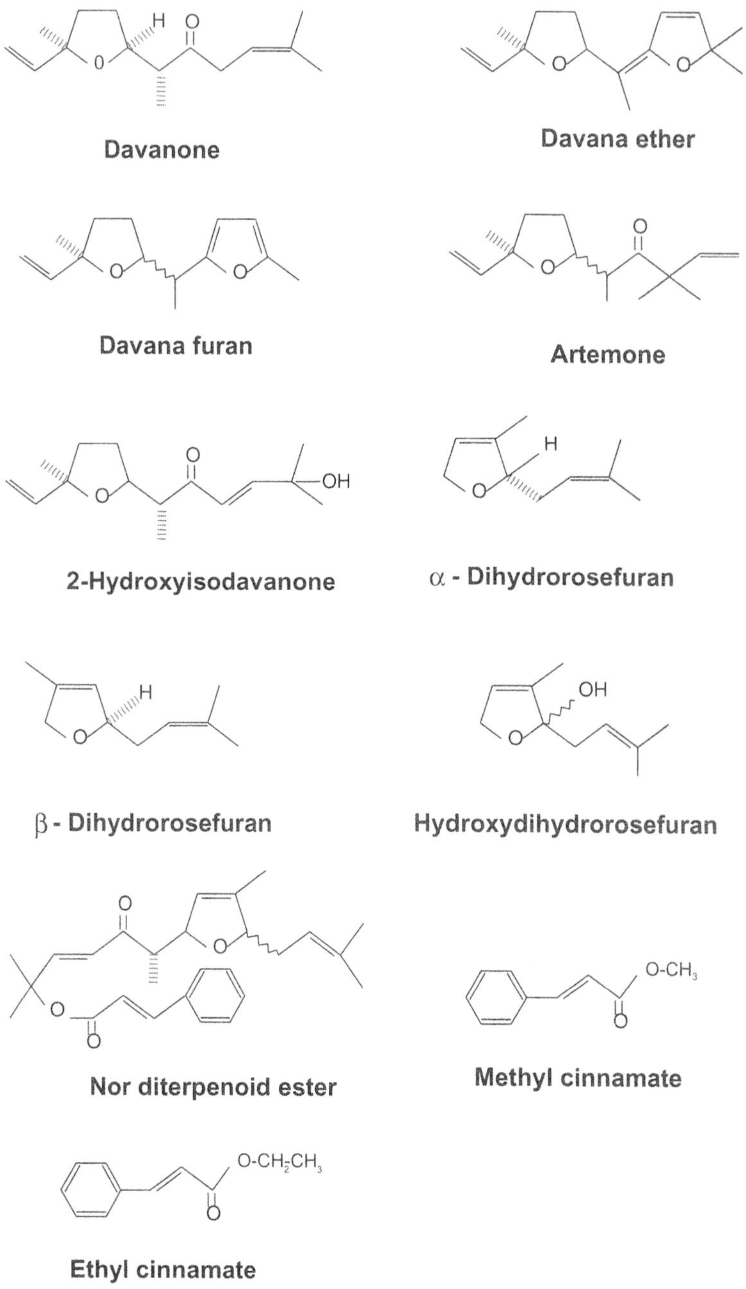

Davanone

Davana ether

Davana furan

Artemone

2-Hydroxyisodavanone

α - Dihydrorosefuran

β - Dihydrorosefuran

Hydroxydihydrorosefuran

Nor diterpenoid ester

Methyl cinnamate

Ethyl cinnamate

Figure 3   Structures of some important constituents of davana oil.

The essential oil of davana, like many other essential oils, has been found to exhibit antifungal and antibacterial properties (Alankara Rao and Rajendra Prasad, 1981, 1981a; Laxmi and Rao, 1991). The dried herb of davana is used as a moth repellant in the preservation of delicate and expensive fabrics (Sastry, 1946).

The fresh herb of davana has been traditionally used as a component of garlands and bouquets, to lend an element of freshness. This property of davana may find application in aromatherapy which, although known since ancient times, in India, China, Egypt and Babylonia, is now becoming increasingly popular (Buchbauer, 1990).

Davana oil is, however, mainly used in fine and expensive perfumes. It is also extensively used in flavouring cakes, pastries, tobacco and alcoholic beverages.

SOME AREAS FOR FUTURE WORK

Davana has been actively studied only since 1980. Therefore, there is considerable scope for further work on all aspects of this plant. However, following are some areas which probably need immediate attention.

1. Very little breeding work has been done on the development of high oil yielding varieties. Therefore, there is considerable scope for increasing the oil yield of davana through breeding.
2. The quality of davana oil is known to be influenced by environmental conditions during crop growth and during drying of the herb before distillation. Systematic studies under controlled environmental conditions are, therefore, necessary to define optimum environmental conditions for obtaining good quality davana oil.
3. Considerable variation is evident in the physico-chemical properties of davana oil reported by various workers. Also, as there is a possibility for adulteration of davana oil there is an urgent need to formulate standards for davana oil for exercising quality control.
4. The demand/supply of davana oil has been found to fluctuate widely over years. This is reflected in export statistics on davana oil (Fig. 1). Market research is, therefore, another very important area which requires immediate attention.

ACKNOWLEDGEMENTS

The author is grateful to Dr. G.R. Mallavarapu for his valuable suggestions on the chemistry of davana oil and for providing the GC profile and the composition of a typical sample of davana oil. The export statistics of davana oil were kindly provided by the Directorate of Commercial Intelligence, Government of India, Calcutta, Regional Officer, CHEMEXIL, Bangalore and Assistant Director,

CHEMEXIL, Bombay, and bibliographical assistance provided by Ms. S. Sharada, are gratefully acknowledged.

REFERENCES

Akhila, A. and Tiwari, R. (1986) Chemistry of davana oil : a review. *CROMAP*, 8(3), 128–138.

Akhila, A. Sharma, P.K. and Thakur, R.S. (1986) A novel biosynthesis of irregular sesquiterpene artemone in *Artemisia pallens*. *Tetrahedron Lett.*, 27, 5885–5888.

Alankararao, G.S.J.G. and Rajendra Prasad, Y. (1981) Investigations on the antibacterial activity of essential oils from *Artemisia pallens* Wall. & *Artemisia vulgaris* Linn. *Indian Perfum.*, 25, 110–111.

Alankara Rao, G.S.J.G and Rajendra Prasad, Y. (1981a) Investigations on the antifungal activity of the essential oils from *Artemisia pallens* Wall. & *Artemisia vulgaris* Linn. *Indian Perfum.*, 25, 112–113.

Anonymous (1947) Davana oil. *Perfum. Essent. Oil Rec.*, 38, 64–65.

Anonymous (1950) Some facts on essential oil industry in India. *Indian Soap J.*, 16, 7–8.

Anonymous (1967) Davana oil. *Perfum. Esent. Oil Rec.*, 58, 363.

Baslas, R.K. (1971) Essential oil from the plants of *Artemisia pallens* L. (Davana oil). *Flavour Ind.*, 2, 370–372.

Benjamin, B.D., Sipahimalani, A.T. and Heble, M.R. (1990) Tissue cultures of *Artemisia pallens*: organogenesis and terpenoid production. *Plant Cell, Tissue, Organ Culture*, 22, 179–182.

Buchbauer, G. (1990) Aromatherapy: do essential oils have therapeutic properties? *Perfum. Flav.*, 15(3), 47–50.

Catalan, C.A.N., Cuenca, M.Del R., Verghese, J., Joy, M.T., Gutierrez, A.B. and Herz, W. (1990) Sesquiterpene ketones related to davanone from *Artemisia pallens*. *Phytochemistry*, 29, 2702–2703.

Chandra, A., Misra, L.N. and Thakur, R.S. (1987) Dihydrofurano-terpenoids of davana oil. *Tetrahedron Lett.*, 28, 6377–6380.

Demole, E.P. (1982) The fragrance of jasmine. In T.H. Theimer, (ed.), Fragrance Chemistry, The Science of the Sense of Smell, Academic, New York, pp. 349–396.

Farooqi, A.A., Devaiah, K.A., Vasundhara, M. and Rao, N.D.D. (1990) Influence of planting season and plant density on growth, yield and essential oil in davana (*Artemisia pallens* Wall.) *Indian Perfum.*, 34, 274–277.

Farooqi, A.A., Rao, N.D.D., Devaiah, K.A. and Rai, S.J. (1990a) Studies on the influence of season on seed setting in davana. *Indian Perfum.*, 34, 286–287.

Farooqi, A.A., Rao, N.D.D., Devaiah, K.A. and Ravi Kumar, R.L. (1990b) Genetic variability in davana (*Artemisia pallens* Wall.) *Indian Perfum.*, 34, 42–43.

Farooqi, A.A., Rao, N.D.D., Devaiah, K.A. and Vasundhara, M. (1990c) Sensitivity of davana (*Artemisia pallens* Wall.) to gamma rays. *Indian Perfum.*, 34, 260–262.

Farooqi, A.A., Devaiah, K.A., Rao, N.D.D., Vasundhara, M. and Raju, B. (1991) Effect of nutrients on growth, yield and essential oil content in davana (*Artemisia pallens* Wall.). *Indian Perfum.*, 35, 63–68.

Farooqi, A.A., Devaiah, K.A. and Vasundhara, M. (1993) Effect of some growth regulators and pinching on growth, yield and essential oil content of davana (*Artemisia pallens* Wall.). *Indian Perfum.*, 37, 19–23.

Fasoulas, A. (1981) Principles and Methods of Plant Breeding. Publ. No.11, Dept. Gen. Plant Breeding, Aristotelian Univ., Thessaloniki, Greece.

Gowda, D.R. and Ramaswamy, M.N. (1965) Davana, *Artemisia pallens* Wall. in India. *Perfum. Essent. Oil Rec.*, **56**, 152–155.

Guenther, E. (1952) The Essential Oils. Vol.V, D. Van Nostrand Company, New York.

Gulati, B.C. (1980) Essential oil of *Artemisia pallens* Wall. (Davana). *Indian Perfum.*, **24**, 101–109.

Gulati, B.C. and Khan, M.N.A. (1980) Essential oil of *Artemisia pallens* Wall. (Davana): a study of minor constituents. *Perfum. Flav.*, **5**(2), 23–24.

Gulati, B.C., Duhan, S.P.S. and Bhattacharya, A.K. (1967) Trial cultivation of davana *Artemisia pallens* Wall. in Uttar Pradesh. *Perfum. Essent. Oil Rec.*, **58**, 523–527.

Husain, A., Virmani, O.P., Popli, S.P., Misra, L.N., Gupta, M.M., Srivastava, G.N., Abraham, Z. and Singh, A.K. (1992) Dictionary of Indian Medicinal Plants. Central Institute of Medicinal and Aromatic Plants, Lucknow.

Janardhana Rao, L., Sreenivas, B. and Reddy, M.G. (1990) Effect of planting dates and intra-row spacing on growth, herbage and oil yield of davana (*Artemisia pallens* Wall.) *Indian Perfum.*, **34**, 250–252.

Klimes, I. and Lamparsky, D. (1986) Analytical results concerning the essential oil of *Artemisia pallens* (Wall.) In E-J. Brunke, (ed.), Progress in Essential Oil Research, Walter de Gruyter, Berlin (cited in Lawerence, 1988).

Kulkarni, R.N. (1991) Three cycles of honeycomb selection for herb yield in davana (*Artemisia pallens* Wall.). *Euphytica*, **52**, 99–102.

Lamparsky, D. and Klimes, I. (1985) Natural flavouring materials – davana oil: some analytical results with respect to biogenetically plausible structural features. In R.G. Berger, S. Nitz and P. Schreier, (eds.), Topics in Flavour Research, H. Eichhorn, Marzling-Hangenheim, pp. 281–304. (cited in Lawrence, 1988).

Lawrence, B.M. (1978) Progress in essential oils. *Perfum. Flav.*, **3**(2), 45–50.

Lawrence, B.M. (1988) Progress in essential oils. *Perfum. Flav.*, **13**(3), 49–56.

Lawrence, B.M. (1995) Progress in essential oils. *Perfum. Flav.*, **20**(1), 47–54.

Laxmi, U.S. and Rao, J.T. (1991) Antimicrobial properties of the essential oils of *Artemisia pallens* and *Artemisia vulgaris*. *Perfum Kosmet.*, **72**, 510–511.

Lewis, Y.S. and Nambudiri, E.S. (1967) Composition of davana oil – some preliminary studies. *Perfum. Essent. Oil Rec.*, **58**, 613–617.

Mallavarapu, G.R. (1995) Essential oils from some aromatic ornamental plants. In K.L. Chadha and S.K. Bhattacharjee, (eds.), Advances in Horticulture, Vol. 12, Ornamental Plants, Malhotra Publishing House, New Delhi, pp. 1035–1052.

Mallavarapu, G.R., Kulkarni, R.N., Baskaran, K., Rao, L. and Ramesh, S. 1999. Influence of plant growth stage on the essential oil content and composition in davana (*Artemisia pallens* Wall.). *J. Agricultural and Food Chemistry*, **47**, 254–258.

Menon, A.K. (1960) Indian Essential Oils, A Review. Council of Scientific & Industrial Research, New Delhi.

Misra, L.N., Chandra, A. and Thakur, R.S. (1991) Fragrant components of oil from *Artemisia pallens*. *Phytochemistry*, **30**, 549–552.

Naegeli, P., Klimes, J. and Weber, G. (1970) Structure and synthesis of artemone. *Tetrahedron Lett.*, **57**, 5021–5024.

Narayana, M.R. and Dimri, B.P. (1990) Cultivation of Davana for its Oil in India. Farm. Bull. No.12, Central Institute of Medicinal and Aromatic Plants, Lucknow.

Narayana, M.R., Khan, M.N.A. and Dimri, B.P. (1978) Davana and its Cultivation in India. Farm Bull. No.12, Central Institute of Medicinal and Aromatic Plants, Lucknow.

Narayana, M.R., Ganesha Rao, R.S., Puttanna, K., Chandrashekara, G. and Prakasa Rao, E.V.S. (1982) Effect of graded levels of N and P with and without FYM on the growth and yield of davana (*Artemisia pallens* Wall.). In C.R. Muthukrishnan, S. Muthuswamy and J.B.M. Md. Abdul Khader, (eds.), Proceedings of National Seminar on Medicinal and Aromatic Plants, Tamil Nadu Agric. Univ., Coimbatore, pp. 180–182.

Pandey, R. (1994) Comparative performance of oil seed cakes and pesticides in the management of root knot disease of davana. *Nematologia Mediterranea*, **22**, 17–19.

Prakasa Rao, E.V.S., Singh, M. and Ganesha Rao, R.S. (1983) Effect of nitrogen and plant spacings on the growth, yield and nutrient uptake in davana (*Artemisia pallens* Wall.). *Intern. J. Trop. Agric.*, **1**, 187–192.

Rai, S.J. and Farooqi, A.A. (1990) Studies on blossom biology of davana (*Artemisia pallens* Wall.). *Indian Perfum.*, **34**, 209–212.

Rajeswara Rao, B.R., Singh, K., Kaul, P. and Bhattacharya, A.K. (1989) Effect of nitrogen on essential oil concentration, yield and nutrient uptake of davana (*Artemisia pallens*). *Indian J. Agric. Sci.*, **59**, 539–541.

Rajeswara Rao, B.R., Singh, K., Kaul, P.N. and Bhattacharya, A.K. (1989a) The effect of plant spacings and application of N and P fertilizers on the productivity and nutrient uptake of davana (*Artemisia pallens* Wall.). *Intern. J. Trop. Agric.*, **7**, 229–236.

Ranganathan, V.N. (1963) Oil of davana from *Artemisia pallens* Wall. *Soap, Perfumery & Cosmetics*, **36**, 64.

Rao, B.S., Mathen, M., Kelkar, N.C. and Singh, J. (1937) Notes on some oils. *Perfum. Essent. Oil Rec.*, **28**, 411–415.

Rekha, K. and Kak, S.N. (1993) Cytological studies in *Artemisia pallens* Wall. ex. DC. *Indian J. Forestry*, **16**, 382–383.

Rojatkar, S.R., Pawar, S.S., Pujar, P.P., Sawaikar, D.D., Gurunath, S., Sathe, V.T. and Nagasampagi, B.A. (1996) A germacranolide from *Artemisia pallens*. *Phytochemistry*, **41**, 1105–1106.

Sarwar, M. and Khan, M.N.A. (1971) Diseases and pests of aromatic plants from south India. *Angew. Bot.*, **45**, 211–216.

Sarwar, M. and Khan, M.N.A. (1975) Collar rot of davana. National Symposium on Recent Advances in the Development, Production and Utilization of Medicinal and Aromatic Plants, Lucknow. February 1975, Section III: Botany, pp. 25 (Abst.).

Sastry, S.G. (1946) Davana oil. *Indian Soap J.*, **11**, 242–244.

Sharief, Md.U. and Jagadishchandra, K.S. (1991) Micropropagation of davana (*Artemisia pallens* Wall.) by tissue culture. In J. Prakash and R.L.M. Pierik, (eds.), Horticulture – New Technologies and Applications, Kluwer Academic, Dordrecht, pp. 255–259.

Shenoy, K.P., Thimma Raju, K.R., Farooqi, A.A. and Devaiah, K.A. (1993) Effect of nitrogen and cycocel on growth and yield of davana (*Artemisia pallens* Wall.). *Indian Perfum.*, **37**, 14–18.

Sipma, G. and Van der Wal, B. (1968) Structure of davanone a new sesquiterpene from davana (*Artemisia pallens* Wall.). *Rec. Trav. Chim. Pay-Bas.*, **87**, 715–720.

Subramoniam, A., Pushpangadan, P., Rajasekharan, S., Evans, D.A., Latha, P.G. and Valsaraj, R. (1996) Effects of *Artemisia pallens* Wall. on blood glucose levels in normal and alloxan-induced diabetic rats. *Journal of Ethnopharmacology*, 50, 13–17.

Sugunakar, G. (1987) Use of davana oil as flavouring agent in India & abroad. *Indian Perfum.*, **31**, 37–39.

Takahashi, T., Muraki, S. and Yoshida, T. (1981) Synthesis and distribution of (–)-mint sulphide, a novel sulphur-containing sesquiterpene. *Agric. Biol. Chem.*, **45**, 129–131.

Thomas, A.F. (1974) Terpenoids derived from linalyl oxide. Part 5. 2-(2-Methyl-2vinylte-trahydrofur-5-yl)-2, 6, 6-trimethyl-2, 6-dihydropyr-3-one, a new sesquiterpenoid isolated from *Artemisia pallens. Helv. Chim. Acta*, 57, 2081–2083.

Thomas, A.F. and Dubini, R. (1974) Terpenoids derived from linalyl oxide. Part 3. The isolation, structure, absolute configuration and synthesis of davanafurans, non-sequiterpenes isolated from *Artemisia pallens. Helv. Chim. Acta*, 57, 2066–2075.

Thomas, A.F. and Dubini, R. (1974a) Terpenoids derived from linalyl oxide, Part 4. The oxidation of davanone. Isolation and synthesis of the davana ethers, sesquiterpenes of *Artemisia pallens. Helv. Chim. Acta*, 57, 2076–2081.

Thomas, A.F. and Ozainne, M. (1974) Terpenoids derived from linalyl oxide. Part 2. The isolation and synthesis of nordavanone a C_{11}-terpenoid from *Artemisia pallens. Helv. Chim. Acta*, 57, 2062–2065.

Thomas, A.F. and Pitton, G. (1971) The isolation, structure and synthesis of davana ether, an odoriferous compound of the oil of *Artemisia pallens* Wall. *Helv. Chim. Acta*, 54, 1890–1891.

Thomas, A.F., Thommen, W., Willhalm, B., Hagaman, E.W. and Wenkert, E. (1974) Terpenoids derived from linalyl oxide. Part 1. The stereochemistry of davanones. *Helv. Chim. Acta*, 57, 2055–2061.

Verghese, J. and Jacob, C.V. (1984) Davana oil. *Pafai J.*, 6(4), 19–23.

Whish, J.P.M. (1996) A flexible distillation system for the isolation of essential oils. *J. Essent. Oil Res.*, 8, 405–410.

8. *ARTEMISIA VULGARIS*

PETER A. LINLEY

The School Of Pharmacy, University of Bradford, West Yorkshire, BD7 1DP, UK.

INTRODUCTION: SYNONYMS AND LOCAL NAMES

Artemisia vulgaris L., most commonly known as Mugwort, is a species of wide distribution throughout Europe, Asia and north America. Several other common names are listed by Grieve (1994) and Bisset (1994) including Felon Herb, Wild Wormwood and St. John's Plant, noting that the latter name should not be confused with St. John's Wort, *Hypericum perforatum*. The historical derivation of these names is suggested by Grieve (1994), the herb having been used over many centuries. Most likely, the name "Mugwort" is linked with the plant's use for flavouring beer prior to the modern use of hops (*Humulus lupulus*). Alternatively, Mugwort, may not relate to either drinking mugs or wort, but from "moughthe", a moth or maggot since the plant has been thought to be useful in repelling moths.

In the United Kingdom *A. vulgaris* has received many local names. Grigson (1975) lists 24 names including Apple-Pie and Mugweed in Cheshire, Green Ginger and Smotherwood in Lincolnshire, Mugwood in Shropshire and Mugger in Scotland.

BOTANY

Habitat

Mugwort is a hardy perennial common throughout temperate regions of the northern hemisphere. It grows readily in hedgerows, roadsides, river banks and waste places such as rubbish tips. Clapham *et al.* (1962) state that geographically the species grows to 70° N. in Norway and 74° N. in Siberia. According to Bisset (1994) the medicinal herb is collected from wild sources in Eastern Europe.

Macroscopical Description

Artemisia vulgaris is a creeping perennial with branching rootstock producing strong, erect stems 60–120 cm in height. The **Stems** are usually glabrous, reddish-purple in colour, angled and longitudinally grooved. The stem has a large central, white pith and a narrow, green outer band of tissues. The **Leaves** are 5–8 cm (length) and 2.5–5 cm (width). The basal leaves have a short petiole and are lyrate-pinnatifid with small ear-like projections (auricles) at the base. The stem-leaves are sessile, clasping and bipinnate or pinnate on the uppermost leaves. The upper surfaces are smooth and dark green whereas the lower surfaces are whitish due to a dense covering of cotton-like hairs (i.e. tomentose). The ultimate leaf segments are 3–6 mm wide, usually lanceolate with the margin entire or toothed. The **Inflorescense** consists of numerous flower heads which are produced on long, leafy stems from each leaf axil. The flower heads are reddish-brown, in dense racemose panicles. Each head is 3–4 mm in diameter and ovoid in shape. The involucral bracts are lanceolate to oblong with a woolly covering, but with a dry, thin appearance (scarious) at the margins. The florets are reddish-brown and all are fertile. The **Fruit** is a glabrous achene about 1 mm in diameter. When fresh, the plant has strong aromatic odour and the taste is described as spicy and bitter.

Microscopical Features

The following description is based on the text of the British Herbal Pharmacopoeia (1996), and illustrations published by Eschrich, (1999).

When viewed after clearing in chloral hydrate solution, the diagnostic features are:

Leaves

The transverse section of the leaf shows a typical dorsiventral arrangement with a single palisade mesophyll. The upper epidermis has sinuous anticlinal walls, few stomata which appear to be anomocytic and T shaped covering trichomes. The lower epidermis has abundant T-shaped trichomes. These consist of a uniseriate stalk (2 or 3 cells) and bear an asymmetrical terminal cell which is elongated at right angles to the stalk. The lower surface has abundant stomata which are difficult to examine due to the large numbers of "woolly" trichomes and epidermal cells with wavy anticlinal walls similar to those of the upper epidermis. Small biseriate glandular trichomes with mulitcellular heads (peltate) are also present.

Inflorescense

The bracts have epidermal cell walls which give a positive reaction for lignin and the epidermises bear trichomes similar to those of the leaf. The epidermises of the corolla are also lignified and have a few biseriate glandular trichomes with multicellular heads. The ovary wall has elongated cells with numerous T-shaped trichomes. The stigma is papillose; the anthers have beaded walls and the pollen is tricolpate with a finely warted exine.

PHYTOCHEMISTRY

The phytochemistry of *A. vulgaris* has received considerable attention and several classes of compounds have been reported. These are summarised in the following sections:

1. Essential Oil

Several reports have been published showing a wide variation in the constituents of the oil, both qualitatively and quantitatively. Nano *et al.* (1976) reported on the steam distillates from the plant's flowers collected in the Piedmont region of Italy. Six oils were examined by conventional packed-column gas chromatography (GC). Some 19 components were identified using a combination of retention times, mass spectrometry, TLC, IR and UV techniques, of which 14 had not been reported previously. The proportion of monoterpene hydrocarbons was highest at the time of flowering, but the sequiterpenes were at their highest proportion in the period before flowering. The most abundant compounds were monoterpenes (mainly myrcene, up to 70%), 1,8-cineole (up to 27%), borneol (up to 19%), bornyl acetate (up to 18%) and camphor (up to 12%), but the amounts varied greatly within the six oils thus demonstrating the variability of the oil even within a restricted locality. Of particular interest, Nano *et al.* (1976) reported the presence of vulgarole 1 which was present in the two oils distilled at the time of flowering in concentrations of 4.5% and 0.8% respectively. This compound received specific investigation by Worner and Schreier (1991a), using a two-column capillary procedure in which the vulgarole on eluting from the first column was "cut" to a second chiral column with MS detection. It was shown that in *A. vulgaris* the enantiomer present was (1S)-(−)-2-endo-acetoxy-3-exo-hydroxybornane and was regarded as an important analytical indicator for the species. Note that vulgarole must not be confused with vulgarol a diterpene found in *Marrubium vulgare* and *Lagochilus inebrians* (Worner and Schreier, 1991).

The oil was further investigated by Hurabielle *et al.* (1981). From the dried herb a yield of 0.1% v/m was obtained. By capillary GC-MS some 24 compounds were identified of which 1:8-cineole and camphor were present in the highest concentration. However, the chromatogram contained a number of unidentified monoterpenes and sesquiterpenes, but all were in trace concentrations only. Using an SE-30 capillary column (30 m) and MS detection, Michaelis *et al.* (1982) examined the oils distilled from *A. vulgaris* blossom. They detected 83 compounds of which about 70 were identified. In this blossom oil the compounds present in the highest concentrations were sabinene (16%), myrcene (14%) and 1:8-cineole (10%). Other compounds ranged down to 0.01% showing the high sensitivity of their instrumentation.

In contrast to earlier reports, Misra and Singh (1986) investigated the leaves of *A. vulgaris* growing in Southern India and found that the hydro-distilled oil contained 56% of α-thujone, but only traces of sabinene, myrcene and 1:8-cineole. Xguyen *et al.* (1992) investigated the oil (0.2% yield) from *A. vulgaris* L. var. indica Maxim grown in Vietnam. This plant gave an oil with a high content of sesquiterpenes, including β-caryophyllene (24%) and β-cubebene (12%). The principal monoterpenes were borneol (8%) and 1:8-cineole (6%). Vulgarole 1 was not detected in this variety or in the Indian plants studied by Misra and Singh (1986).

Milhau *et al.* (1997) investigated *in vitro* antimalarial activity of several essential oils including a distillate from *A. vulgaris*. The oil contained α-thujone (36%) and camphor (26%) with lesser amounts of 1:8-cineole (8%), β-thujone (6%) and camphene (6%); it showed only low antiplasmodial potency.

Naef-Mueller *et al.* (1981) reported the presence of irregular monoterpenes in a steam-distilled oil from Moroccan *A. vulgaris*. Although the oil contained thujone (and iso-thujone) (35%) and camphor (30%), the authors identified 21 non-head-to-tail (irregular) isoprenoid monoterpenes, several being reported for the first time. These were grouped into two classes, based on the compounds artemisia triene 2 and santalina triene 3.

Woerner *et al.* (1991) extracted *A. vulgaris* with pentane – dichloromethane (2:1 v/v), fractionated the extract by gel permeation chromatography and liquid chromatography before analysis of the fractions by GC-MS and GC-FTIR. From this procedure, the authors identified 54 compounds not previously reported in the species and which may be summarised as follows:

Type of Compound	*Number Identified*
Acids	15
Alcohols	15
Carbonyls	5
Esters	7
Lactones	4
Miscellaneous	8

The paper gave only approximate data on the amount of each constituent, but the qualitative results demonstrated the complex phytochemistry of the species.

Two sesquiterpenes were identified by Marco *et al.* (1991) from a methanolic extract. These were eudesmane acids 4, 5, which, after chromatographic purification, were isolated in low yields of 12 mg and 8 mg respectively from 2 kg of fresh plant material. A third compound, a eudesmane dialcohol, 6 which had previously been reported was also present). Recently, Pino *et al.* (1999) have reported that oil from *A. vulgaris* grown in Cuba contained caryophyllene oxide (31%) as its main constituent. Sesquiterpene lactones including vulgarin 7 (tauresmin), psilostachyin 8 and psilostachyin C 9 have also been reported in the oil (Stephanovic *et al.*, 1972).

2. Phenolic Compounds and Coumarins

Phenolic compounds in *A. vulgaris* have been investigated, but the reports have identified known compounds. The flavanols reported by Hoffman and Herrmann, (1982) included the glycosides of quercetin 10 as its 3-O-glucoside and 3-O-rhamnosidoglucoside (rutin). The same sugar residues were associated with isorhamnetin.

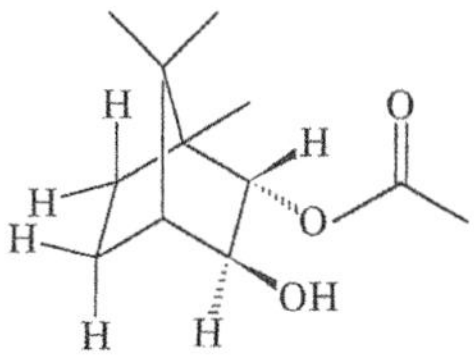

1

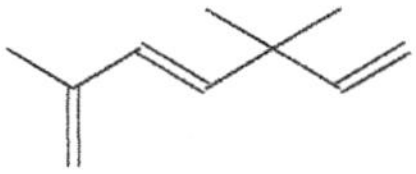

2

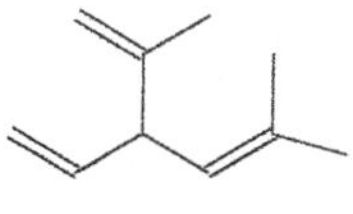

3

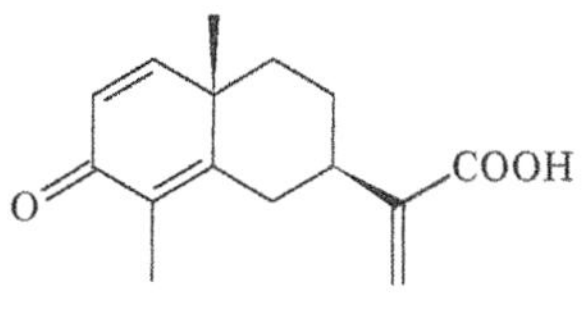

4

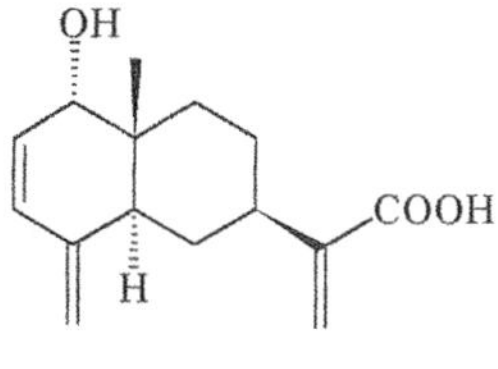

5

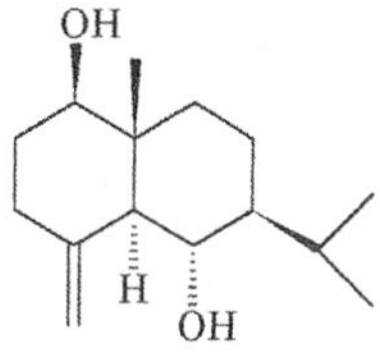

6

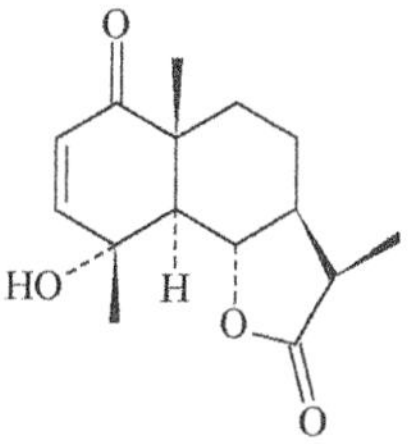

7

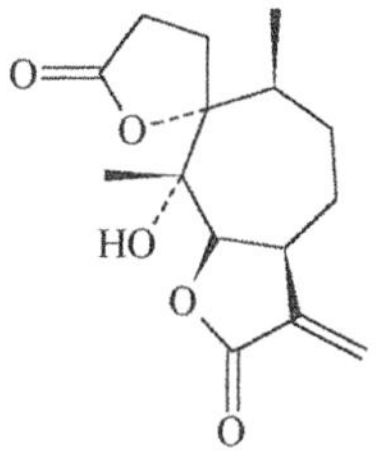

8

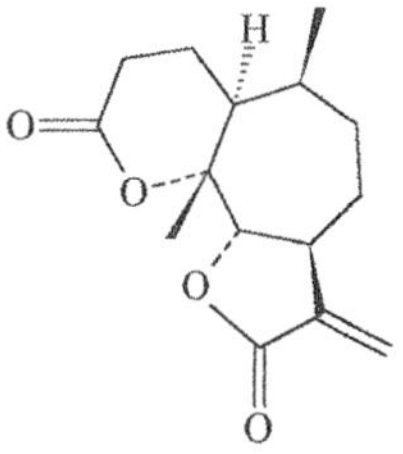

9

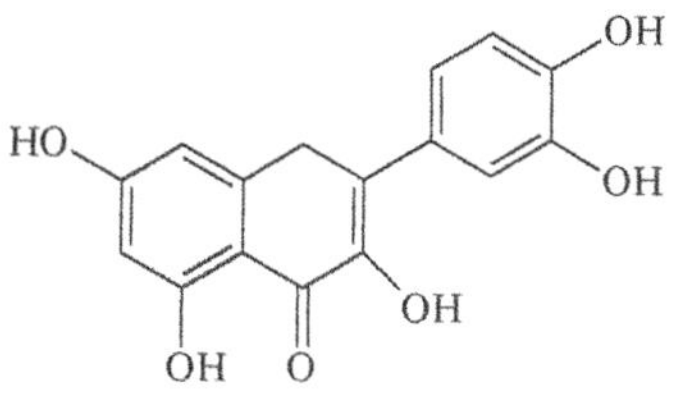

10

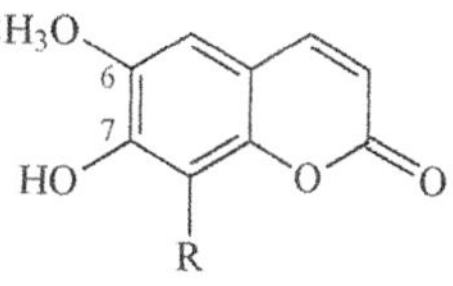

11, R=H
12, R=OMe

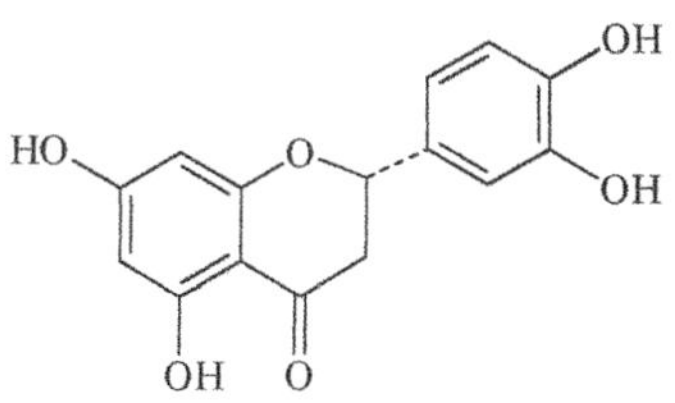

13

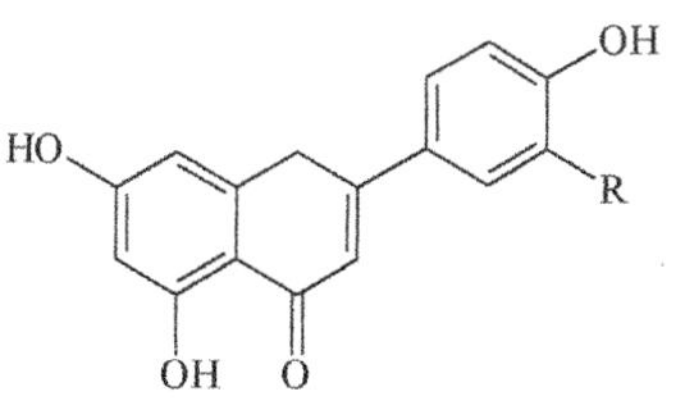

14, R=OH
15, R=H

Similarly, the coumarins reported by Ikhsanova *et al.* (1986) included coumarin, aesculetin, aesculen, umbelliferone and scopoletin (6-methoxy-7-hydroxycoumarin) **11**. Banthorpe and Brown (1989), found that callus cultures of *A. vulgaris* yielded significant amounts of scopoletin and isofraxidin **12**, the latter at about 5 times the level present in the foliage of the plant. Murray and Stefanovic (1986), confirmed the structure of a coumarin previously reported by Stefanovic *et al.* (1982) as 6-methoxy-7,8-methylenedioxycoumarin.

Plants originating in the Korean Republic were examined by Lee *et al.* (1998) for their flavonoid content. Twenty known flavonoids were isolated and identified; the principal compounds were eriodictyol **13** and luteolin **14**.

All the flavonoids were tested for oestrogenic activity using a yeast transformed with both a human oestrogenic receptor expression plasmid and a reporter plasmid. Two compounds, eriodictyol **13** and apigenin **15**, were able to induce the transcription of the receptor gene in the transgenic yeast.

3. Polyacetylenes

Drake and Lam (1974) reported the polyacetylenes in *A. vulgaris*, using techniques which had not been available in previous investigations such as that by Bohlmann *et al.* (1957). Eight compounds were isolated and identified, three of which had been reported in other species of *Artemisia*. Of particular interest was the presence of three compounds present in extracts of the flowers; no polyacetylenes had been detected in a previous study (Stavholt and Sorensen, 1950). The isomers of one compound, dehydromatricaria ester, were investigated during the growing season. It was shown that the proportion of the cis-isomer increased relative to the trans-isomer. Bohlmann *et al.* (1957) had found the reverse effect.

MEDICINAL AND OTHER USES

Mugwort has been used for a wide range of medical conditions and several non-medical uses. Examples of the latter are summarised by Bown (1995). The herb was used by Roman soldiers in their sandals to protect their feet during marches. Grigson (1975) refers to the plant's title as the "Mother of Herbs" and the fact that it was one of nine herbs used in Anglo-Saxon and earlier times to ward off poisons and evil. Pipe smoking of the dried herb ("Gypsy's tobacco") packed in acorn cups was noted by Mabey (1996).

Despite the many indications quoted for the use of Mugwort, these do not appear to have been supported by clinical investigations, and as quoted by Bisset (1994), its therapeutic use cannot be substantiated. The medicinal uses of the Mugwort quoted in readily accessible herbal compendia are shown in Table 1.

Table 1 Medicinal uses of mugwort (*A. vulgaris*) cited in some well known herbals.

Medicinal Use	Reference				
	Bisset, 1994	Williamson and Evans, 1988	Culpepper, 1981	Bown, 1995	Grieve, 1998
Amenorrhoea	*	*			
Anorexia/appetite loss		*	*		
Anthelmintic	*	*			
Antibacterial (essential oil)	*	*			
Antidote to opium (fresh plant juice)			*		
Antifungal (essential oil)	*				
Choloretic	*	*			
Diaphoretic		*			*
Diuretic			*		*
Dyspepsia		*		*	
Dysmenorrhoea	*				
Emmenagogue		*			*
Menstrual complaints				*	
Nausea (Chinese medicine)		*			
Nervine (anti-epileptic)					*
Repellent (insects)	*				
Roundworm		*		*	
Threadworm		*		*	
Tonic/stimulant					*
Vomiting(Chinese Medicine)		*			

TOXIC EFFECTS

There are no reports of toxicity following oral ingestion of the herb although the German Commision E monograph cited in Bisset (1994) states that an abortifacient action has been described and Williamson and Evans (1994) note that it should not be used in pregnancy. There is one report of contact dermatitis described by Kurz and Rapaport (1979). In animals, Forsyth (1968) records one incidence of poisoning in cattle attributed to *A. vulgaris*.

The literature contains many references to Mugwort pollen due to its allergenic effects. For example, Nilsen and co-workers have published a series of papers which identified and characterized allergens from the pollen using a variety of immunological techniques (Nilsen, 1990a,1991a) and have purified allergen Ag7 by concanavalin A affinity chromatography (Nilsen, 1990b). They have also reported the structural analysis of the glycoprotein allergen in Mugwort pollen (Nilsen *et al.*, 1991b).

REFERENCES

Banthorpe D.V. & Brown G.D. (1989) Two unexpected coumarin derivatives from tissue cultures of Compositae species. *Phytochemistry*, **28**, 3003–3007.

Bisset, N.G. (1994) Herbal drugs and phytopharmaceuticals. (Edited and translated from the German *Teedrogen* by M. Wichtl). Medpharm, Stuttgart and CRC Press London, pp. 88–90.

Bohlmann F., Inhoffen E. & Herbst P. (1957) *Chem. Ber.*, **90**, 124.

Bown, D. (1995) *RHS Encyclopedia of herbs and their uses*, Dorling Kindersley: London, pp. 88.

British Herbal Pharmacopoeia (1996) British Herbal Medicine Association.

Clapham, A.R., Tutin, T.G. & Warburg, E.F. 1962. *Flora of the British Isles*, 2nd edition, Cambridge University Press pp. 859.

Drake D. & Lam J. (1974) Polyacetylenes of *Artemisia vulgaris*. *Phytochemistry*, **13**, 455–457

Forsyth A.A. (1968) *British Poisonous Plants*. Ministry of Agriculture, Fisheries and Food Bulletin, HMSO London, **161**, 97.

Grieve, M. (1998) *A modern herbal*. Tiger Books International, Twickenham, U.K. pp. 556–558.

Grigson, G. (1975) *The Englishman's Flora*, Paladin: St. Albans, pp. 410–413.

Hoffmann B. & Herrmann K. (1982) Flavonol glycosides of wormwood (*Artemisia vulgaris* L.), tarragon (*Artemisia dracunculus* L.) and absinthe (*Artemisia absinthium* L.). 8. Phenolics of spices. *Z. Lebensm. Unters. Forsch.*, **174**, 211–215.

Hurabielle M., Malsot M. & Paris M. (1981) Chemical study of two oils from wormwood: *Artemisia herb-alba* Asso and *Artemisia vulgaris* L. of chemicotaxanomic interest. *Riv. Ital. EPPOS*, **63**, 296–299.

Ikhsanova M.A., Berezovskaya T.R. & Serykh E.A. (1986) Coumarins of *Artemisia vulgaris*. *Khim. Prir. Soedin*, pp. 110 (CA **104**, 203928, 1986)

Kurz G. & Rapaport M.J. (1979) *Contact Dermatitis*, **5**, 407 (cited in Bisset, 1994).

Lee S.J., Chung H.Y., Maier C.G.A., Wood A.R., Dixon R.A. & Mabry T.J. (1998) Estrogenic flavonoids from *artemisia vulgaris* L. *J. Agric. Food Chem.*, **46**, 3325–3329.

Mabey R. (1996) *Flora Britannica*, Sinclair-Stevenson, U.K., pp. 370.

Marco J.A., Sanz, J.F. & Del Hierro P. (1991) Two eudesmane acids from *Artemisia vulgaris*. *Phytochemistry*, **30**, 2403–2404.

Michaelis K., Vostrowsky O., Paulini H., Zintl R. & Knobloch K. (1982) Essential oil from blossoms of *Artemisia vulgaris* L. *Z. Naturforsch., C: Biosci.* **37C**, 152–158.

Milhau G., Valentin A., Benoit F., Mallie M., Bastide J-M., Pelissier Y. & Bessiere J-M. (1997) *In vitro* antimalarial activity of eight essential oils. *J. Essent. Oil Res.*, **9**, 329–333.

Misra L.N. & Singh S.P. (1986) Alpha-Thujone, the major component of the essential oil from *Artemisia vulgaris* growing wild in Nilgiri hills. *J. Nat. Prod.*, **49**, 940.

Murray R.D.H. & Stefanovic M. (1986) 6-Methoxy-7,8-methylenedioxycoumarin from *Artemisia dracunculoide* and *Artemisia vulgaris*. *J. Nat. Prod.*, **49**, 550–551.

Naef-Mueller R., Pickenhagen W. & Willhalm B. (1981) New irregular monoterpenes in *Artemisia vulgaris*. *Helv. Chim. Acta.*, **64**, 1424–1430.

Nano G.M., Bicchi C., Frattini C. & Gallino M. (1976) On the composition of some oils from *Artemisia vulgaris*. *Planta Med.* 30, 211–215.

Nilsen B.M. & Paulsen B.S. (1990a) Isolation and characterization of a glycoprotein allergen, Art v II, from pollen of mugwort (*Artemisia vulgaris* L.). *Mol. Immunol.*, **27**, 1047–1056.

Nilsen B.M., Paulsen B.S, Clonis Y. & Mellbye K.S. (1990b) Purification of the glycoprotein Ag7 from mugwort pollen by concanavalin A affinity chromatography. *J. Biotechnol.*, **16**, 305–316.

Nilsen B.M., Grimsoen A. & Paulsen B.S. (1991a) Identification and characterization of important allergens from mugwort pollen by IEF, SDS-PAGE and immunoblotting. *Mol. Immunol.*, **28**, 733–742.

Nilsen B.M., Sletten K., Paulsen B.S., O'Neill M. & Van Halbeek H. (1991b) Structural analysis of the glycoprotein allergen Art v II from the pollen of mugwort (*Artemisia vulgaris* L.). *J. Biol. Chem.*, **266**, 2660–2668.

Pino, J.A., Rosado, A., & Fuentes, V. (1999) Composition of the essential oil of *Artemisia vulgaris* L. herb from Cuba. *J. Ess. Oil Res.*, **11**, 477–478.

Stavholt, K. & Sorensen, N.A. (1950) Naturally-occurring acetylene compounds V. Dehydromatricaria ester (methyl *n*-decenetriynoate) from the essential oil of *Artemisia vulgaris*. *Acta Chem. Scand.* **4**, 1567–1574.

Stephanovic, M., Jokic, A. & Behbudi, A. (1974) *Glas. Hem. Drus. (Beograd)*, **37**, 463. (CA **80**, 80080 (1974), cited in Bisset (1994).

Stefanovic M., Dermanovic M. & Verencevic M. (1982) Chemical investigation of the plant species of *Artemisia vulgaris*. *Glas. Hem. Drus. Beograd*, **47**, 7 (CA **96**, 177968e).

Williamson, E.M. & Evans, F.J. 1994. *Potter's new cyclopaedia of botanical drugs and preparations*. The C.W. Daniel Company Ltd., Saffron Walden, U.K. pp. 194–195.

Woerner M. & Schreier P. (1991) Multidimensional gas chromatography/mass spectrometry (MDGC/MS): a powerful tool for the direct chiral evaluation of aroma compounds in plant tissues III. (1S)-(-)-2-endo-acetoxy-3-exo-hydroxybornane (vulgarole) in mugwort (*Artemisia vulgaris* L.) herb. *Phytochem. Anal.* **2**, 260–262.

Woerner M. Pflaum M. & Schreier P. (1991) Additional volatile constituents of *Artemisia vulgaris* L. herb. *Flavour Fragrance J.* **6**, 2257–260.

Xguyen X.D., Vu V.N., Hoang T.H. & Leclercq P.A. (1992). Chemical composition of the essential oil of *Artemisia vulgaris* L. var. indica Maxim. from Vietnam. *J. Essent. Oil Res.*, **4**, 433–434.

9. *ARTEMISIA* SPECIES IN TRADITIONAL CHINESE MEDICINE AND THE DISCOVERY OF ARTEMISININ

HONGWEN YU AND SHOUMING ZHONG

East West Biotech Ltd., Langston Priory Mews, Kingham, Oxon., OX7 6UP, UK.

QING HAO-AN ANTIMALARIAL HERB

A herb, named Qing Hao (usually pronounced ching how) in Chinese, sweet Annie or sweet wormwood in English, and properly known as *Artemisia annua* L. has become well known in western countries during the last 20 years. Herbal companies, which deal with traditional Chinese medicine (TCM), receive several inquiries concerning this herb every day. A question commonly asked by those about to travel to Africa or S.E. Asia is "Can I take the herb called Qing Hao to prevent malaria during my trip?" Unfortunately, the answer has disappointed many people because although this herb is used for the treatment of malaria in TCM, usually combined with other herbs, it is not recommended for the prevention of the disease or as a deterrent to mosquitoes. However, the leaves of Qing Hao were burned as a fumigant insecticide to kill mosquitoes in ancient China but this practice no longer continues today since the development and marketing of more efficient mosquito-repellant devices.

THE DISCOVERY OF ARTEMISININ

Qing Hao is a herb commonly used in China with a long history of use as an antipyretic to treat the alternate chill and fever symptoms of malaria and other "heat syndromes" in the traditional Chinese medical system. The name Qing Hao first appeared in a silk book excavated from the tomb at Mawangdui belonging to the Han dynasty; it was entitled "Wu Shi Er Bin Fang" (Prescriptions for Fifty-two ailments) and dated from as early as 168 BC, and described the use of Qing Hao for the treatment of haemorrhoids. In 340 AD, Qing Hao was recorded for the first time as a treatment for fevers in a medical book, "Zhou Hou Bei Ji Fang" (Handbook of Prescriptions for Emergency Treatment). In this work, the author, Ge Hong, recommended that, to reduce fevers one should soak a handful of Qing Hao in one sheng (approx. 1 L) of water, strain the liquor and drink it all. Li Shi-Zhen, the author of the famous materia medica "Ben Cao Gang Mu" (1596 AD), based on the former medical text records and his own experience, stated that malaria with chills and fever could be treated with Qing Hao (Klayman, 1985 and references therein). Qing Hao has been recommended for use alone or with other herbs for malaria as well as for other conditions (see below).

Qing Hao is the aerial part of *Artemisia annua* L. of the family Asteraceae (formerly Compositae), and it is now officially listed in the Chinese pharmacopoeia (Anon. 1992). In the early literature, Qing Hao referred to two species, *A. annua* and *A. apiacea* but nowadays Qing Hao is usually taken to be *A. annua*.

In 1967, the Chinese government began a systematic examination of plant species used in TCM in order to discover new drugs, especially for malaria. The many plants tested included Qing Hao, but no activity was seen when hot-water extracts of this herb were tested in mice infected with the rodent malaria parasite *Plasmodium berghei*. However, when cold ether extracts of Qing Hao were tested, encouraging activity was observed (Klayman, 1985).

Since then, Chinese researchers in many institutions have devoted considerable resources to investigate the active components responsible for the anti-malarial activities of the crude drug Qing Hao and this has resulted in the discovery of the active compound, Qinghaosu, which means "a principle from Qing Hao" and which was also referred to as arteannuin and artemisinine; today, the accepted western name is artemisinin (since the compound is a terpene, not an alkaloid as the suffix-ine implies). Artemisinin was first isolated and identified in 1972 as a novel compound with its chemical structure representing a new type of sesquiterpene lactone with an unusual endoperoxide moiety (see chapter 12).

Following intensive investigations into the pharmacology, pharmacokinetics and toxicology of artemisinin and its natural and synthetic derivatives, artemisinin was found to have a rapid schizonticidal action (i.e. it kills the forms of malaria parasites infecting red blood cells), with low toxicity (see chapter 13). Of particular significance was its remarkably potent therapeutic effect against malaria caused by drug-resistant *Plasmodium falciparum*. In clinical studies with 2099 patients infected with *Plasmodium vivax* or *P. falciparum*, artemisinin showed very good therapeutic efficacy in almost all patients without obvious side effects (Anon. 1979). Moreover, artemisinin was also found to be effective in the treatment of chloroquine resistant *falciparum* malaria and its often fatal complication, cerebral malaria (see chapter 14).

The discovery of a new potent antimalarial from a TCM herb was so significant that it attracted great attention from the World Health Organsation and a number of research establishments around the world. Although artemisinin has been synthesised by a few research groups in several countries, the high cost made synthetic artemisinin an uneconomic choice for large-scale production of the drug so that it is extracted from its natural source (see chapter 12). Consequently, many taxonomy-orientated scientists started screening dozens of other *Artemisia* species in an attempt to find new source plant for artemisinin but *Artemisia annua* remains the only species that is known to produce artemisinin.

Several hundred years ahead of modern medicine, practitioners of TCM recorded the presence of a biological clock in human body, the diagnosis of diabetes, the function of hormones, described the circulation of the blood and the application of inoculation (Temple 1986). The discovery of artemisinin from a TCM herb is an excellent example of the contribution that TCM may make to modern medical science.

TRADITIONAL CHINESE MEDICINE

A large population in South East Asia has used TCM for thousands of years, and it is still practised in parallel with western medicine in the healthcare system in China owing to its unique theoretical basis and proven therapeutic efficacy. The theory and practise of TCM is complex and space does not permit a detailed consideration here; however, a brief account is given below in order to help the reader appreciate the use of *Artemisia* species in TCM.

The underlying principle of TCM lies in the comprehensive and dynamic application of the theories of Yin-Yang and the Five Elements. Everything in the universe is considered to have the opposite but inter-related aspects, Yin and Yang. When the Yin and Yang are balanced in a human body, the person is in a healthy condition, not only less susceptible to diseases, but also likely to have a better quality of life; conversely, illness is the result of an imbalance of Yin and Yang in the body. The aim of treatment in TCM is always to restore the balance of the Yin and Yang. The theory of Five Elements (metal, wood, water, fire and earth), relates the properties of the five elements to universally interdependent and mutually restraining relationships of matters. This theory is used in TCM to explain the correlation and pathological influences among five viscera, (heart, liver, spleen, lung and kidney), and to guide the diagnosis and the treatment.

Traditional Chinese medicine consists of acupuncture, massage, Qi-Gong (Chinese deep-breathing excercises), medicated diet and herbal treatment with herbal treatment holding a dominant position. The theory of the herb property is one of the most important parts of the TCM. According to this theory, herbs have four properties – hot, warm, cool and cold, and five tastes – sour, bitter, sweet, pungent and salty. Basically, herbs with a hot or warm property have the nature of Yang and herbs with a cool or cold property have the nature of Yin. This forms the very basis of the theory of TCM materia medica. In line with the Yin-Yang concept in TCM diagnosis, the use of a Yin herb or a Yang herb to re-balance the Yin and Yang in the different organs in the body and to treat the illnesses accordingly can be appreciated. Herbs in TCM are also uniquely considered to have selective therapeutic effects on certain parts of the body and this is called channel tropism. This resembles the concept of drug targeting in modern pharmaceutical research although the origin of the "magic bullet" concept is unlikely to have been derived from channel tropism! Both the herbal property and the channel tropism represent the different pharmacological functions of each herb.

Herbs with the property of cold or cool are able to clear heat, to purge the pathogenic fire and to detoxify, and can be used to treat heat syndromes. For example, Artemisia herb (*A. annua*) and Coptis root (*Coptis chinensis*) are cool in nature and are used to treat internal heat caused by bacterial infections or by non-infectious disease. However, these two herbs are not necessarily inter-changeable in their usage because they are different in other respects. Herbs with the property of warm and hot are able to warm internally, to expel cold and to tonify Yang, and are thus used to treat cold syndromes. For example, ginger (*Zingiber officinale*), and cinnamon bark, (*Cinnamomum cassia*) are hot in nature and are commonly used to warm the interior and expel cold. Herbs with a sour taste possess astringent actions and can be used to treat sweating and diarrhoea, such as Schisandra fruit (*Schisandra chinensis*), for excessive sweating and Cherokee rosehip (*Rosa laevigata*), for chronic diarrhoea. Herbs with sweet taste usually possess a tonifying (or strenghening) function, such as wolfberry fruit, (*Lycium barbarum*) used for tonifying the liver and kidney and the Chinese date, (*Ziziphus jujuba*) used for tonifying the blood.

TCM is characterised by its unique understanding of the human body's physiology and pathology, and by its intricate methods of diagnosis and treatment of diseases. TCM regards the human body as an organic whole, whose component parts are physiologically interconnected, so a local lesion may affect the entire body and indeed it may well be a result of the imbalance of some other parts of the body. Disorders of internal organs may be reflected on the surface of the body. When a Chinese practitioner sees a patient, he/she will usually gather all the relevant information through four methods of diagnosis (inspection, listening and smelling, inquiring and pulse-feeling), then differentiate symptoms and signs into "Zheng" of a certain nature (syndrome). He/she will make a diagnosis based on "Eight Principal Syndromes", i.e. Yin and Yang, interior and exterior, cold and heat, deficiency and excessiveness, but it is important to appreciate that each syndrome will be further differentiated according to the particular symptoms present.

It is normal practice in TCM for some different diseases to be treated with similar herbs while diseases which appear to be similar (from a western point of view), may be treated with different herbs depending on the differentiation of syndromes. For example, a patient suffering from a cold who has symptoms of severe chilliness, slight fever and a tongue with thin, white fur, which indicates the exterior wind-cold syndrome, should be treated with herbs which are pungent in taste and warm in property. If a patient with a cold has symptoms of high fever, mild chilliness and a tongue with thin yellow fur, which indicates the exterior wind-heat syndrome, he should be treated with those herbs which are mild, diaphoretic, pungent in taste and cool in property.

ARTEMISIA ANNUA IN CHINESE TRADITIONAL MEDICINE

Yeung (1985), in a short monograph on Qing Hao gives *A. apiacea* Hance as a synonym for *A. annua* and describes the taste and property of the herb as bitter, pungent and cold. Its functions are antimalarial, to reduce the heat caused by deficiency of Yin, and to clear the summer heat. The medicinal uses of Qing Hao are given as malaria, febrile diseases, tidal fever, low grade fever and summer heat stroke. Although Qing Hao may be used as a cooling herb for the relief of symptoms, TCM places great emphasis on treating the underlying cause of an illness and as explained above, diagnosis is often much more precise than it is in western medicine. This helps to explain why complex combinations of Chinese herbs are used; additional herbs (which may be referred to as "minister", " assistant" or "guide" herbs are added to the principal (or "emperor" herb) in order to complement or modify its action so that the TCM prescription is tailored for the needs of the individual patient.

An example of a prescription for the treatment of malaria using TCM is the classical formula Qing Hao Bie Jia Tang (decoction of Carapax Trionycis and Qing Hao) which is found in the "Wen Bing Tiao Bian" (Detailed analysis of febrile diseases) of 1798. It is composed of the following: Herba Artemisia Annuae, Carapax Trionycis (fresh water turtle shell), Folium Mori (*Morus alba* leaf), Pollen (species not specified), Rhizoma Anemarrhenae (*Anemarrhenae asphodeloides* rhizome) and Cortex Moutan (*Paeonia suffruticosa* bark). Herba Artemisia Annuae is used to clear the heat from the channels so as to lower the fever and Carapax Trionycis has the effect of nourishing Yin and lowering the fever. The combination of the two expels heat without injuring Yin and these are the principal ingredients but Herba Artemisia Annuae may also act as the "guiding herb". Cortex Moutan may also cool the blood and help Herba Artemisia Annuae to eliminate the heat while Rhizoma Anemarrhenae serves to benefit Yin and help Carapax Trionycis to lower the fever so that they are considered to be minister herbs. Folium Mori and Pollen are assistant herbs in the formula.

Although it is well established that artemisinin itself is a highly effective antimalarial, to date there have been no published reports of clinical trials in which *A. annua* herb has been used alone for the treatment of malaria. One reason for this is that, as illustrated above, TCM is individualised for each patient so that different combinations of herbs are used. Currently however, there is interest in Africa in growing the herb and using it as a locally available and inexpensive antimalarial, but before this practise can be recommended, studies demonstrating the efficacy of *A. annua* will be needed. However, studies in which a crude ethanolic extract of *A. annua* was formulated with oil in a soft gel capsule and administered to mice and tried clinically in man have been reported (Yao-De *et al.*, 1992). In mice infected with *Plasmodium berghei* the soft gel capsule was found to be superior to a tablet formulation of the extract; for the capsule the ED_{50} value was 11.9 g/kg (with reference to the raw herb) and 35.1 mg/kg (with respect to the artemisinin content) while the corresponding values for the tablet were 46.8 g/Kg and 124 mg/kg respectively. Artemisinin itself was found to have an ED_{50} of 122 mg/kg. The soft gel capsule was administered to 103 patients with malaria (*P. falciparum* or *P. vivax*) and compared with the tablet formulation which was given to 41 malaria patients. Both formulations were effective in reducing fever and clearing parasites at doses equivalent to 73.6 g raw herb (for the capsule), and 80.8 g (for the tablet), given over 3 days, but recrudescent rates were high with both dosage forms although they were reduced by increasing the duration of treatment. Recrudescence (which arises because not all of the parasites have been killed by the drug and is not due to drug resistance), may also occur when artemisinin and its derivatives are used to treat malaria (see chapter 13), but the high rate of recrudescence reported in the above study suggests that crude extracts alone may not give an acceptable cure rate and additional drug therapy may need to be given.

OTHER SPECIES OF *ARTEMISIA* USED IN TRADITIONAL CHINESE
MEDICINE

There are 184 artemisia species and 44 varieties found in China. Most of them are
distributed in the north, northeast and north west of China, with some in eastern,
central and southern areas. Most species of *Artemisia* contain polyacetylenes,
flavonoids and terpenoids but some species also contain other types of constituents
such as cyanogenic glycosides and coumarins.

According to a national survey of the medicinal plants of China, more than sixty
Artemisia species are used in different areas for certain ailments such as inflamma-
tion, liver and stomach disorders and gynaecological problems (Table 1). The leaves
of more than ten species are used for the preparation of moxas (see below). A
number of *Artemisia* species are used as choleretic, anti-inflammatory and diuretic
agents in the treatment of hepatitis. Two of these are *A. scoparia* and *A. capillaris*
and are known by the same Chinese name as Yin Chen. Both species contain essential oils, flavones and coumarins. A flavone, capillarisin, the major constituent of
A. capillaris, together with two new stereoisomeric constituents, capillartemisin A
and B, showed choleretic effects in experiment studies. The coumarin derivative scoparone isolated from both species had a preventative effect on carbon tetrachloride
or galactosamine-induced hepatotoxicity in hepatocyte cell cultures, (Hikino, 1985,
Kiso *et al.*, 1984), and also had anti-inflammatory and analgesic effects (Yamahara
et al., 1982). Decoctions of Yin Chen have been used in many clinical studies for
icteric hepatitis with good success. In one study, 32 patients with icteric hepatitis
were treated with 30–45 g of Yin Chen to be made into a decoction taken daily. In
all cases, fever subsided quickly, jaundice disappeared, and the liver returned to
normal size. The mean length of treatment was seven days (Huang, 1959). Other
species used locally for the same purpose include *A. anethifolia, A. anethoides* and
A. demissa.

More than ten *Artemisia* species are used in TCM for certain gynaecological problems. According to the theory of TCM, a number of conditions, such as amenor-
rhea, menstrual pain and prolonged menstrual bleeding, are usually related to Qi
(vital energy) and blood deficiency, Qi stagnation or blood stasis due to cold. Thus
those herbs with the acrid and warm property can be used to treat such ailments
with a good clinical response and various species of *Artemisia* which have the same
property are often the principal ingredients of TCM formulae for the illnesses mentioned above (Table 1).

Along with acupuncture, the application of moxibution is used in TCM. It is
carried out by applying an ignited moxa cone or moxa stick on the acu-points to
elicit heat stimulation. It exerts an effect by warming and regulating the channels,
promoting the circulation of Qi and blood. Moxibustion is often combined with
acupuncture for the treatment and prevention of many diseases including pain relief
and injury management.

There are number of herbal materials used to make moxa cones or sticks but the
dried leaves of mugwort, (*Artemisia vulgaris* leaf) is the most common one. The
dried leaves are pounded into fine pieces (moxa wool), which can then be easily

Table 1   Sixty *Artemisia* species used in traditional chinese medicine.

Medicinal use	Botanical name	Parts used
Stomachic	A. absinthium	whole herb
(aids digestion)	A. adamsii	whole herb
	A. macrocephala	whole herb
	A. rupestris	whole herb
Choleretic and anti-icteric	A. anethifolia	young plant
(stimulates bile flow)	A. anethoides	young plant
	A. capillaris	young plant
	A. demissa	young plant
	A. hedinii	whole herb
	A. mattfeldii	whole herb
	A. scoparia	young plant
	A. selengensis	whole herb
	A. tanacetifolia	whole herb
Antipyretic	A. angustissima	whole herb
	A. annua	whole herb
	A. anomala	whole herb
	A. caruifolia	whole herb
	A. caruifolia var. schochii	whole herb
	A. dracunculus	whole herb
	A. japonica	whole herb
	A. littoricola	whole herb
	A. manshurica	whole herb
	A. vestita	whole herb
Antitussive	A. conaensis	whole herb
	A. dubia	whole herb
	A. dubia var. subdigitata	whole herb
	A. halodendron	young plant
	A. lagocephala	leaves
Antirheumatic	A. brachyloba	whole herb
	A. eriopoda	whole herb
	A. integrifolia	whole herb
	A. ordosica	whole herb
Emmenagogic	A. atrovirens	leaves
(regulates menstruation)	A. indica	leaves
	A. keiskeana	whole herb
	A. lactifolia	whole herb
	A. lancea	leaves
	A. lavandulaefolia	leaves
	A. leucophylla	leaves
	A. mongolica	leaves
	A. princeps	leaves
	A. sylvatica	whole herb
	A. vulgaris	leaves

Table 1   Sixty *Artemisia* species used in traditional chinese medicine. (*Continued*)

Medicinal use	Botanical name	Parts used
Antidotal	*A. anomala var. tomentella*	whole herb
(clears heat and detoxifies)	*A. deversa*	whole herb
	A. orientali-hengduangensis	whole herb
	A. roxburghiana	whole herb
	A. sacrorum	whole herb
	A. sacrorum var. messerschmidtina	whole herb
	A. sieversiana	whole herb
	A. tournefortiana	whole herb
Antiparasitic	*A. frigida*	whole herb
	A. moorcroftiana	whole herb
	A. smithii	whole herb
	A. emeiensis	whole herb
Moxibustion	*A. argyri*	leaves
	A. argyri var. gracilis	leaves
	A. igniaria	leaves
	A. verbenacea	leaves
	A. verlotorum	leaves
	A. vulgaris	leaves

shaped into moxa cones or prepared as moxa sticks for use. A good example of the therapeutic application of the mugwort moxa stick is its use to stimulate acupuncture point BL 67, (Zhi Yin), to promote the turning of a foetus in breech presentation. This technique is taught in every TCM university in China and a recent randomised, controlled clinical study demonstrated its effectiveness (Cardini and Huang, 1998).

Pure moxa sticks contain only mugwort as the main ingredient, while medicated moxa sticks may contain some or all of the following materials: Cortex Cinnamon (cinnamon bark), Rhizoma Zingiberis (ginger rhizome), Flos Caryophylli (flowers of *Dianthus* sp.), Radix Angelicae Pubescentis (*Angelica pubescens* root), Herba Asari (*Asarum sieboldii* herb), Radix Angelicae Dahuricae (*Angelica dahurica* root), Rhizoma Atractylodis (*Atractylodes macrocephala* rhizome), Myrrha (myrrh), Olibanum (frankincense) and Pericarpium Xanthoxyli (*Xanthoxylum bungeanum* fruit peel). Although there is no report of any chemical effects, apart from heat, of moxibustion in classical TCM, it is likely that essential oil containing herbs in moxa sticks play a part in the therapeutic results of moxibustion.

While over sixty *Artemisia* species are used for medicinal purposes, the active chemicals of most of the species are still unknown. The demand for further research on this genus can be justified with special emphasis upon their chemistry, taxonomy, and clinical verification of claimed therapeutic efficacy.

REFERENCES

Anon. (1979) *China Pharmaceutical Bulletin*, **2**, 49.

Anon. (1992) *Pharmacopoeia of the People's Republic of China*, English edition of the Chinese 1990 Pharmacopoeia. Gaungdong Science and Technology press, Guangzhou, China, p. 91.

Cardini, F. and Huang, W. (1998) Moxibustion for correction of breech presentation – A randomized controlled trial. *The Journal of the American Medical Association*, **280**, 1580–1584.

Huang, Y. (1959) *Fu Jian Traditional Chinese Medicine*, 7, 42.

Hikino, H. (1985) Antihepatoxic constituents of Chinese drugs. *China Pharmaceutical Bulletin*, **20**, 415–417.

Kiso, Y., Ogasawara, S., Hirota, K., Watanabe, N., Oshima, Y., Konno, C. *et al.* (1984) Validity of oriental medicines. LV. Liverprotective drugs. 10. Antihepatotoxic principles of Artemisia capillaris buds. *Planta Medica*, **50**, 81–85,

Klayman, D.L. (1985) Qinghaosu (Artemisinin): An antimalarial drug from China. *Science*, **228**, 1049–1054.

Temple, R. (1986) In: Stephens, P. (Ed.). *China – Land of Discovery and Invention*, Part 5, Medicine and Health, UK, pp. 123–135.

Yamahara, J., Matsuda, H., Sawada, T., Mibu, H. and Fujimira, H. (1982) Biologically active principles of crude drugs. Pharmacological evaluation of *Artemisia capillaris* flos. *Yakugaku Zasshi*, **102**, 285–291.

Yao-De, W., Qi-Zhong, Z. and Jie-Sheng, J. (1992) Studies on the antimalarial action of gelatin capsule of *Artemisia annua. Chung Kuo Chi Sheng Ching Hsueh Yu Chi Sheng Chung Ping Tsa Chih*, **10**, 290–294.

Yeung, H. (1985) *Handbook of Chinese herbs and formulas*. Volume 1. Institute of Chinese Medicine, Los Angeles, CA 90025, pp. 430.

10. CULTIVATION OF *ARTEMISIA ANNUA* L.

J.C. LAUGHLIN[1], G.N. HEAZLEWOOD[2] AND B.M. BEATTIE[2]

[1]*Agricultural Consultant (Medicinal Crops), 1/14A Sherburd Street, Kingston, Tasmania 7050, Australia.*

[2]*Department of Primary Industry & Fisheries, PO Box 303, Devonport, Tasmania 7310, Australia.*

INTRODUCTION

The genus *Artemisia* includes a large number of species and some have been cultivated as commercial crops with a wide diversity of uses. Some better known examples include antimalarial (*A. annua* – annual or sweet wormwood), culinary spices
(*A. dracunculus* – French tarragon), liquor flavouring (*A. absinthium* – absinthe),
garden ornamental (*A. abrotanum* – southernwood) and insect repellent (*A. vulgaris*
– mugwort). However this review will concentrate on the cultivation of *A. annua*
because of its contemporary importance as a source of new and effective antimalarial drugs.

During World War II and in the years immediately following, the world wide incidence of malaria was dramatically reduced. On the one hand the *Anopheles* mosquito vector was successfully controlled by the advent of the insecticide DDT and on
the other the organisms causing human malaria – the single celled *Plasmodium*
species: *falciparum, vivax, malariae* and *ovale* – were effectively controlled by the
use of synthetic derivatives of quinine. The specific statistics for India illustrate this
dramatic reduction. In 1961 the incidence of malaria had fallen to about 100,000
reported cases, however by 1977 the number of reported cases in India had risen to
at least 30 million (Harrison, 1978). The reasons for this dramatic increase were the
dual factors of the development of resistance to DDT by the *Anopheles* mosquito
and the development of resistance to quinine and quinine analogues by the
Plasmodium. Although the problem of malarial drug resistance in the *Plasmodium*
varies throughout the world – with the worst in South East Asia – the overall situation is that malaria is currently the most serious and devastating tropical disease in
the world. At least 300 million clinical cases occur worldwide with up to 2.7 million
deaths annually (Nussenzweig and Long, 1994).

It was because of this background of a massive increase in the incidence of
malaria and the serious reduction in effectiveness of standard antimalarial drugs
that the possibilities extended by extracts from *Artemisia annua* were eagerly taken
up. *Artemisia annua* L. is a member of the family *Asteraceae* (formerly *Compositae*)
and is considered to be a native of Asia with its origin in China (McVaugh, 1984)
where it forms part of the native plant population of the steppes of Chahar and
Suiyuan Provinces (40° N, 109° E) (Wang, 1961). *A. annua*, or *Qing Hao* (= green

159

herb) as it is referred to in the Chinese literature and quite often in the general literature, has been used as a traditional herbal treatment for malaria in China for centuries with the first recorded use dating back almost 2000 years (Anon, 1979). This traditional use of *A. annua* utilised extraction of the plant with hot or boiling water either alone or more commonly as part of a mixture with other herbs. Chinese scientists studied the effectiveness of the herbal infusion method utilising *A. annua* in the late 1960's and found that it was not effective (Klayman, 1989). However extraction of *A. annua* leaf with diethyl ether gave a compound with marked antimalarial activity in animal experiments. This compound was isolated and identified as artemisinin by Chinese scientists in 1972 who also refer to it by its Chinese name of *qinghaosu* (= extract of green herb) (Anon, 1982). Artemisinin is more specifically an uncommon sesquiterpene lactone endoperoxide of the cadinane series (Luo and Shen, 1987). With the possible exception of *A. apiacea* (Liersch *et al.*, 1986) *A. annua* appears to be the only species in the plant kingdom which contains artemisinin.

The apparent lack of effectiveness of infusions of *A. annua* in hot or boiling water raises the question of why the plant should have established its reputation as a febrifuge. One explanation offered is that *A. annua* was commonly used as an infusion in conjunction with other herbs. Extracts from these herbs such as saponins may have acted as detergents and released artemisinin into the water. Flavonoids, which may also have been present in the other herbs, could also have potentiated the antimalarial activity of artemisinin (Phillipson, 1994, S.C. Vonwiller, personal communications). One particular mixture recorded in an old Chinese herbal (Wen Bing Tiao Bian, 1798) utilised *Artemisia annua, Amyda sinensis, Rehmania glutinosa, Anemarrhea aspholeloides* and *Paeonia suffruticosa*. A much earlier Chinese herbal (Ge Hong, 340 AD) prescribed the top growth of *A. annua* for the treatment of fevers.

Since the isolation and identification of artemisinin from *A. annua* as the compound which exerts the antimalarial effect, detailed studies have been carried out on other related compounds in the plant and on ways in which artemisinin can be modified to have more effective medicinal qualities. Artemisinic acid – also known as *qinghao* acid – a precursor of artemisinin which is present in some parts of the plant at concentrations up to tenfold that of artemisinin (Laughlin, 1993) is one of the more important compounds of potential commercial importance. Arteether (Davidson, 1994) and artemether (Roche and Helenport, 1994) the ethyl and methyl derivatives respectively of artemisinin are two of the more important conversion products of enhanced antimalarial efficacy which are being developed commercially.

Despite the fact that chemical synthesis of artemisinin can be accomplished (Schmid and Hofheinz, 1983, Xu *et al.*, 1986, Rabindranathan *et al.*, 1990, Avery *et al.*, 1992) the cost and complexity of the synthesis is such that extraction of artemisinin from *A. annua* plant material is a much more economic method of production. Although micro propagation of *A. annua* can be easily accomplished (Whipkey *et al.*, 1992) it is unlikely that the *in vitro* production of artemisinin will be a commercially viable proposition (Ferreira and Janick, 1995a). Because of this, the production of artemisinin from field cultivated plants will be the preferred option for the foreseeable future. A number of useful abbreviated general outlines of *A. annua* cultivation have been recorded (Simon and Cebert, 1988, Simon *et al.*, 1990, Laughlin, 1993, 1994, Ferreira *et al.*, 1997). This review will give a detailed assessment of the various contributions which have been made to the knowledge of cultivation of *A. annua*. The review will follow the growth cycle of the plant from seed drilling or transplanting through to vegetative growth, final harvest and post harvest operations.

BOTANICAL DESCRIPTION

A. annua is an annual shrub of the family *Asteraceae* (formerly *Compositae*). The height ranges from 30–250 cm depending on strain, region of production and a range of agronomic factors particularly plant population density. The leaves are bipinnatifid, glabrous with segments linear and dentate. The inflorescence is a terminal compound panicle. The flower heads are small and typically yellow in colour, heterogamous, 2–3 mm wide, globular, bracts linear, oval-acuminate or oval. The marginal flowers are female with a corolla 4 lobed and the disk (central) flowers are hermaphrodite, with the corolla 5 lobed and stamens 5. Glandular trichomes containing a strongly volatile oil are present in both flower heads and leaves of *A. annua*. The fruit is an obovoid achene often light grey in colour, smooth and about 0.5 mm long (Keys, 1976). A very detailed botanical description of *A. annua* has been recorded by Ferreira *et al.* (1997).

CLIMATIC RANGE OF PRODUCTION

Although *A. annua* is generally considered to have its origins in the temperate regions of China (40° N) where it forms a part of the natural steppe vegetation at an elevation of 1000–1500 metres above sea level (Wang, 1961) the range of the plant is much wider. In China it is fairly generally distributed and extends as a native into southern Siberia, Vietnam and northern India (Keys, 1976). Outside of Asia *A. annua* has been introduced and grows wild in a wide range of countries in Europe (Hungary, Bulgaria, Rumania, France), North America (USA) and South America (Argentina) (Klayman, 1989, 1993). In addition, it has been introduced into experimental cultivation in Vietnam, Thailand, Burma, Madagascar, Malaysia, USA, Brazil, Australia (Tasmania) and in Europe into Holland, Switzerland, France and as far north as Finland.

The climatic range of *A. annua* is of considerable importance in determining areas for potential production. Although *A. annua* originated in relatively temperate latitudes (Wang, 1961) it appears that the plant can grow effectively at much lower tropical latitudes with seed selections which are either native to these areas (Woerdenbag *et al.*, 1994) or which have been adapted by breeding (Magalhaes and Delabays, 1996). However other workers have concluded that cultivation of *A. annua* would be unadapted to the tropics because the plant is a short day annual which would flower without achieving sufficient biomass (Anon, 1996, Ferreira *et al.*, 1997). These conclusions may well need to be reassessed in the light of other evidence from late flowering strains of *A. annua*. The high artemisinin concentrations (0.5–1.5%) in the leaves of some of these strains could allow high artemisinin yields in tropical latitudes even though the leaf biomass may not be as high as some strains of *A. annua* grown in temperate latitudes. In Vietnam (near Hanoi), at a latitude of 21° 02′ N, a field experiment achieved adequate leaf dry matter (5.3 t/ha) before flowering with high artemisinin yields equivalent to about 45 kg/ha (Woerdenbag *et al.*, 1994). Also in Madagascar at 18° 52′ S adequate leaf dry matter (4.7 t/ha) was also achieved before flowering with a high artemisinin yield of about 40 kg/ha (Magalhaes *et al.*, 1996). In this study the field experiment was carried out at an elevation of ca. 1500 metres above sea level. The beneficial influence that higher altitudes may have on the production of *A. annua* at tropical latitudes may be a principle which could be applied to other parts of tropical Africa and elsewhere where similar areas could be investigated. The possibility that drugs derived from *A. annua* may soon be needed in the battle against malaria in Africa should be an incentive to seek out these niche areas of production (see **Time of establishment**).

SITE AND SOIL SELECTION

The ideal site selection for *A. annua* cultivation would depend on the scale of operation and the location of commercial extraction plants. The concentration of artemisinin in *A. annua* (see **Germplasm**) is relatively low and the total fresh plant yield from which the chemical is extracted can be as high as 100t/ha and greater. Generally it would seem that the extraction and processing plant should be as close as possible to the area of production, or vice versa depending on circumstances, to minimise transport costs. If ultimately mechanical harvesting is a feasible proposition (Mediplant, 1995) then the selection of relatively flat locations would be appropriate. *A. annua* has been grown in a wide diversity of soils and its effectiveness in colonising waste areas suggests that it is very adaptable. Apart from an intolerance of some *A. annua* selections to acid soils below pH 5.0–5.5 (see **Nutrition**) many soil types could be utilised.

SEED SELECTION

Germplasm

Since the identification of artemisinin as the compound in *A. annua* which conveys the antimalarial effect there have been a number of investigations to select strains with high artemisinin. These studies have shown wide variability in artemisinin content which can range from 0.01% (Trigg, 1989) to about 1.0% (Delabays *et al.*, 1993) and approaching 1.5% (Debrunner *et al.*, 1996). The economics of commercial development of artemisinin derived drugs and their use in areas of most need hinge on plant raw material with high artemisinin content. Because of this, investigations have been carried out to select seed from high artemisinin producers coupled with other desirable agronomic characteristics. These include good seed and plant vigour, high leaf to stem ratio with high dry matter leaf yield, disease resistance and desirable time of flowering appropriate to the region of production. The seeds of *A. annua* will maintain their vigour for at least three years if stored under dry, cool conditions (Ferreira *et al.*, 1997).

Germplasm comparison and selections have been carried out on the basis of (i) plants which have been introduced and established in the investigating countries and (ii) promising high artemisinin lines introduced from countries where *A. annua* is native (e.g. China and Vietnam). However commercial competitiveness in the possession of high artemisinin lines has limited the widespread availability of these lines at the present time. Likewise the general access to hybrids which have incorporated the high artemisinin (1.1%) but low vigour of Chinese clones with the low artemisinin (0.04–0.22%) but high vigour of a range of European clones (Delabays *et al.*, 1993) are also generally unavailable. Similarly the more recent hybrids between Chinese and Vietnamese selections with even higher artemisinin (1.0–1.5%) are only available to a limited extent (Debrunner *et al.*, 1996). The cost and time involved in carrying out accurate assays for artemisinin have also been other limitations on the screening and selection of *A. annua* lines. Rapid and economical methods of artemisinin assay developed by Charles *et al.* (1990) and Ferreira and Janick (1995a, 1996a) and the technique of appraising plantlets growing *in vitro* in the laboratory for artemisinin content (Pras *et al.*, 1991) are all strategies by which the process of screening *A. annua* germplasm could be speeded up. Germplasm assessment and studies of a range of other agronomic factors which have a bearing on the successful cultivation of *A. annua* have been carried out in Australia (Tasmania) (Laughlin, 1993, 1994, 1995), Brazil (Magalhaes, 1994, 1996), India (Singh *et al.*, 1986, 1988), Japan (Kawamoto *et al.*, 1999), Madagascar (Magalhaes *et al.*, 1996), Netherlands (Woerdenbag *et al.*, 1990), Switzerland (Delabays *et al.*, 1992, 1993, Debrunner *et al.*, 1996), and in the United States (Simon and Cebert, 1988, Simon *et al.*, 1990, Morales *et al.*, 1993, Ferreira *et al.*, 1995).

Alternative selection strategies

Artemisinic acid (*qinghao acid*) a precursor of artemisinin is present in *A. annua* at concentrations up to tenfold that of artemisinin (Laughlin, 1993) and can be converted to artemisinin with an efficiency up to 40% (Roth and Acton, 1989, Haynes and Vonwiller, 1991). *A. annua* also contains an oil which is used in perfumery and as an anti-microbial (Laurence, 1990). At the present time the volume of oil traded is relatively small but if the cultivation of *A. annua* expands, the greater availability of the oil may make it a more commercially attractive commodity. Various strains of *A. annua* which are high in specific oil components of commercial interest have been identified in studies in the United States (Charles *et al.*, 1991). The screening of *A. annua* germplasm for both artemisinic acid and oil content as well as artemisinin may be a very useful technique in that it may be possible to extract both oil and artemisinic acid in the one operation (Laughlin, 1994) (see **Oil production**).

PLANT CULTURAL TECHNIQUES

Plant Establishment

Natural stands

In China A. *annua* traditionally has been harvested from wild natural self seeded stands. Although no specific crop production statistics are available, because of a confidentiality policy of Chinese authorities, it is believed that the bulk of Chinese production still comes from wild stands. These stands are the source of much of the artemisinin derived drugs used in China and probably the bulk of those drugs exported elsewhere (WHO, 1994) although some selected lines of A. *annua* are cultivated as a row crop in Szechwan Province (Ferreira *et al.*, 1997). Ideally the harvesting of raw material for medicinal drug production from wild stands is not a good policy (Fritz, 1978, Franz, 1983). The plant material in wild stands is typically very variable in its content of the required medicinal constituents and this has an impact on the economics of drug extraction. Added to this the continual encroachment and elimination of wild stands will ultimately limit the source of genetic variability which is vital to the development of improved seed lines (Chatterjee, 1993). Another negative factor against utilisation of wild stands is that transport distances often become uneconomic with a crop such as A. *annua* with a relatively low artemisinin content and a large bulk of material required.

Where plants traditionally have been harvested from the wild for centuries there is most likely a general conservative reluctance in many quarters to believe that plantation grown crops could have equal quality characeristics. Because A. *annua* has been grown in China for its medicinal qualities for many centuries the belief in the superiority of the wild grown plant may be very deeply entrenched. A comparison of the quality of wild grown and plantation grown A. *annua* plants in key production areas in China would be a useful manoeuvre to help overcome this barrier. No such comparisons of wild and plantation grown A. *annua* appear to have been published elsewhere. However the analogy of wild and plantation grown *Gentiana lutea* L. plants which have both medicinal and liquor industry uses may supply a useful model for this type of comparison (Franz and Fritz, 1978).

Transplanting

Because supplies of high quality A. *annua* seed have been generally limited up to the present time most experimental programs have utilised transplanting as the preferred method of establishment (Acton *et al.*, 1985, Liersch *et al.*, 1986, Singh *et al.*, 1988, Simon and Cebert, 1988, Delabays, 1993, Laughlin, 1993, Ferreira *et al.*, 1995). These transplants have taken the form of cuttings (Ferreira *et al.*, 1995) but in most other studies the investigators have used some form of cellular tray system such as Speedlings® (Fig. 1). This system involves germinating A. *annua* seed in shallow (5–6 cm deep) seed trays using a sterilised potting mix such as 2 parts of sand, 2 parts of peat and 1 part vermiculite which has had the pH adjusted to pH

Figure 1 A plantlet of *A. annua* after eight weeks of growth, initially in seed trays and then in cellular Speedling® trays. The dense root mass allows rapid growth after transplanting.

6.0–7.0 and a low to moderate quantity of complete nutrient fertiliser added after sterilisation. The seeds are sprinkled uniformly onto the surface of this mixture and covered very lightly with vermiculite (1–2 mm) and germinated in a greenhouse. The surface must be kept moist and not let dry out. When the plantlets are about 2 cm high (4–5 true leaves) – which takes about 3 weeks – they are pricked out, taking care not to damage the roots, and transplanted into the cellular trays using the same mixture as above or whatever alternative container system may be used. The plantlets are then grown on in the greenhouse in these containers for about an additional 4 weeks or until the young plants have about 10 true leaves and are about 10–14 cm high. After this the plants are hardened off outside or in a shade house for 3–5 days and then transplanted to the field. There is no published information on the optimum size for transplanting *A. annua* plantlets but the key point is that they should be robust enough to cope with mechanical transplanting systems but not so tall that they become spindly and susceptible to wind damage. Until the transplanted seedlings start to actively grow away the soil should be kept moist. It would be a useful exercise to compare the optimum age and size of transplants in each environment in which *A. annua* is introduced.

Figure 2   A comparison of *A. annua* transplanted (background) and direct seed drilled (foreground) seven weeks after establishment at Cressy Research Station, Tasmania, Australia. At final harvest three months later both methods gave similar yields of leaf dry matter.

Seed drilling

If supplies of seed were freely available, the direct drilling of seed would be the most economical method of plant establishment provided the length of the growing season and other environmental factors were suitable and weeds could be controlled (see **Herbicides** and **Plant population density**). In Australia (Tasmania) direct drilling of seed gave leaf dry matter and artemisinin yields very similar to the yields from transplanting in 2 out of 3 field experiment (Laughlin, 1993). In these experiments seed drilling and transplanting took place in mid October (spring). Both treatments matured at the same time and were harvested at the early bud stage in late February (summer) four and a half months later (Fig. 2). There is little information on the possibility of autumn sowing of seed and the over-wintering survival of *A. annua* seedlings in geographical zones where this technique may be appropriate (Ferreira *et al.*, 1997). However the performance of self sown seeds of a Yugoslavian strain of *A. annua* which germinated in the field in autumn (April) in Tasmania (Forthside 41° 12′ S) may add some information. This seed germinated in late April and the young seedlings survived the winter very well. They grew away strongly in spring but did not flower until late February in the following year (J.C. Laughlin, unpublished data). Climatic and day length conditions at Forthside are shown in Table 1. In Germany *A. annua* has been direct drilled successfully in field experiments with drilling operations taking place in the third week of June and full bloom in late August (Liersch, 1986) Also in Vietnam, *A. annua* was direct drilled in January with floral induction in October (Woerdenbag *et al.*, 1994) (see **Time of establishment**).

Direct drilling of *A. annua* seed can be carried out by various methods ranging from the basic hand application or simple hand pushed single row seeders to sophisticated combined multi row seed and fertiliser drills. In all of these operations the soil needs to be cultivated down to a fine tilth and consolidated by rolling where appropriate. Because *A. annua* seed is so small (10–15,000 per gram) it needs to be diluted either by some inert material or by mixing with an appropriate neutral fertiliser. In Tasmanian experiments the technique of mixing a 50:50 blend of fine ground limestone and superphosphate with *A. annua* seed gave successful establishment (Laughlin, 1993). This technique is useful in many situations ranging from basic hand sowing to machine drilling and has the advantage of not requiring a sophisticated drill with a specialised small seed facility. Although the 50:50 fertiliser limestone blend is neutral, the seed should only be mixed through it immediately before drilling. Ideally the drilled area should be irrigated soon after sowing if soil moisture is less than optimal. In this technique of mixing seed and fertiliser it is important to have some detailed knowledge of (i) laboratory germination and (ii) field emergence and survival in order to use a seed sowing rate that avoids the problems of either very sparse stands or very thick stands which may necessitate hand or machine thinning.

Depth of sowing is also critical with a small seeded crop such as *A. annua*. In Tasmanian studies, a depth of drilling of 5 mm below the surface gave good emergence and establishment (J.C. Laughlin, unpublished data). However it is important, with a shallow drilling depth such as this, that the soil be cultivated down to a fine tilth to avoid seeds being sown on the surface and thus be in danger of losses in dry weather. For this reason also it is important to irrigate soon after drilling so that young emerging seedlings do not suffer moisture stress as well as obviating any possibility of fertiliser "burn" (see **Irrigation**). An alternative technique to mixing seed and fertiliser has been the drilling of a finely broken up mixture of seed and other inert floral part of *A. annua* (Galambosi, 1985). Ultimately the success of any techniques of direct seed drilling depend on a detailed knowledge of local soil and environmental conditions such as rainfall and temperature.

Time of Establishment

The choice of time of establishment is an important decision which must allow a sufficient period for rapid early growth and the production of a vigorous vegetative framework before the commencement of flowering. In continental climates good growth and yields were obtained from establishment in late spring in Switzerland (Delabays *et al.*, 1993) and early summer in Germany (Liersch *et al.*, 1986) and in the USA (Charles *et al.*, 1990). In a cool temperate maritime climate in Australia (Tasmania) a field experiment compared transplanting in the spring months of October and November with early summer in December. Although all transplants flowered at the same time, the leaf dry matter yield from October transplants were double and fourfold those from November and December transplants respectively (Laughlin, 1993). Although the concentration of artemisinin was the same for these three times of establishment the concentration of artemisinic acid decreased by 25% and 50% respectively from November and December transplanting compared to the concentration in October transplants (Laughlin, 1993). Later field experiments in Tasmania compared winter transplanting in July and August with spring

Table 1 The long term average monthly meteorological data for Forthside Research Station, near Devonport, Tasmania, Australia.

| Month | Temperature (C°) | | Rain | Wind | Sun | Day Length | |
	Min	Max.	mm	km	Hrs.	Hrs.	Min.
Jan.	11.0	20.6	46.1	6411	253.1	14	49
Feb.	11.6	21.2	50.2	5390	223.1	13	41
Mar.	10.4	19.7	54.4	5391	197.9	12	25
Apr.	6.2	14.8	80.3	6248	198.2	11	02
May	6.5	14.0	102.8	4970	121.6	9	51
June	4.4	12.1	100.3	4775	107.0	9	13
July	3.5	11.1	122.8	5311	115.2	9	28
Aug.	4.0	11.7	114.2	5592	127.5	10	28
Sep.	5.0	14.1	90.3	6161	161.5	11	47
Oct.	6.3	15.3	84.2	6515	227.9	13	07
Nov.	8.5	17.2	74.1	6241	225.7	14	25
Dec.	9.8	18.9	76.0	6648	244.1	15	08

transplanting in September and October. There were no differences in either leaf dry matter, artemisinin or artemisinic acid yield between any of these times of transplanting (Laughlin, 1993). This experiment showed that there was no point in transplanting earlier than October in Tasmania with the associated problem of weed control. It also showed that although July and August transplants had 2 and 3 months of growth respectively by October, when day length had increased to 13:30 h, flowering did not occur until February of the following year when day length had decreased to 13:30 h again. In addition to absolute day length the factor of decreasing day length *per se* also may be of some significance in triggering flowering.

In the USA (Indiana) a greenhouse experiment at a constant temperature of 27° C with cuttings of a Chinese selection showed that *A. annua* was a short day plant and that floral induction was initiated when the photoperiod was 13:31 h. In a related field experiment bud formation occurred two weeks later at a day length of 12:57 h (Ferreira *et al.*, 1995). Temperature × photoperiod interactions were not studied in this greenhouse investigation. However floral induction and flowering seem to show a very marked shift from this model in warmer tropical and subtropical climates. In India at Lucknow, (26° 52′ N) seedlings of the same European selection which were used in the Tasmanian field experiment described above (Laughlin, 1993) were transplanted in the relatively cool winter period of mid December (Singh *et al.*, 1988). In this experiment, bud formation (preflowering) occurred at a day length of 11:16 h with full flowering and maximum dry matter yield of leaves + flowers and artemisinin about 6 weeks later on 26 March at a day length of 12:15 h (Table 2). Later work on establishment times concluded that the optimum transplanting time on the north Indian plains was mid October (Bagchi *et al.*, 1997). Another field experiment in Vietnam with an *A. annua* seed source native to Langson in Northern Vietnam was carried out near Hanoi and it studied the pattern of dry matter yield and artemisinin content over a growing season (Woerdenbag *et al.*, 1994). In this experiment the seed was direct drilled into field plots at a density of 25 plants/m² (200 mm × 200 mm) on 10 January. Maximum dry matter yield of leaves (5.3t/ha), maximum artemisinin concentration (0.86%) and maximum artemisinin yield (45.4 kg/ha) occurred at the vegetative stage on 15 June at a day length of 13:24 h (Table 3). The plants remained vegetative until the preflowering stage on 15 October when there was a dry matter leaf yield of 3.8t/ha and artemisinin concentration of 0.42% at a day length of 11:41 h (Table 3). The conclusion from this study is that with a strain of *A. annua* adapted to the tropical Vietnamese environment an adequate biomass at a high concentration of artemisinin was achieved before the onset of flowering.

The adaptability of Vietnamese seed selections from the northern latitudes near Hanoi is illustrated by their performance in the climatic environment of Tasmania (Table 1). Tissue cultured plants from Vietnamese selections were grown-on in the greenhouse in mid December 1994. These plants were transferred to the field to a number of locations near Devonport in mid February 1995 and grew vegetatively throughout the autumn. They survived the winter without any problem and continued good vegetative growth throughout the following spring with some achieving

Table 2 Dry leaf yield, artemisinin content and yield of *A. annua* strains at different growth stages (Adapted from Singh *et al*., 1988).

Date of observation	Stage	European strain			Day length		*Temperature (C°)	
		Leaf yield (t/ha)	Artemisinin (%)	(kg/ha)	Hrs.	Min.	Max.	Min.
26 Oct. 1985	Seed sown	–	–	–	11	19	32.8	19.8
14 Dec. 1985	Transplanted	–	–	–	10	33	24.8	9.1
2 Feb. 1986	Vegetative	0.57	0.012	0.068	11	00	26.4	11.5
6 Feb. "	"	1.33	0.015	0.199	11	5	"	"
14 Feb. "	Preflowering	1.78	0.021	0.374	11	16	"	"
14 Mar. "	50% flowering	2.00	0.090	1.800	11	57	32.9	16.3
26 Mar. "	100% flowering	4.80	0.079	3.73	12	15	"	"

* Temperatures are recorded as monthly means (Takahashi and Arakawa, 1981).

Table 3   Development of *Artemisia annua* plants in Vietnam, sown on 10 January 1992, during a vegetation period (Adapted from Woerdenbag et al., 1994).

Date	Stage	Aerial tissue (fresh wt.) t/ha	Leaf (dry wt.) t/ha	Artemisinin Conc. (% dry leaf)	Artemisinin Yield (kg/ha)	Day length Hrs.	Min.	*Temperature C° Max.	Min.
10 Jan.	seed sown	†NA	NA	NA	NA	10	56	20.2	13.4
15 Feb.	vegetative	"	"	"	"	11	27	20.4	14.4
15 Mar.	"	"	"	"	"	12	0	23.1	17.1
15 Apr.	"	"	0.4	"	"	12	37	27.5	20.5
15 May	"	"	NA	"	"	13	8	32.0	23.6
15 June	"	60	5.3	0.86	46	13	24	33.2	25.3
15 July	"	70	5.1	0.63	32	13	18	32.9	25.4
15 Sep.	"	75	4.6	0.55	25	12	17	31.1	24.3
15 Oct.	Preflowering	78	3.8	0.42	16	11	41	29.0	21.4
10 Nov.	Full bloom	80	a	0.20	NA	11	14	25.6	17.8
20 Nov.	Cease bloom	85	b	0.16	NA	11	5	25.6	17.8

* Temperatures are recorded as monthly means (Takahashi and Arakawa, 1981).

† NA = Not available. *a* = No leaves only flowers. *b* = No leaves only ceased flowers and fruit.

a height of two metres. Flowering did not commence until late February 1996, but only on some plants. None of the flowers from these plants set viable seed (J.C. Laughlin, unpublished data).

In Madagascar at Antananarivo (Latitude 18° 52'S) at an elevation of ca. 1500 metres above sea level *A. Annua* plantlets resulting from a cross between plants of Chinese and Vietnamese origin were transplanted into field plots on 12 March at a spacing of 50×70 cm. Mature plants were harvested on 4 August and leaf dry matter yield of 4.7 t/ha and an artemisinin yield of 41.3 kg/ha were obtained (Magalhaes *et al.*, 1996). In Penang, Malaysia (5° 30' N) a seed line of *A. annua* obtained from the latitude of Hanoi in Vietnam was set out in field plots to study the performance at low latitudes and about sea level. Three week old plantlets were transplanted into field plots and flowering occurred at 14 weeks with maximum artemisinin (0.39%) one week before flowering (Chan *et al.*, 1995). Although detailed yields of leaf dry matter were not studied in this experiment plants grew to one metre tall. Even at these low latitudes (0–10°) it would be a useful exercise to explore the yield potential *A. annua* may have in lower temperature areas at higher altitudes. It may also be possible, in these tropical climates with very rapid growth, to consider the possibility of two following crops as has been suggested for Vietnam (Woerdenbag *et al.*, 1994) and Brazil (Magalhaes, 1996).

Plant Nutrition

Major elements

There are only a small number of recorded studies of the vegetative growth responses of *A. annua* to the specific major elements nitrogen, phosphorus and potassium or of their effects on the concentration of artemisinin and related compounds. Good yields of total plant and leaf dry matter relative to a wide range of published yield figures, have been obtained in Mississippi USA where a complete fertiliser mixture containing 100 kg N, 100 kg P and 100 kg K/ha was broadcast and worked uniformly through the soil (WHO, 1988). Similarly in Australia (Tasmania) good to high plant and dry leaf yields have been obtained with a mixed fertiliser containing 60 kg N, 60 kg P and 50 kg K/ha pre-drilled in bands 150 mm apart and about 50 below seed and 75 mm below transplants (Laughlin, 1993). The technique of banding fertiliser is to be generally recommended in soils where phosphorous fixation is a problem. The manoeuvre of pre-drilling fertiliser in 150 mm rows prior to drilling seed or transplanting plantlets allows very simple and inexpensive drilling equipment to be used and obviates the need for sophisticated and expensive drills which place fertiliser and seed in the one operation. In this technique the fertiliser bands can never be more than 75 mm (half the row width) laterally displaced from the plant row when seed is drilled at random in a second operation parallel to the fertiliser and closer to the surface (Laughlin, 1978).

Although there are no specific experimental data on field responses of *A. annua* to phosphorous or potassium there is some specific evidence of the response of *A. annua* to nitrogen. Trials in the USA (Indiana) compared different rates of nitrogen fertiliser supplying zero, 67 and 135 kg N/ha and obtained the highest total plant yield with 67 kg N/ha (Simon *et al.*, 1990). In India, sand culture studies with an American strain of *A. annua* (Washington) also showed that nitrogen deficiency was associated with a large decrease in artemisinin (Srivastava and Sharma, 1990). Similar hydroponic studies in Brazil concluded that the omission of nitrogen or phosphorus drastically reduced plant growth and dry matter production (Figueira, 1996). Later field trials with nitrogen fertiliser compared 0, 32, 64 and 97 kg N/ha applied as urea (Magalhaes *et al.*, 1996). In this trial the dry matter leaf yield increased up to the highest rate of N but the concentration of artemisinin was reduced by 22% so that maximum economic yield of artemisinin was given by 64 kg N/ha. This represented a 50% increase in artemisinin yield above zero N.

Because nitrogen is a very mobile element it can easily be leached out of the root zone specially in areas of high or concentrated rainfall periods. This leaching effect may well be very significant in tropical and sub-tropical regions and in these situations the method and timing of nitrogen fertiliser application may be very important. Banding of nitrogen near the seed or plant row may give less leaching than broadcasting and uniform mixing. Split applications of nitrogen or slow release nitrogen may be other means whereby leaching is minimised, especially if the growing season is a long one. For example in the growth pattern of *A. annua* in Vietnam near Hanoi the plants remained vegetative from seed drilling in January until September (Woerdenbag *et al.*, 1994). The optimum time of harvest in this experiment was mid June and at this stage the mean total rainfall between January and June was 533 mm with more than twice this quantity possible in very wet years (Takahashi and Arakawa, 1981). In situations such as this the splitting of the total quantity of nitrogen into two, three or more doses may be an appropriate method to adopt. Ideally some form of leaf or tissue analysis to determine the critical concentration of N at which a vegetative growth or artemisinin (or artemisinic acid) concentration response would be obtained could be the ultimate to aim for because soil analyses for N (unlike P and K) are often unreliable.

It has been shown that some strains of *A. annua* are sensitive to soil pH below 5.0–5.5 (Laughlin, 1994) (see **Soil acidity**). A comparison of different forms of nitrogen such as urea and the neutral calcium ammonium nitrate with the more acidic ammonium sulphate and ammonium nitrate may be useful trials to carry out with *A. annua*. Ammonium sulphate has been compared with ammonium nitrate in a field experiment on sandy soil in Switzerland (Magalhaes *et al.*, 1996). When 90 kg N/ha was applied both forms of nitrogen increased leaf dry matter yield and artemisinin yield by about 50%. However a hydroponic nutrient culture experiment suggested that, under some circumstances nitrogenous fertiliser in the nitrate form may give higher yields of artemisinin (mg/plant) than the ammonium form (Magalhaes *et al.*, 1996). More work needs to be done on this aspect on a range of soil types in the field.

Micro elements

In China a range of growing media and nutrient treatments were tested for their effect on the synthesis of artemisinin. There were no effects on artemisinin from any of these treatments (Chen and Zhang, 1987). However Indian sand culture experiments with the American strain Washington showed that both copper and boron deficiencies were related to large decreases in artemisinin (Srivastava and Sharma, 1990). Similarly in Brazilian studies the omission of any of the elements N, P, K, Ca, Mg, or S from nutrient solutions limited artemisinin and artemisinic acid production (Figueira, 1996). Where experience has shown that boron or copper deficiencies can occur with other crops, particularly on lighter soils in specific regions, standard soil tests for these elements would be a wise precaution if the cultivation of *A. annua* was to be attempted.

Soil Acidity

Cultivated crops show well defined patterns for those that are tolerant or intolerant of acid or alkaline soils. However there are only a few studies of the effect of soil pH on the vegetative growth of *A. annua* and on artemisinin concentration. In Australia (Tasmania) a field experiment studied the effect of zero and 10t/ha of fine ground limestone (calcium carbonate) on the growth of Chinese and Yugoslavian strains of

Figure 3 The effect of lime (calcium hydroxide) on the growth of a Yugoslavian strain of *A. annua* at the bud stage. The rates of lime shown above (left to right) gave soil pH values of 5.0, 5.2, 5.3, 5.4, 6.0, 7.4 and 8.2 respectively. Plant height and leaf dry matter yields markedly declined below pH 5.3 and above 7.4.

A. annua on a red krasnozem soil of pH 5.0 in the top 500 mm (Laughlin, 1993, 1994). The 10t/ha limestone treatment increased soil pH from 5.0 to 5.5. The leaf dry matter yield of the Yugoslavian strain was increased from 1.0t/ha to 6.5t/ha while the Chinese strain increased from 4.5t/ha to 8.0t/ha. The concentrations of neither artemisinin nor artemisinic acid were affected by the change in soil pH. These results suggested that there were large differences in strain (genotype) susceptibility to soil pH. The responses of the Chinese and Yugoslavian strains of *A. annua* to a wide range of soil pH were later studied in a greenhouse pot experiment using the same krasnozem soil as the above field experiment. Fine ground calcium hydroxide at the equivalent of zero, 1, 2.5, 5, 10, 20 and 40 t/ha was uniformly mixed through the soil to give mean soil pH values of 5.0, 5.2, 5.3, 5.4, 6.0, 7.4 and 8.2 respectively. Both strains of *A. annua* grew well in the soil pH range 5.4 to 7.4 but the Chinese strain was much more tolerant of both very high (8.2) and very low (5.0) pH conditions than the Yugoslavian strain (Fig. 3) (Laughlin, 1993, 1994). In Indian pot culture experiments, with *A. annua* grown on soils of widely varying pH, the oil yields at pH 4.9 and 9.9 were respectively about 75% and 25% of those grown on soils of pH 7.9–8.9 (Prasad *et al.*, 1998).

The response of plants to soil pH is not generally to the applied limestone *per se* but to the increased or decreased availability of other nutrients. In the above study on krasnozem soil in Tasmania the results may imply that the Yugoslavian strain was intolerant of high levels of manganese or aluminium at low soil pH as is the case with a number of other crops on this soil type. The work of Srivastava and Sharma (1990) which drew an association between boron and copper and artemisinin concentration may also have implications for soil amelioration practices by lime application. On some light soils the application of lime can lower the availability of boron. It may be useful for further studies to more closely examine the response of *A. annua* to the combined effects of copper, boron and lime application. An alternative to the sometimes costly strategy of ameliorating soil pH by lime application may be the selection of strains of *A. annua* which are not only adapted to the local environment but also tolerant of extremes of soil pH below 5.5 and above 7.5.

Plant Population Density

Plant population density and its components of inter and intra-row spacings (Holliday, 1960, Willey and Heath, 1969, Ratkowsky, 1983) are of considerable importance in determining yield and the practicability of both weed control (see **Herbicides**) and harvesting (see **Methods of harvest**). If inter-row cultivation were to be used to control weeds before the rows close then inter-row spacings of 0.5 to 1.0 metres may be appropriate. Similarly wide intra-row spacings may also be appropriate. However, if effective herbicides were available, then yield per unit area could be increased by using much higher plant population densities. In some earlier studies, low densities of 1 plant/m^2 (Maynard, 1985, WHO, 1988) and 2.5 plants/m^2 (Delabays *et al.*, 1993) were used and gave yields of 1–4 t/ha of dried leaf. In other studies higher densities have been used. Simon *et al.* (1990) in the USA compared 3,

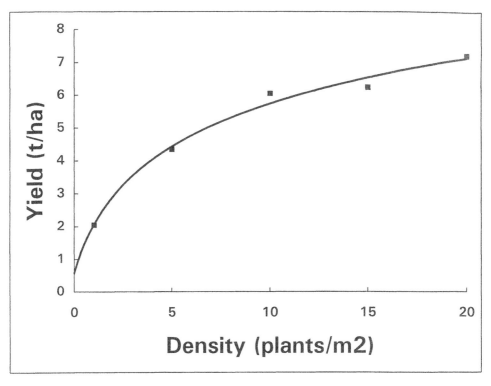

Figure 4   Relationship between plant population density (ρ) and total dry leaf yield (t/ha)
■ — ■ ($1/\omega = 0.003764 + 0.00131\rho$), ω = wt/plant (Adapted from Laughlin, 1993).

7 and 11 plants/m^2 and obtained the highest biomass at the highest density. In Australia (Tasmania) a field experiment with a Yugoslavian strain compared 1, 5, 10, 15 and 20 plants/m^2 at a November transplanting and found that leaf dry matter yield increased up to a density of 20 plants/m^2 (Fig. 4). However yield at 10 plants/m^2 (Fig. 5) was about 90% of the maximum of 6.8t/ha (Laughlin, 1993). High densities of 25 plants/m^2 were also used in a field experiment in Vietnam to give a maximum leaf dry matter yield of 5.3t/ha (Woerdenbag *et al.*, 1994). In the above Australian study (Laughlin, 1993) plant population density had no effect on the concentration of either artemisinin or artemisinic acid in the Yugoslavian strain of *A. annua* which was used and therefore yields reflected leaf dry matter yield (Fig. 6). However plant density may have different effects on the concentration of anti-malarial constituents in other climatic regions and with other strains of *A. annua* this area of study may be worth investigating. On the north Indian plains it has been recommended that *A. annua* should be cultivated at a high plant density of about 22 plants/m^2 (Ram *et al.*, 1997). The effect of variation in rectangularity (the ratio of inter- to intra-plant spacings) at constant plant population density (Chung, 1990) has not been studied with *A. annua* and may also be worth investigating.

Figure 5 Experimental plots of *A. annua* in left foreground at Forthside Research Station near Devonport, Tasmania, Australia. The density was 10 plants/m2 and height about 2 metres at maturity.

Herbicides

Weeds are a constant problem for crop production throughout the world and any system of *A. annua* cultivation must give careful thought to weed control. In small areas of cultivation in developing countries hand control of weeds may be appropriate and if this system is used, row spacings wide enough to allow easy access, should be used (see **Plant population density**). Similarly, if inter-row cultivation by hand pushed or tractor drawn implements is to be used, careful thought must be given to row spacing to allow easy access while the crop is small and before the rows close. This system may also be the only practical one available even in developed economies where stringent regulations governing the registration of herbicides for new crops demand lengthy lead times and investigations (Ferreira *et al.*, 1997).

If *A. annua* is to be established from seed, weed control in the early stages of growth is even more critical than with transplants. The young seedlings which develop from the tiny *A. annua* seed (see **Seed drilling**) are very small and can easily be choked out by weeds in the first days and weeks of growth. In this situation weed control with herbicides is certainly the most convenient and efficient method. Some detailed experiments on the use of herbicides with *A. annua* have been carried out. Application of 2.2 kg active ingredient (a.i.)/ha napropamide before transplanting

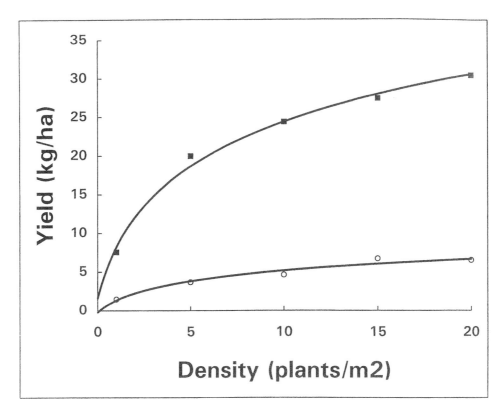

Figure 6   Relationship between plant population density (ρ) and artemisinin yields (kg/ha) ■ — ■ ($1/\omega = 5.852 + 1.301\rho$) and artemisinic acid 0–0 ($1/\omega = 1.406 + 0.291\rho$), ω = wt/plant (Adapted from Laughlin, 1993).

gave good weed control without phytoxicity in the USA (Simon and Cebert, 1988). More detailed field studies have been carried out in the USA where a range of herbicides was tested (Bryson and Croom, 1991). Chloramben was very effective when applied at 2.2 kg a.i./ha before emergence also trifluralin at 0.6 kg a.i./ha incorporated before transplanting followed by fluazifol at 0.2 + 0.2 kg a.i./ha broadcast after emergence and acifluorfen at 0.6 kg a.i./ha after emergence. All of these treatments gave good weed control without any significant reduction in leaf yield or concentration of artemisinin (Bryson and Croome, 1991). Arteether, a derivative of artemisinin, has been shown to be a very effective growth inhibitor of dicotyledenous weeds (Bagchi *et al.*, 1997).

Diseases and Pests

Very few significant plant diseases or pests of *A. annua* have been reported in the literature to date. In a wide selection of experiments in the USA ranging from harvest at the green vegetative stage to seed production and utilisation as a dried flower arrangement no obvious plant pathological symptoms were observed (Simon and Cebert, 1988). The only pests observed in these trials were caterpillars but with no visible feeding injury. In Australia (Tasmania) across a similar wide range of experiments, the only disease observed in some trials was a very low incidence (<1% of plants) of *Sclerotinia* stem infection on the lower third of the plant (J.C. Laughlin, unpublished data). The symptoms took the form of conspicuous white fungal patches on the surface of the main stem. The possibility of *Sclerotinia* stem infection and appropriate control measures should be considered when *A. annua* is grown in plantations at high plant densities. Under these conditions the build up in localised humidity could induce the infection (Laughlin and Munro, 1983). In Saudi Arabia *Orobanche cernua* was identified as a root parasite of *A. annua* with the potential to cause yield losses (Elhag *et al.*, 1997).

Irrigation

Whether *A. annua* is transplanted as seedlings or directly drilled into the field as seed it is important that soil moisture be adequate. In many situations frequent light irrigations are necessary to ensure good safe establishment and the frequency of these water applications will depend very much on soil type and climate. Another compelling reason to ensure that soil moisture does not become low at establishment or at later stages is the possibility of fertiliser "burn". This problem is caused by a detrimental osmotic effect from high concentrations of soluble mobile elements such as nitrogen and sometimes potassium (Simon *et al.*, 1990, Laughlin and Chung, 1992). Water stress at critical periods of growth can cause yield reductions in many plants. With *A. annua*, water stress during the two weeks before harvest gave a reduced leaf artemisinin concentration (Charles *et al.*, 1993). More work on the irrigation requirements of *A. annua* needs to be done to define critical growth stages in a range of environments, especially with relation to nitrogen application and its effect on leaf dry matter yield and the concentration of both artemisinin and artemisinic acid.

Growth Regulators

The effects of growth regulators on the yield and artemisinin concentration of *A. annua* have had very little study. In India triacontanol at 1.0 and 1.5 mg/litre significantly increased artemisinin levels, plant height and fresh weight. Chlormquat at 100 and 1500 mg/litre also increased artemisinin levels. When applied at 1500 and 2000 mg/litre plant height was reduced (Shukla *et al.*, 1992). The application of GA_3 (25 or 50 mg/litre) has also significantly increased plant fresh weight, artemisinin concentration and oil yield (Farooqi *et al.*, 1996). The possibilities of increasing artemisinin levels and reducing plant height have important practical implications. If ultimately *A. annua* may be grown in plantations at high densities to maximise leaf yield (Laughlin, 1993, Woerdenbag *et al.*, 1994) the plant height is generally increased and is typically between two and three metres tall (Fig. 7). Plants of this height may be a problem for machine harvesting *in situ* by leaf stripping (see **Method of harvest**) and a significant reduction in

Figure 7   A semi-commercial crop of *A. annua* at Forthside Research Station near Devonport, Tasmania, Australia. The density was 20 plants/m² and height about 2.5 metres at maturity.

height may be a practical advantage. Growth retardants such as cycocel (CCC), dimethylsulphoxide (DMSO and maleic hydrazide (MH) could show useful effects in this regard. Related to this technique of leaf stripping, plant desiccants such as diquat and paraquat may also be worth investigating for their impact on ease of leaf removal and artemisinin content.

A study of the endogenous plant growth substances in *A. annua* may give useful leads as to which of the large range of plant growth substances available may give the best effect when applied exogenously. Another potentially important area in which growth regulators may play a significant role is in seed set and seed germination. Although some strains of *A. annua* may grow vegetatively very well when tested in a new environment the seed set and/or germination may be poor (see **Time of establishment**). This aspect may be very important in establishing cultivation of *A. annua* from direct seeding. A sustainable source of seed would be an important pre-requisite for economical production rather than having to rely on the importation of seed from another country or region (see **Seed drilling**). The growth substances uracil and 5-bromouracil have been effective in the seed set of *Papaver somniferum* (Theuns, 1975) and could well be investigated with *A. annua*.

Distribution of Artemisinin in *A. annua*

The distribution of artemisinin in the *A. annua* plant varies between its constituents of leaf, main stem, lateral branches, roots, flowers and seeds. The concentration of artemisinin within these constituents also changes over time as the plant develops during the vegetative phase and again as the plant moves into the flowering stage of growth. These changes in concentration over time are sometimes reflected in changes in the plant profile from the top to the bottom of the plant. The relative differences in concentration of artemisinin between main stem, lateral branches, roots, leaves, flowers and seed appear to be fairly similar between ecotypes. There is evidence that artemisinin is localised in special cell structures of *A. annua*. Duke and Paul (1993) showed that biseriate glandular trichomes were present along both sides of the leaf midrib as well as on abaxial surfaces of the leaf and on the stem. Later work (Duke *et al.*, 1994) concluded that these glandular trichomes were the site of sequestration of artemisinin. Similarly, Ferreira and Janick (1995b) showed that biseriate glandular trichomes were common in the bracts, receptacles and florets of the capitulum of *A. annua* and that sequestration of artemisinin also occurred in these structures. These workers also hypothesised that artemisinin may also be synthesised within the glandular trichomes.

Distribution between plant components

In experiments in the USA it was found just prior to flowering that about 89% of the total plant artemisinin was in the leaves with only about 10% in the lateral branches (Charles *et al.*, 1990). In terms of specific concentration of artemisinin these workers found 0.15% in the upper leaves, 0.04% in side shoots, trace amounts in the main stem and none in the roots. After flowering 0.04% of artemisinin was found in seeds. A similar pattern of distribution was found by Acton *et al.* (1985). Later greenhouse and field experiments in which plants were sequentially harvested from the vegetative stage to flowering and seed set showed a similar relative distribution of artemisinin in the vegetative stage (Ferreira *et al.*, 1995). These workers also showed that the concentration of artemisinin in the inflorescences was 4–11 fold that in leaves and that the content of artemisinin in the seed was mainly associated with floral remnants and debris.

Changes in artemisinin over time

Most researchers have found, in sequential harvesting studies, that the concentration of artemisinin in the leaves increases as the plant develops in the vegetative phase. In some experiments artemisinin concentration peaked just before flowering (Acton *et al.*, 1985, Liersch *et al.*, 1986, El-Sohly, 1990, Woerdenbag *et al.*, 1990, 1994, Laughlin, 1993). Other workers agree that artemisinin concentration increases during the vegetative phase but have found that peak artemisinin was achieved later during full flowering (Singh *et al.*, 1988, Pras *et al.*, 1991, Morales *et*

Figure 8   The small (2–3 mm diameter) yellow flowers of *A. annua* at full bloom.

al., 1993, Ferreira *et al.*, 1995). The differences between the two times of peak artemisinin may be attributable to climatic conditions, ecotype, cultural practices or a combination of these factors. However, as the time of peak artemisinin concentration is of considerable importance to optimum yield of artemisinin it would be a wise precaution for this point to be determined experimentally for any new area of production rather than to rely only on published figures. The difference between the time of peak artemisinin concentration in the data of Singh *et al.* (1988) and Laughlin (1993) illustrates this point. The same European strain was used in both studies but with peaks at different stages: namely full bloom (Singh *et al.*, 1988) and late vegetative (Laughlin, 1993) respectively. Although a knowledge of the time of peak artemisinin concentration is one of the necessary conditions to maximise yield of artemisinin it is not sufficient in itself. It must be coupled with a knowledge of the way that the dry matter yields of leaves and flowers change over time (see **Time of harvest**). In the data of Laughlin (1993) although the concentration of artemisinin peaked at the late vegetative stage and decreased by about 25% at flowering the dry matter yield of leaves and flowers about doubled over the same period. Because of this, the yield per unit area of artemisinin was at a maximum at full bloom (Fig. 8).

Changes down the plant profile

The variation of artemisinin concentration down the plant profile varies considerably between strain origin. In the USA a field experiment (plant density unspecified) with an accession of *A. annua* of Chinese origin sampled leaves from the top, middle and bottom thirds of the plant. The concentration and yield of artemisinin in the top third of the plant was about double that from the middle and lower thirds when sampled in the vegetative phase just before flowering (Charles *et al.*, 1990). In Australia (Tasmania) a similar field experiment – grown at a density of 10 plants/m² – showed different patterns with a Yugoslavian and Chinese strain. The leaves of the Yugoslavian strain were harvested at a range of times from early vegetative to full bloom for each quarter of the stem from top to bottom of the plant. At the vegetative stage just prior to flowering the concentration of artemisinin was very similar in each quarter. The earliest time of harvest with the Chinese strain was early bud and at this stage the concentration from the top to bottom quarters increased from 0.12%, 0.17%, 0.19% to 0.22% (Laughlin, 1995). In another study in the USA the leaves of six Chinese clones of *A. annua* were taken from the top, middle and bottom thirds of the stems at both vegetative and flowering stages (Ferreira and Janick, 1996b). Artemisinin was evenly distributed. It is apparent that there are large differences in the profile of artemisinin distribution in the various *A. annua* strains. Plant population density may be an important factor in the difference between the results of Laughlin (1995) and those of Charles *et al.* (1990) and Ferreira and Janick (1996b); however these profile distribution patterns should be explored in developing harvesting strategies in new areas of *A. annua* production (see **Time of harvest** and **Method of harvest**).

Artemisinic Acid: Precursor of Artemisinin

In addition to artemisinin some studies also have shown that its precursor artemisinic acid (= arteannuic acid = *qinghao* acid) may be a viable supplementary or alternative method of obtaining artemisinin (Roth and Acton, 1989, Jung *et al.*, 1990, Vonwiller *et al.*, 1993, Laughlin, 1993, 1995). Methods of conversion of artemisinic acid to artemisinin with efficiencies of about 18% (Roth and Acton, 1989) and 40% (Vonwiller *et al.*, 1993) have been developed. The concentration of artemisinic acid in *A. annua* is strongly influenced by strain selection. In European and Chinese selections the concentration of artemisinic acid can be up to tenfold that of artemisinin (Roth and Acton, 1989, Jung *et al.*, 1990, Laughlin, 1993, 1995). However in a Vietnamese strain Woerdenbag *et al.* (1994) found that, in the vegetative stage, the opposite was the case with the concentration of artemisinin being up to tenfold that of artemisinic acid.

Changes in artemisinic acid over time

There are only a limited number of studies which have related the concentration of artemisinic acid and dry matter yield of leaves to plant developmental stages by sequential harvesting from the vegetative stage to flowering. In an Australian (Tasmanian) study a Yugoslavian strain of *A. annua* was transplanted in the field at the optimum time in October. The mean concentration of artemisinic acid in the leaves from the whole plant peaked in mid February – just before early bud and about the same time as artemisinin – and then decreased about threefold at full bloom (Laughlin, 1993). In another Australian field experiment a Chinese selection of *A. annua* which was transplanted in October gave a similar pattern of artemisinic acid accumulation which peaked just before the early bud stage. However in this case, the concentration of artemisinic acid only decreased by about 30% at full bloom (J.C. Laughlin, unpublished data). In a study of the changes in artemisinic acid over a vegetation period in Vietnam, plants were sampled from the time of maximum leaf yield (mid June) to just before flowering (mid October). The mean leaf concentration of artemisinic acid decreased almost threefold (0.16% to 0.06%) over this period and to 0.02% at full bloom (10 November) (Woerdenbag *et al.*, 1994).

The data from all three experiments suggest that the concentration of artemisinic acid peaks at about the late vegetative or early bud stages and then decreases. The extent of the decreases vary greatly between strain selections and as with artemisinin the final yield of artemisinic acid hinges on the relative changes in the dry matter yield of leaves and flowers as well. The maximum yield of artemisinic acid in the Vietnamese study (Woerdenbag *et al.*, 1994) and in the Australian study with the Yugoslavian strain (Laughlin, 1993) would have been in the vegetative stage. In the Australian study this was because artemisinic acid concentration decreased threefold while leaf + flower dry matter only doubled between the late vegetative and full bloom stages of growth. In contrast to artemisinin, the time of maximum artemisinic acid yield was influenced more by the decrease in acid concentration than by the increase in leaf + flower dry matter. (see **Time of harvest, Method of harvest** and **Oil production**).

Oil Production

A. annua is the source of an essential oil which is located mainly in the leaves and flowers. In eastern Europe where *A. annua* has been established as a weed for a lengthy period the oil has been extracted commercially in Hungary, Rumania and Bulgaria and has found a limited market for perfumery and as an anti-bacterial (Laurence, 1990). In Hungary a field experiment studying the potential of *A. annua* as a source of oil found that the concentration of oil was at a maximum at full bloom and obtained oil yields of 20–40 kg/ha (Galambosi, 1982). In Bulgaria other experiments showed that all parts of *A. annua* contained the oil but it was mainly concentrated in the flowers with a maximum concentration of 3.2% in dried flowers at full bloom (Georgiev *et al.*, 1981). In the Ukrainian SSR a late maturing selection gave an oil yield of 111 kg/ha with a high perfume quality (Kapelev, 1984). In later more detailed experiments it was confirmed that oil was mainly concentrated in the leaves and flowers with only trace amounts in the main stem, side branches and roots (Charles *et al.*, 1991) with the highest concentration at flowering (Simon *et al.*, 1990). In excess of sixty individual chemical constituents have been reported in the oil distilled from *A. annua* with a marked variation in composition between the selection used (Charles *et al.*, 1991). Artemisia ketone made the major contribution (68.5%) in a Chinese selection (Charles *et al.*, 1991) but it formed a maximum of only 4.4% in a Vietnamese strain (Woerdenbag *et al.*, 1994).

Up to the present time the volume of *A. annua* oil traded has been relatively small. However with the increased interest in *A. annua* for the production of antimalarial drugs the potential volume of oil available is much greater and may induce greater interest from the perfumery or general chemical industries. The extraction of oil from *A. annua* by steam distillation can denature many of the non-volatile sesqui-terpenes (Ferreira *et al.*, 1997). However there may be strategies whereby a dual purpose use of *A. annua* for antimalarial drugs and oil could be carried out. In a field experiment in Australia (Tasmania), European, USA and Chinese selections of *A. annua* were harvested at the late vegetative and full bloom stages of growth. Whole plants were chipped into small segments and oil extracted by steam distillation for one hour. A comparison of distilled and non-distilled plants showed that while artemisinin was either completely denatured or showed only minute traces in the distilled plants, artemisinic acid was unaffected by distillation (J.C. Laughlin, unpublished data). These results were similar for all selections of *A. annua* and at both times of harvest. Depending on the concentration of artemisinic acid relative to artemisinin and on the concentration of oil it may be economic to consider a dual purpose production with artemisinin obtained by conversion from artemisinic acid (see **Artemisinic acid, Time of Harvest** and **Method of harvest**).

Time of Harvest

The optimum time of harvest of *A. annua* will depend on the key target compound which is required. On the assumption that artemisinin is the main objective then maximum yield of artemisinin was at full bloom in most cases (Singh *et al.*, 1988, Pras *et al.*, 1991, Morales *et al.*, 1991, Laughlin, 1993, Ferreira *et al.*, 1995). However this was not the case with a Vietnamese selection in a trial in that country where maximum yield was at the vegetative stage (Woerdenbag *et al.*, 1994). If artemisinic acid were the main objective then the limited amount of data available at present suggest that the late vegetative stage would give maximum yield (Laughlin, 1993, 1995, Woerdenbag *et al.*, 1994). In the situation where oil was the main objective or a dual purpose objective of oil and artemisinic acid then full bloom would be the preferred time. In the latter case of dual purpose use it would be important to use a selection of *A. annua* which did not show a large decrease in artemisinic acid between the late vegetative stage and full bloom (Laughlin, 1993).

Methods of Harvest

Hand harvest

There is little published detailed information on the method of harvesting *A. annua*. In China and Vietnam, the two countries where the plant has been traditionally grown, it is reported that the crop is cut at the base, left to dry and then the leaves are removed by some form of shaking (P. Kager, personal communication).

Mowing

One of the earliest attempts at mechanical harvesting was that of Dr E. Croom of the University of Mississippi, USA who grew a crop of two hectares of *A. annua* for the production of one kilogram of artemisinin for World Health Organisation experiments (Maynard, 1985). In this project the crop, which was grown in rows three feet apart (ca. 90cm), was successfully cut at the base with a tractor drawn mower (see **Post harvest treatment**).

Leaf stripping

Because artemisinin and artemisinic acid and oil are all concentrated mainly in the leaves and flowers the ideal harvesting technique would be some form of mechanical stripping of these components. Some preliminary trials of this method were carried out in Australia (Tasmania) using a standard "Mather and Platt" green bean harvester. Although obvious blockages occurred in these rudimentary trials with a completely unmodified machine the system showed some promise and should be explored further with suitable modifications (Laughlin, 1994,1995). Other techniques which should be explored and which could facilitate the possibility of leaf stripping as a viable method of harvest may be the use of leaf desiccants and growth (height) retardants (see **Growth regulators**). The importance of testing this method of harvest is because of the high percentage of main stem and lateral branch components with low artemisinin which must be handled in some way in alternative methods of harvest. At the late vegetative stage the fresh leaf contribution to total above ground plant yield was only 26.9% in a Vietnamese study (Woerdenbag *et al.*, 1994) and 29% in Tasmania (Laughlin, 1995). In the Tasmanian study where maximum artemisinin yield occurred at full bloom the composite fresh yield of leaves and flowers contributed 46% to total plant yield (Laughlin, 1995).

Forage harvesting

In this technique the fresh plant is cut close to the base, or at any required height, mechanically chopped into small segments and blown into a trailing or following bin (Fig. 9). The second stage in this method involves kiln or some alternative form of drying before the separation of leaf (plus flowers) from stem segments by sieving. This system has been given a very preliminary screening in Tasmania (J.C. Laughlin, unpublished data) and although feasible it involves the obvious problem of cartage, handling and drying (see **Post-harvest treatment**). In any system of harvesting such as this it would be important that the drying and post-harvest handling installation be close to the area of production. The harvested plant mass in this system can heat up quickly and could result in losses of artemisinin. There are no published studies on this aspect but it is an area which merits investigation.

Figure 9 Harvesting *A. annua* with a forage harvester. The experimental plots in Figure 5 were first cut with a mower 10 cm above ground level, wilted for seven days and then picked up with a forage harvester and blown into a trailing bin.

Cutting and field wilting

In this method plants are cut at or near the base and then left to lose moisture for a period of time appropriate to the climatic region of production. The next stage is to pick up the wilted plants with a forage harvester (Fig. 9) which also has the facility for breaking the stem into pieces and blowing them into a trailing or following bin. This system has been successfully tested in Tasmania (J.C. Laughlin, unpublished data) and the advantage is the large loss of moisture (up to 50% or more) and hence the reduction in cartage and drying costs and time. The adaptability of this system will depend on the climatic conditions in the region of production, mainly temperature, rainfall, humidity and possibly ultra violet effects.

Post Harvest Treatment

Shade drying

In the USA, where two hectares were initially harvested with a tractor drawn mower, whole plants were transported to a large shaded enclosure and hung up for drying (Maynard, 1985). In India (Lucknow) where *A. annua* has been introduced for experimental purposes the plants are harvested by cutting at the base, mechanically chipped into small segments and dried in the shade prior to leaf removal by sieving (Singh *et al.*, 1988). In more recent field studies in Madagascar (Antananarivo) whole plants were also dried under cover before leaf removal (Magalhaes *et al.*, 1996).

Field (sun) drying

The wilting of *A. annua* in the field after harvest with exposure to direct sunlight is believed to reduce artemisinin. Large decreases occurred when plants were dried in the open in India at Lucknow (A. Singh, personal communication). In comparisons of the effect of sun and shade drying in Oregon, USA, leaf (or branch) samples were dried (i) in open sunshine, (ii) in open sunshine protected by paper bags, and (iii) air dried under cover at ambient temperature. Air drying under cover gave the highest artemisinin concentration with sun drying the lowest and dried in bags intermediate between the two (Charles *et al.*, 1993, Feibert, 1994). In this study the maximum air temperature was 30° C and the maximum sample temperatures were: sun 42.2° C, shaded 22.8° C and bagged 35.6° C. In another similar comparison of drying methods in Australia (Tasmania) whole plants were cut off at the base and either (i) sun dried in the open, (ii) shade dried or (iii) leaves were removed and oven dried at 35° C immediately after harvest. In this study the sun dried treatments were left for 7, 14 and 21 days and the shade dried for 21 days. The maximum air temperature in this trial was 22° C. There were no differences in artemisinin or artemisinic acid concentration between any of these treatments (J.C. Laughlin, unpublished data). Maximum air temperatures may be the simple explanation for the differences between these two studies.

However, given that sun drying would be most economical, if feasible, more detailed comparisons need to be carried out in a range climatic zones, especially sub-tropical, to confirm or otherwise the suspected effects of sun drying. To date no data are available for the effect of sun drying on artemisinic acid in hot continental or tropical environments and the effect of drying whole intact plants may give different results from drying stripped leaves or detached branches. Wilting and drying *A. annua* plants in the sun appears to have no detrimental effects on oil content (Charles *et al.*, 1991). This is another incentive to obtain more data on the effect of sun drying on artemisinic acid because of the possibilities of dual purpose harvesting of these two products (see **Oil production**).

Oven (kiln) drying

Oven or kiln drying may be an expedient necessity in some areas where *A. annua* could be cultivated. Although air drying of *A. annua* has been commonly practised in experimental programs there are not many examples in the literature of detailed results from oven drying. In the study of Charles *et al.* (1992), drying in a forced fan oven at 30° C, 50° C and 80° C was included in the comparison with sun and air drying. Plants were dried at these temperatures for 12, 24, 36 and 48 hours respectively. Although generally the results of this study suggested that oven drying gave lower artemisinin than shade drying the variability was such that further studies are well warranted. In Australia (Tasmania) leaves of *A. annua* were dried in a forced fan oven at 35° C, 45° C, 55° C and 65° C for 48 hours. On a relative scale the artemisinin concentration of leaves after these drying temperatures were 100, 97, 90 and 75 respectively. No comparison with air drying was made in this study (J.C. Laughlin, unpublished data).

In a study in the USA, shoots of *A. annua* were oven dried at 40° C and compared with indoor air drying and freeze drying. Indoor air drying gave the highest artemisinin concentration (0.13%), oven drying next (0.10%) and freeze drying lowest (0.02%) (Ferreira *et al.*, 1992). In this study some supplementary investigation was given to the effect of microwave drying. Drying fresh leaves for 2 minutes at 100% power virtually eliminated artemisinin while 5 minutes at 50% power reduced artemisinin concentration by about 50%. In this study (Ferreira *et al.*, 1992) air drying gave 30% more artemisinin than oven drying at 40° C. There may be situations where the cost of drying and the reduction in artemisinin may be acceptable penalties for the speed and efficiency of production flow. The answer to this scenario would depend on the particular cost structure in the region of production. Here again it would be important to know what effect oven drying may have on the concentration of artemisinic acid in that this may be an alternative route to artemisinin.

Pelleting and crude extract production

In a number of situations the extraction of artemisinin or artemisinic acid may take place at a considerable distance from the site of *A. annua* production: possibly in another country. In this event not only land but sea or air transport may be necessary

Figure 10   Pellets of *A. annua* dry leaf material 5–10 mm long × 4 mm diameter.

and the volume and weight of the *A. annua* plant material could be a problem. A crude extract with an organic solvent such as hexane is a possible solution to this problem in that the volume of the extract would only be 5–10% of the original *A. annua* plant material. If this method were to be used, pelleting of the *A. annua* plant material (Fig. 10) may be necessary to allow efficient movement of the plant material during the counter-current extraction process. It may be possible that the physical texture of the *A. annua* leaf material could be suitable for extraction without pelleting but this could only be determined by experiment. In preliminary pelleting trials in Australia (Tasmania), in which the only heat involved was in the grinding and compression phases of the operation, there was no reduction in either artemisinin or artemisinic acid concentration of the final pellets compared with the original *A. annua* leaf material (J.C. Laughlin, unpublished data). However the feasibility of extraction from original unpelleted *A. annua* plant material was not attempted in this study and experiments to establish this possibility would be well worthwhile.

Cost of Production

The cost of production of *A. annua* in both the cultivation and post-harvest phases will vary widely with region and method of production used. However an obvious necessity is the availability of strains of *A. annua* with a high artemisinin and/or artemisinic acid content which are adapted to the region of production. This is a vital necessity in order to make the production of antimalarial drugs profitable for the producer and affordable to the end user.

ACKNOWLEDGEMENTS

The assistance of the Tasmanian Department of Primary Industry and Fisheries was a valuable contribution towards the preparation of this review. We express appreciation to Mrs Helen Sims for the typing of the review.

REFERENCES

Acton, N., Klayman, D.L. and Rollman, I.J. (1985) Reductive electrochemical H PLC assay for artemisinin (qinghaosu). *Planta Med.*, 51, 445–446.

Anon. (1979) Qinghaosu Antimalarial Coordinating Research Group. Antimalarial studies on qinghaosu. *Chin. Med. J.*, 92, 811–816.

Anon. (1982) China Cooperative Group on Qinghaosu and its derivatives as Antimalarials. Chemical studies on qinghaosu (Artemisinin). *J. Trad. Chin. Med.*, 2, 3–8.

Anon. (1996) The wrong resistance. Editorial, *New Scientist*, 151, (No. 2038, 13 July), pp. 3.

Avery, M.A., Chong, W.K. and Jennings-White, C. (1992) Stereoselective total synthesis of (+)-artemisinin, the antimalarial constituent of *Artemisia annua* L. *J. Am. Chem. Soc.*, 114, 974–979.

Bagchi, G.D., Jain, D.C. and Kumar, S. (1997) Arteether: a potent plant growth inhibitor from *Artemisia annua. Phytochemistry*, 45(6), 1131–1133.

Bagchi, G.D., Ram, M., Sharma, S. and Kumar, S. (1997) Effect of planting date on growth and development of *Artemisia annua* under subtropical climatic conditions. *Journal of Medicinal and Aromatic Plant Sciences*, 19(2), 387–394.

Bryson, C.T. and Croom, E.M. Jr. (1991) Herbicide inputs for a new agronomic crop, annual wormwood (*Artemisia annua*). *Weed Technol.*, 5, 117–124.

Chan, K.L., Teo, K.H., Jinadasa, S., and Yuen, K.H. (1995) Selection of high artemisinin yielding *Artemisia annua. Planta Med.*, 61, 185–287.

Charles, D.J., Simon, J.E., Wood, K.V. and Heinstein, P. (1990) Germplasm variation in artemisinin content of *Artemisia annua* using an alternative method of artemisinin analysis from crude plant extract. *J. Nat. Prod.*, 53, 157–160.

Charles, D.J., Cebert, E. and Simon, J.E. (1991) Characterization of the essential oil of *Artemisia annua* L. *J. Ess. Oil Res.*, 3, 33–39.

Charles, D.J., Simon, J.E. and Shock, C.C., Feibert, E.B. and Smith, R.M. (1993) Effect of water stress and post-harvest handling on artemisinin content in the leaves of *Artemisia annua* L. In J. Janick and J.E. Simon, (eds.), *New Crops*, Wiley, New York, pp. 640–642.

Chatterjee, S.K. (1993) Introduction and domestication of new crops. Recommendations from a workshop of the first world congress on medicinal and aromatic plants for human welfare: WOCMAP I, Maastricht, Netherlands, July 19–25, 1992, Acta Hort., 331, 16.

Chen, F.I. and Zhang, G.H. (1987) Studies on several physiological factors in artemisinin synthesis in *Artemisia annua* L. *Plant Physiol. Comm.*, 5, 26–30.

Chung, B. (1990) Effect of plant population density and rectangularity on the growth and yield of poppies (*Papaver somniferum*). *J. Agric. Sci., Cambridge*, **115**, 239–245.

Davidson, D.E. (1994) Role of arteether in the treatment of malaria and plans for further development. *Trans. R. Soc. Trop. Med. Hyg.*, **88 (Suppl. 1)**, 51–52.

Debrunner, N., Dvorak, V., Magalhaes, P. and Delabays, N. (1996) Selection of genotypes of *Artemisia annua* L. for the agricultural production of artemisinin. In F. Pank (ed.) *Proceedings of an International Symposium: Breeding Research in Medicinal and Aromatic Plants*. Quedlinburg, Germany, June 30th – 4 July 1996, pp. 222–225.

Delabays, N., Blanc, C. and Collet, G. (1992) La culture et la selection d' *Artemisia annua* en vue de la production d' artemisinine. *Revue Suisse Vitic. Arboric. Hort.*, **24**, 245–251.

Delabays, N., Benakis, A. and Collet, G. (1993) Selection and breeding for high artemisinin (qinghaosu) yielding strains of *Artemisia annua*. *Acta Hort.*, **330**, 203–207.

Duke, S.D. and Paul, R.N. (1993) Development and fine structure of glandular trichomes of *Artemisia annua* L. *Int. J. Plant Sci.*, **154**, 107–118.

Duke, M.V., Paul, R.N., El-Sohly, H.N., Sturtz, G. and Duke, S.D. (1994) Localization of artemisinin and artemisitene in foliar tissue of glanded and glandless biotypes of *Artemisia annua* L. (1994) *Int. J. Plant Sci.*, **155**, 365–372.

Elhag, H., Abdel-Sattar, E., El-Domiaty, M. El-Olemy, M. and Mosa, J.S. (1997) Selection and micropropagation of high artemisinin producing clones of *Artemisia annua* L. Part II. Follow up of the performance of micropropagated clones. *Arab Gulf Journal of Scientific Research*, **15**(3), 683–693.

El-Sohly, H.N. (1990) A large scale extraction technique of artemisinin from *Artemisia annua*. *J. Nat Prod.*, **53**, 1560–1564.

Farooqi, A.H., Shukla, A., Sharma, S. and Khan, A. (1996) Effect of plant age and GA$_3$ on artemisinin and essential oil yield in *Artemisia annua* L. *Journal of Herbs, Spices and Medicinal Plants*, **4**(1), 73–80.

Feibert, E. (1994) Sweet wormwood (*Artemisia annua*) research at Ontario. *Hort. Science*, **29**(5), 557 (Abstr. 683).

Feibert, E., Shock, C.C. and Saunders, M. (1992) Sweet wormwood (*Artemisia annua*) research at Ontario in 1991. In *Oregon State University Agricultural Experiment Station Special Report*, **924**, 125–130.

Ferreira, J.F., Charles, D., Simon, J.E. and Janick, J. (1992) Effect of drying methods on the recovery and yield of artemisinin from *Artemisia annua* L. *Hort. Science*, **27**, 650 (Abstr. 565).

Ferreira, J.F., Simon J.E. and Janick, J. (1995) Developmental studies of *Artemisia annua*: flowering and artemisinin production under greenhouse and field conditions. *Planta Med.*, **61**, 167–170.

Ferreira, J.F. and Janick, J. (1995a) Production and detection of artemisinin from *Artemisia annua*. *Acta Hort.*, **390**, 41–49.

Ferreira, J.F. and Janick, J. (1995b) Floral morphology of *Artemisia annua* with special reference to trichomes. *Int. J. Plant Sci.*, **156**(6), 807–815.

Ferreira, J.F. and Janick, J. (1996a) Immunoquantitative analysis of artemisinin from *Artemisia annua* using polyclonal antibodies. *Phytochemistry*, **41**, 97–104.

Ferreira, J.F. and Janick, J. (1996b) Distribution of artemisinin in *Artemisia annua*. In J. Janick and J.E. Simon (eds.) *Proceedings of the Third National New Crops Symposium* Indianapolis, USA October 1995, **Vol. 3** pp. 578–584.

Ferreira, J.F., Simon, J.E. and Janick, J. (1997) *Artemisia annua*: Botany, Horticulture, Pharmacology. *Horticultural Reviews*, **19**, 319–371.

Figueira, G.M. (1996) Mineral nutrition, production and artemisinin content in *Artemisia annua* L. *Acta Hort.*, **426**, 573–577

Franz, Ch. (1983) Preface, Third International Symposium on Spices and Medicinal Plants. Hamburg, Germany 19 August – 4 September 1982. *Acta Hort.*, **132**, 13.

Franz, Ch. and Fritz, D. (1978) Cultivation aspects of *Gentiana lutea* L. *Acta Hort.*, **73**, 307–314.

Fritz, D. (1978) Welcoming address, First International Symposium on Spices and Medicinal Plants. Freising – Weihenstephan, Germany 31 July – 4 August 1977. *Acta Hort.*, **73**, 15–16.

Galambosi, B. (1982) Results of cultural trials with *Artemisia annua. Herba Hungarica*, **21** (2/3), 119–125.

Georgiev, E., Genov, N., Christova, N. (1981) Changes in the yield and quality of essential oil from wormwood during growth. *Rasteniev dni Nauki*, **18** (7), 95–102. (Bulgarian with English summary).

Harrison, G. (1978) *Mosquitos, Malaria and Man*. John Murray-London.

Haynes, R.K. and Vonwiller, S.C. (1991) The development of new peroxide antimalarials. *Chemistry in Australia*, **58**(2), 64–67.

Holliday, R. (1960) Plant population and crop yield. *Nature*, **186**, 22–24.

Jung, M.H., El Sohly, N., and McChesney, J.D. (1990) artemisinic acid: a versatile chiral synthon and bioprecursor to natural products. *Planta Med.*, **56**, 624.

Kapelev, I.G. (1984) Brief results from the introduction of essential oil plants of the wormwood genus. *Byulleten Gosudarstvennogo Nikitskogo Botanicheskogo Sada*, **54**, 60–65 (Russian with English summary).

Kawamoto, H., Sekine, H. and Furuya, T. (1999) Production of artemisinin and related sesquiterpenes in Japanese *Artemisia annua* during a vegetation period. *Planta Med.*, **65**, 88–89.

Keys, J.D. (1976) *Chinese Herbs*. pp. 216–217. Swindon Book Company, London.

Klayman, D.L. (1989) Weeding out malaria. *Nat. History*, Oct, 18–26.

Klayman, D.L. (1993) *Artemisia annua*: From weed to respectable antimalarial plant. In A.D. Kinghorn and M.F. Balandrin (eds.), *Human Medicinal Agents from Plants*. Am. Chem. Soc. Symp. Ser. Washington DC, pp. 242–255.

Laughlin, J.C. (1978) The effect of band placed nitrogen and phosphorus fertiliser on the yield of poppies (*Papaver somniferum* L.) grown on krasnozem soil. *Acta Hort.*, **73**, 165–172.

Laughlin, J.C. (1993) Effect of agronomic practices on plant yield and antimalarial constituents of *Artemisia annua* L. *Acta Hort*, **331**, 53–61.

Laughlin, J.C. (1994) Agricultural production of artemisinin – a review. *Trans. R. Soc. Trop. Med. Hyg.*, **88** (**Suppl. 1**), 21–22.

Laughlin, J.C. (1995) The influence of distribution of antimalarial constituents in *Artemisia annua* L. on time and method of harvest. *Acta Hort.*, **390**, 67–73.

Laughlin, J.C. and Chung, B. (1992) Nitrogen and irrigation effects on the yield of poppies (*Papaver somniferum* L.). *Acta Hort.*, **306**, 466–473.

Laughlin, J.C. and Munro, D. (1983) The effect of *Sclerotinia* stem infection on morphine production and distribution in poppy (*Papaver somniferum* L.) plants. *J. Agric. Sci., Cambridge*, **100**, 299–303.

Laurence, B.M. (1990) Progress in essential oils. *Perfumer and Flavorist*, **15**, 63–64.

Liersch, R., Soicke, C., Stehr, C. and Tüllner, H.V. (1986) Formation of artemisinin in *Artemisia annua* during one vegetation period. *Planta Med.*, **52**, 387–390.

Luo, X.D. and Shen, C.C. (1987) The chemistry, pharmacology and clinical application of qinghaosu (artemisinin) and its derivatives. *Med. Res. Rev.*, **7**, 29–52.

Magalhaes, P.M. (1994) A experimentacao agricola com plantas medicinais e aromaticas. *Atualidades Cientificas*, **3**, 31–56 (CPØBA/UNICAMP, Campinas, Brazil).

Magalhaes, P.M. (1996) Selecao, melhoramento, e nutricao da *Artemisia annua* L. para cultivo em regiao intertropical. Ph. D. Thesis, University of Campinas, Brazil.

Magalhaes, P.M. and Delabays, N. (1996) The selection of *Artemisia annua* L. for cultivation in intertropical regions. In F. Pank (ed.) *Proceedings of an International Symposium on Breeding Research on Medicinal and Aromatic Plants*. Quedlinburg, Germany 30 June – 4 July 1996, pp. 185–188.

Magalhaes, P.M., Raharinaivo, J. and Delabays, N. (1996) Influences de la dose et du type d'azote sur la production en artemisinine de l'*Artemisia annua* L. *Revue Suisse de Viticulture, Arboriculture et Horticulture*, **28**(6), 349–353.

Maynard, L., (1985) Malaria Cure. *Pharmacy Report, The University of Mississippi School of Pharmacy*, **7**(1), 10–13.

McVaugh, R (1984) Flora Novo-Galiciana: a descriptive account of the vascular plants of Western Mexico. In W R Anderson (ed.), *Flora, Vol 12 (Compositae)*, University of Michigan Press, Ann Arbor.

Mediplant, (1995) *Rapport D'Activite 1995 Mediplant (Centre de Recherches sur les Plantes Medicinales et Aromatiques* Conthey, Switzerland), pp. 5.

Morales, M.R., Charles, D.J. and Simon, J.E. (1993) Seasonal accumulation of artemisinin in *Artemisia annua* L. *Acta Hort.*, **344**, 416–420.

Nussenzweig, R.S. and Long, C.A. (1994) Malaria vaccines: multiple targets. *Science*, **265**, 1381–1383.

Phillipson, J.D. (1994) Natural products as drugs. *Trans. R. Soc. Trop. Med. Hyg.*, **88** (**Suppl. 1**), 17–19.

Pras, N.J., Visser, J.E., Batterman, S., Woerdenbag, H.J., Malingre, T.M. and Lugt, C.B. (1991) Laboratory selection of *Artemisia annua* L. for high artemisinin yielding types. *Phytochem. Anal.*, **2**, 80–83.

Prasad, A., Ram, M., Gutpa, N. and Kumar, S. (1998) Effect of different soil characteristics on the essential oil yield of *Artemisia annua*. *Journal of Medicinal and Aromatic Plant Sciences*, **20**(3), 703–705.

Ram, M., Gupta, M., Dwivedi, S. and Kumar, S. (1997) Effect of plant density on the yield of artemisinin and essential oil in *Artemisia annua* cropped under low input cost management in North-Central India. *Planta Med.*, **63**(4), 372–374.

Ratkowsky, D.A. (1983) *Non-linear regression modeling: A unified practical approach*. Marcel Dekker, New York.

Ravindranathan, T., Kumar, M.A., Menon, R.B. and Hiremath, S.V. (1990) Stereoselective synthesis of artemisinin+. *Tetrahedron Lett.*, **31**, 755–758.

Roche, G. and Helenport, J.P. (1994) The view of the pharmaceutical industry. *Trans. R. Soc. Trop. Med. Hyg.*, **88** (**Suppl. 1**), 57–58.

Roth, R.J. and Acton, N. (1989) A simple conversion of artemisinic acid into artemisinin. *J. Nat. Prod.*, **52**, 1183–1185.

Schmid, G., and Hofheinz, W. (1983) Total synthesis of qinghaosu. *J. Am. Chem. Soc.*, **105**, 624–625.

Shock, C.C. and Stieber, T. (1988) Preliminary observations on sweet wormwood production in the Treasury Valley. In *Oregon State University Agricultural Experiment Station Special Report*, **832**, 125–130.

Shukla, A., Farooqi, A.H., Shukla, Y.N., and Sharman, S. (1992) Effect of triacontanol and chlormquat on growth plant hormones and artemisinin yield in *Artemisia annua* L. *Plant Growth Regulation*, **11**, 165–171.

Simon, J.E. and Cebert, E. (1988) *Artemisia annua*: A production guide. In J.E. Simon and L.Z. Clavio, (eds.), *Third National Herb Growing and Marketing Conference*, Purdue Univ. Agr. Exp. Sta. Bull., **552**.

Simon, J.E., Charles, D., Cebert, E., Grant, L, Janick, J. and Whipkey, A. (1990) *Artemisia annua* L: a promising aromatic and medicinal. In J. Janick and J.E. Simon (eds.) *Advances in New Crops*. Timber Press, Portland, Oregon, USA, pp. 522–526.

Singh, A.V., Kaul, V.K., Mahajan, V.P., Singh, A. Misra, L.N., Thakur, R.S. *et al.* (1986) Introduction of *Artemisia annua* in India and isolation of artemisinin a promising anti-malarial drug. *Indian J. Pharm. Sci.*, 48, 137–138.

Singh, A., Vishwakarma, R.A. and Husain, A. (1988) Evaluation of *Artemisia annua* strains for higher artemisinin production. *Planta Med.*, 54, 475–476.

Srivastava, N.K. and Sharma, S. (1990) Influence of micronutrient imbalance on growth and artemisinin content in *Artemisia annua*. *Indian J. Pharm. Sci.*, 52, 225–227.

Takahashi, K. and Arakawa, H. (1981) Climates of Southern and Western Asia. In H.E. Landsberg, (ed.), *World Survey of Climatology* Vol. 9, Elsevier Scientific Publishing Company, Amsterdam – Oxford – New York., p. 42 (Hanoi), p. 139 (Lucknow).

Theuns, H.G. (1975) Literature survey on the influence of plant growth regulators and other substances on *Papaver species*. *Internal report Laboratory of Organic Chemistry, Division of Chemistry of Natural Products, University of Utrecht, Netherlands*.

Trigg, P.I. (1989) Qinghaosu (artemisinin) as an antimalarial drug. *Economic and Medicinal Plant Research*, 3, 19–55.

Vonwiller, Simone C., Haynes, R.K., King, S. and Wang, H. (1993) An improved method for the isolation of *qinghao* (artemisinic) acid from *Artemisia annua*. *Planta Med.*, 59, 562–63.

Wang, C.W. (1961) The forests of China, with a survey of grassland and desert vegetations. In *Harvard Univ. Maria Moors Cabot Foundation No. 5*, Harvard Univ., Cambridge, MA.

Whipkey, A., Simon, J.E., Charles, D.J. and Janick, J. (1992) In vitro production of artemisinin from *Artemisia annua* L. *J. Herbs Spices Med. Plants*, 1, 15–25.

Willey, R.W. and Heath, S.B. (1969) The quantitative relationships between plant population and crop yield. *Advances in Agronomy*, 21, 281–321.

WHO (World Health Organization), (1988) The development of artemisinin and its deriva-tives. *World Health Organization mimeographed document* WHO/TDR/CHEMAL ART 86.3.

WHO (World Health Organization), (1994) The role of artemisinin and its derivatives in the current treatment of malaria. *World Health Organization publication*.

Woerdenbag, H.J., Lugt, C.B. and Pras, N. (1990) *Artemisia annua* L: a source of novel anti-malarial drugs. *Pharm. Weekblad Sci.*, 12, 169–181.

Woerdenbag, H.J., Pras, N., Chan, N.G., Bang, B.T., Bos, R. van Uden, W. *et al.* (1994) Artemisinin related sesquiterpenes and essential oils in *Artemisia annua* during a vegetation period in Vietnam. *Planta Med.*, 60, 272–275.

Xu, X.X., Zhu, J., Huang, D.Z. and Zhou, W.S. (1983) The stereocontrolled synthesis of qinghaosu and deoxyqinghaosu from artennuic acid. *Huaxue Xuebao*, 42, 940 (Cited by Luo and Shen [1987]).

11. VARIATION AND HERITABILITY OF ARTEMISININ CONTENT IN *ARTEMISIA ANNUA* L.

NICOLAS DELABAYS[1], CHARLY DARBELLAY[2] AND NICOLE GALLAND[3]

[1]*Federal Agricultural Research Station, 1960 Nyon, Switzerland.*

[2]*Mediplant, Centre des Fougères, 1964 Conthey, Switzerland.*

[3]*University of Lausanne, 1015 Lausanne, Switzerland.*

INTRODUCTION

Artemisinin, a sesquiterpene lactone endoperoxide produced by the medicinal herb *Artemisia annua* L. (Asteraceae), has very potent antimalarial properties (Klayman, 1985; White, 1994). Since its isolation from the plant by Chinese scientists in 1972 (Anon., 1979), numerous pharmacological and clinical studies have been carried out with this molecule and some of its derivatives in order to assess their efficacy against malaria (Tang and Eisenbrand, 1992; Hien and White, 1993; Benakis, 1996). As artemisinin is very difficult to synthesize, only its extraction from cultivated plants can assure viable production (Woerdenbag *et al.*, 1990). Thus, during the last few years, several cultivation trials have been carried out with *Artemisia annua*, especially in India (Singh *et al.*, 1988), the United States (Simon *et al.*, 1990), Madagascar (Raharinaivo, 1993), Tasmania (Laughlin, 1993), Switzerland (Delabays *et al.*, 1994), Vietnam (Woerdenbag *et al.*, 1994) and Brazil (Magalhaes, 1996).

An important step in the domestication of a new species concerns its genetic improvement (Delabays, 1992a). With respect to many characteristics, wild plants usually display a wide genetic variability (Briggs and Walters, 1988). For instance, a variation is often observed on the content of secondary metabolites in the plants, allowing successful genetic improvement on such characteristics (Khanna and Shukla, 1991). With *Artemisia annua*, the main breeding goal is to enhance its artemisinin content. In order to achieve this, a description of the genetic variability expressed by the species on this trait is necessary. Then, in order to choose the best breeding program, as well as to predict the gain that can be expected, the heritability of the artemisinin content must be estimated (Gallais, 1990).

VARIABILITY IN ARTEMISININ CONTENT

A large variation in artemisinin content has been observed in the leaves of different samples of *Artemisia annua*. Thus, concentrations varying from 0.02 to 1.38% in the dry leaves have been reported in the literature (Table 1).

Table 1   Contents in artemisinin in different samples of leaves of *Artemisia annua* (% dry weight) reported in the literature.

Origin of the plant	Artemisinin content(%)	References
Europe	0.02	Singh *et al.*, 1988
Connecticut (USA)	0.06	Charles *et al.*, 1990
Argentina	0.10	Liersch *et al.*, 1986
China	0.14	Charles *et al.*, 1990
Dakota (USA)	0.21	Charles *et al.*, 1990
Spain	0.24	Delabays *et al.*, 1993
China	0.79	Anon., 1980
Vietnam	0.86	Woerdenbag *et al.*, 1994
China	1.07	Delabays *et al.*, 1993
Hybrid (Switzerland)	1.38	Debrunner *et al.*, 1996

This variation obviously has several causes. Besides the utilization of diverse methods for extraction and analysis, the collecting and the preparation of the samples, especially the separation of the leaves from the stems, are very variable too. Also, it is well known that the artemisinin content of the plant varies during the season (Woerdenbag *et al.*, 1994); obviously, a part of the variation reported is due to harvesting at different stages of growth and at different times. Moreover, environmental factors such as temperature (Chen and Zhang, 1987) or availability of nutrients (Magalhaes *et al.*, 1996) can also influence the artemisinin content in the plant. However, the variation observed is so considerable that a genetic basis is certainly involved. In order to quantify its significance, it is necessary to grow different strains of *A. annua* in the same environment and use the same method of analysis to determine the artemisinin content.

THE GENETIC BASIS OF ARTEMISININ VARIATION IN *ARTEMISIA ANNUA*

In 1989 and 1990, plants of very diverse origins were grown in comparative trials in Switzerland and great variability in their artemisinin content was found with amounts from 0.05 to 1.07% in the dry leaves (Delabays *et al.*, 1993). Also, in a preliminary trial, we crossed 2 lines presenting very different artemisinin concentrations. The plants of the progeny presented very variable contents, but generally they were intermediate to the concentrations found in the parents (Delabays, 1992b). Moreover, the distribution in artemisinin content observed in the second generation revealed an increase in variance, suggesting a classical segregation of the genetic components. However, classification of the individual plants, according to their artemisinin content, could no longer be made: the variation was continuous (Delabays, 1992b). This observation pointed out the necessity of using a model of quantitative genetics to describe the genetic background governing this trait in the species.

The continuous variation observed with the artemisinin content is quite logical if one thinks of the nature of such a characteristic. The amount of a secondary metabolite accumulated in a plant is the result of a very complex process. It involves the whole biosynthetic pathway of the molecule and its accumulation in the plant, often in a specialized structure. Concerning the biosynthesis of artemisinin, authors agree that it starts through the classical metabolic pathway of the terpenoids, from mevalonic acid to farnesyl pyrophosphate (Kudakasseril *et al.*, 1987; Akhila *et al.*, 1987). However, there is divergence of opinion concerning the complete pathway, especially about the possible role played by artemisinic acid as a precursor (Nair *et al.*, 1985; Nair *et al.*, 1986; El-Feraly, *et al.*, 1986; Jung *et al.*, 1990; Woerdenbag *et al.*, 1990, Nair and Basile, 1993; Woerdenbag *et al.*, 1990; Sangwan *et al.*, 1993; Kim and Kim, 1992; Wallart *et al.*, 1994). Obviously, several steps are involved in the biosynthesis of artemisinin, each one likely to be genetically regulated.

Concerning the localization of the molecule in the plant, it is well established today that it is accumulated in the glandular trichomes present on the surface of the leaves as well as on the corolla and receptacle of the florets (Hu *et al.*, 1993, Duke *et al.*, 1994, Ferreira and Janick, 1995).

In quantitative genetics, what one seeks to determine is that part of the total variation that can be put down to genetic factors; this is known as heritability.

BROAD-SENSE HERITABILITY

The broad-sense heritability (H) of a characteristic is defined as the proportion of the variation observed that can be attributed to genetic factors (Falconer, 1989; Nyquist, 1991). It represents the ratio of the genetic variance (σ_g^2) to the total variance expressed by the phenotype on the feature (σ_p^2): H = σ_g^2/σ_p^2.

Concerning the heritability of the artemisinin content in *Artemisia annua*, data reported in the literature are very scarce. In fact, only Ferreira (1994) has proposed an estimate of broad-sense heritability for this trait. In a study carried out with 24 clones of *A. annua* grown in two environments (in the open and in a glasshouse), the partition of the variance observed in their artemisinin contents revealed a broad-sense heritability of 0.98. Such estimates must be interpreted with caution as they apply only to the plant material used in the study, as well as to the environments, sites and years, in which they have been calculated. Prior to a generalization, several experiments need to be compared.

A study was carried out in 1993 and 1994 in Switzerland, with clones of very diverse origins (Delabays, 1997). Figure 1 presents the contents of artemisinin measured in the dry leaves of these different genotypes. Very variable concentrations were observed. The clones CA and CN, natives of Asia (from China and Vietnam respectively), contained high amounts of artemisinin. In comparison, clones CC and CD, coming respectively from North America (Virginia) and Europe (Spain), contained lower concentrations. A particularly low content has been observed in clone CB, isolated from a strain which came from the Botanical Garden of Göttingen (Germany). The case of clone CE is a little different: this genotype has been isolated in the progeny resulting from the cross between clones CA and CC.

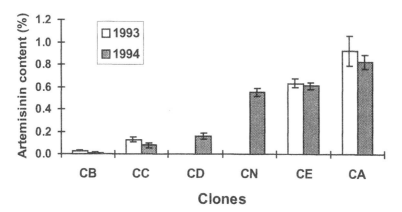

Figure 1 Artemisinin contents (means and standard deviations) in the dry leaves of different clones of *Artemisia annua* grown in 1993 and 1994.

Obviously, genetic differences appear between these different clones which are relevant to their artemisinin content. In order to quantify this observation, especially to estimate the heritability of the artemisinin content, an analysis of variance has been carried out (Delabays, 1997). Table 2 presents the analyses obtained from 6 clones cultivated in 1994 (A), as well as with 4 clones grown in 1993 and 1994 (B). It also shows the models used for the estimation of the heritability (Nyquist, 1991). These estimates are presented in Table 3.

The differences in artemisinin concentrations observed between the genotypes are highly significant. Moreover, the heritabilities estimated are also very high, always equal to or greater than 0.95, that is to say similar to those reported by Ferreira (1994). Thus, it appears that an important part of the variation in the artemisinin content of the different clones can be attributed to genetic factors. These estimates refer to a series of genotypes coming from different parts of the world (Delabays, 1997). They may be considered as indicative of the variation offered by the species as a whole.

In order to check the possible presence of genetic variability within strains or populations, we have grown them along with several replicates of a clone, propagated by means of *in vitro* techniques and used as a control. The variation observed between the different replicates of the clone cannot be attributed to genetic factors as such plant material is homogenous; this variation results only from environmental factors, such as soil, climate, watering or competition. It also includes the sampling error and the variation depending on the precision of the analytical method. By comparing this variation with the one observed in the populations grown in parallel, the genetic component present in the latter can be estimated. This method, called the "difference method", has been described among others by Nyquist (1991). Once the genetic component is determined, the heritability, that is to say the ratio of the genetic variance (σ_g^2) to the total variation (phenotypic variance: σ_p^2) can be calculated (Falconer, 1989).

Table 2 Analysis of variance and expected mean square for the estimation of the broad-sense heritability of the artemisinin content in *Artemisia annua* (H_g: genotypic heritability; H_{ind}: heritability for an individual; y: number of years; r: number of replications; n: number of individuals within plots; df: degrees of freedom; MS: mean square; F: ratio of the MS of the source of variation studied (blocks, lines etc.) and the MS of the error (this ratio indicates if a source of variation is significant); ***: p<0,001).

A. For the 6 clones grown in 1994.

Sources of variations	df	Artemisinin content		Expected mean square
		MS	F	
Blocks (b)	2	0.033		
Genotypes (g)	5	9.425	409.8***	$\sigma^2_i + n\sigma^2_\varepsilon + bn\sigma^2_g$
Error (ε)	10	0.023		$\sigma^2_i + n\sigma^2_\varepsilon$
Plants (i)	90	0.004		σ^2

$$H_g = \frac{\sigma^2_g}{\sigma^2_g + \sigma^2_\varepsilon/r + \sigma^2_i/rn} = \frac{MS_g - MS_\varepsilon}{MS_g}$$

$$H_{ind.} = \frac{\sigma^2_g}{\sigma^2_g + \sigma^2_\varepsilon + \sigma^2_i} = \frac{MS_g - MS_\varepsilon}{MS_g + (r-1)\,MS_\varepsilon + (rn-r)\,MS_i}$$

B. For the 4 clones grown in 1993 and 1994.

Sources of variations	df	Artemisinin content		Expected mean square
		MS	F	
Blocks (b) within years	4	0.017		
Years (y)	1	1.302		
Genotypes (g)	3	21.592	863.7***	$\sigma^2_i + n\sigma^2_\varepsilon + bn\sigma^2_{gy} + ybn\sigma^2_g$
Genotypes x years (gy)	3	0.443	17.7***	$\sigma^2_i + n\sigma^2_\varepsilon + bn\sigma^2_{gy}$
Error (ε)	12	0.025		$\sigma^2_i + n\sigma^2_\varepsilon$
Plants (i)	120	0.005		σ^2_i

$$H_g = \frac{\sigma^2_g}{\sigma^2_g + \sigma^2_{gy}/y + \sigma^2_\varepsilon/yr + \sigma^2_i/yrn} = \frac{MS_g - MS_{gy}}{MS_g}$$

$$H_{ind.} = \frac{\sigma^2_g}{\sigma^2_g + \sigma^2_{gy} + \sigma^2_\varepsilon + \sigma^2_i} = \frac{MS_g - MS_{gy}}{MS_g + (a-1)\,MS_{gy} + (ry-y)\,MS_\varepsilon + (rny-ry)\,MS_i}$$

Table 4 shows the variances observed within different samples of *Artemisia annua* grown between 1990 and 1996 as well as the heritabilities estimated with the "difference method" (Delabays, 1997). The variations in artemisinin content observed in the different strains tested, with values of variance generally superior to that measured with the control clones, demonstrate that most of them contain a

Table 3 Estimations of the genotypic and individual broad-sense heritabilities for the artemisinin content in *Artemisia annua* (Delabays, 1997).

Analysis	Broad-sense heritability for artemisinin content	
	Genotype	Individual
1994 (6 clones)	0.997	0.986
1993 and 1994 (4 clones)	0.979	0.949

genetic component. For example, with the Vietnamese population G, evaluated over 3 years, the ratio σ_g^2/σ_p^2 varied between 0.20 and 0.74, and those of population H, received from Madagascar and tested in 1994 and 1995, between 0.57 and 0.70. Here again, such estimates must be interpreted with caution, but obviously, an important genetic variability is present within most of the strains tested.

NARROW-SENSE HERITABILITY

In quantitative genetics, different components are distinguished in the genetic variance, especially an additive part (σ_A^2) and a dominance one (σ_D^2). The additive part of the genetic variation is correlated with the degree of resemblance between relatives (Falconer, 1989; Nyquist, 1991) and the ratio of this additive component to the total variation, σ_A^2/σ_p^2, represents what is called the narrow-sense heritability (h^2). Knowledge of this value is important in order to choose the best selection program and also to predict the genetic gains that can be expected through breeding (Gallais, 1990).

No data concerning the narrow-sense heritability of the artemisinin content in *Artemisia annua* has been reported in the literature. To estimate it, related lines have to be grown and analyzed in parallel, as the additive variance can be calculated from the covariance observed between them (Falconer, 1989; Gallais, 1990). Thus, we crossed 5 clones of *A. annua* with very different artemisinin contents in a diallel set, which means that each clone was hybridized with each of the others. Except for one clone which was male sterile, we produced reciprocal offspring, that is to say that for each crossing seeds were collected from both parents. All together, we obtained 16 progenies that were then grown in a comparative trial (Delabays, 1997). Figure 2 presents the contents of artemisinin found in these different progenies.

The structure of the hybridizations (diallel) allows, through the analysis of the related progenies, the estimation of several genetic components of the variation observed. In particular, the specific and general combining ability (sca and gca) of the clones tested can be estimated. These terms were first defined by Sprague and Tatum, (1942): the general combining ability concerns the mean performance of a parental line in crossing and the specific combining ability designates the case of a hybrid line whose value is different to the one expected, based on the mean performance of the two parents. Griffing (1956) described different models for the analysis of diallel mating systems, based especially on the presence or not of reciprocals. As a

Table 4 Variation in the artemisinin content (% dry weight) of leaves collected from plants of differents populations of *Artemisia annua* grown between 1993 and 1996, in comparison with several replicates of clones used as control (Po.: populations; d.: number of days between the planting of the seedlings and the collection of the sample; n.: number of plants of the sample; n.: number of plants analyzed, m: mean artemisinin content of the n samples; min.: minimum artemisinin content; max.: maximum artemisinin content; var: variance of the artemisinin content of the n samples; H: heritability).

Year	Ac.	d.	n	m.	min.	max.	var.	Control	n	m.	var.	H
1993	G	46	125	0.33	0.13	0.53	0.005	CD	21	0.45	0.004	0.20
1994	G	52	86	0.51	0.29	0.90	0.012	CM	10	0.83	0.006	0.50
	H	52	86	0.56	0.24	0.80	0.014	CM	10	0.85	0.006	0.57
1995	G	77	67	0.84	0.59	1.24	0.023	CR	12	0.91	0.006	0.74
	H	77	42	1.06	0.42	1.43	0.040	CR	13	1.00	0.012	0.70
	SA	77	80	0.97	0.39	1.42	0.040	CR	13	1.03	0.008	0.80
	I	77	13	0.68	0.56	0.88	0.008	CR	6	0.97	0.010	-0.25
1996	SB	80	324	1.04	0.55	1.55	0.035	CR	24	0.88	0.014	0.66

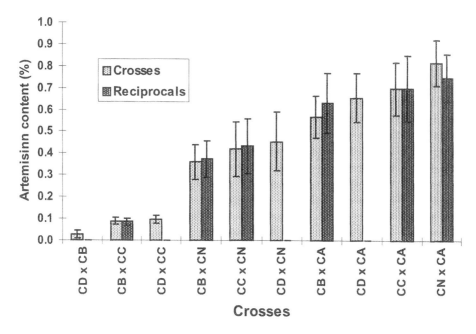

Figure 2 Artemisinin contents (means and standard deviations) in the dry leaves of the 16 progenies obtained from a diallel cross between 5 clones of *Artemisia annua*.

first step, we have analyzed the results obtained with the 4 clones for which we had reciprocal progenies (Table 5).

A highly significant difference appears between the lines. In order to check the possible presence of reciprocal effects on the artemisinin content of the plants, method 3, model II, described by Griffing (1956) has been applied to the data (Table 6).

Table 5 Analysis of variance for the artemisinin content among twelve progenies resulting from a diallel cross set, with reciprocal, between 4 clones of *Artemisia annua* (Delabays, 1997).

Sources of variations	df	Artemisinin content	
		MS	F
Blocks (b)	2	0.011	
Progenies (g)	11	2.019	183.5***
Error (ε)	22	0.011	
Plants (i)	180	0.009	

*** : p<0.001.

Table 6 Analysis of reciprocal effects on the concentration in artemsinin in progenies issued of a complete diallel cross set between 4 clones of *Artemisia annua* (Delabays, 1997).

| Sources of variations | df | Artemisinin content | |
		MS	F
Reciprocal effect	6	0.00072	1.18 ns
Error (ε)	180	0.00061	

ns: not significant.

No maternal effects were detected. Thus, for the analysis of the diallel trial, reciprocal progenies have been mixed and the crossings involving clone CD, male-sterile, have been added. Method 4, model II, reported by Griffing (1956), has been used for the analysis (Table 7 and 8).

The differences between the lines is confirmed, as well as the presence of general and specific combining ability variations. The estimation of the variance components is calculated as follows (Griffing, 1956):

$$\sigma^2_{sca} = MS_{sca} - MS_{\epsilon}$$

$$\sigma^2_{gca} = (MS_{gca} - MS_{sca})/3$$

Table 7 Analysis of variance for the content in artemisinin among the progenies resulting from the diallel cross set, without reciprocals, obtained with 5 clones of *Artemisia annua* (Delabays, 1997).

| Sources of variations | df | Artemisinin content | |
		MS	F
Blocks (b)	2	0.011	
Lines (g)	9	9.005	191.60***
Error (ε)	18	0.047	
Plants (i)	330	0.011	

*** :p<0.001.

Table 8 Analysis for the general and specific combining ability (GCA and SCA) among 5 clones of *Artemisia annua* for the artemisinin content (Delabays, 1997).

| Sources of variations | df | Artemisinin content | | Expected mean square |
		MS	F	
GCA	4	0.4619	5'.71*	$\sigma^2 + \sigma^2_{asc} + 3\,\sigma^2_{agc}$
SCA	5	0.0809	31.16***	$\sigma^2 + \sigma^2_{asc}$
Error (ε)	18	0.0026		σ^2

* : p<0,05; *** : p<0,001.

Moreover, taking into account the family connections between the lines and the corresponding covariances that can be expected, the additive (σ^2_A) and the dominance (σ^2_D) genetic variance can be estimated from the general and specific combining ability observed (Griffing, 1956; Gallais, 1989). With this strongly outcrossing species, in which the selfing coefficient has been estimated at only 0.01 (Delabays, 1997), the following relations apply (Gardner, 1963):

$$\sigma^2_A = (4/1 + F)\, \sigma^2_{gca}$$

$$\sigma^2_D = (2/1 + F)^2\, \sigma^2_{sca}$$

Table 9 presents the genetic variances (σ^2) as well as the broad- and narrow-sense heritabilities (H and h^2) estimated for the artemisinin content in the diallel trial (Delabays, 1997).

The large genetic component of the variability observed for the artemisinin content is confirmed. Moreover, it appears that the additive part (σ^2_A) is relatively important. Actually, narrow-sense heritabilities superior to 0.60 have been estimated. This finding suggests that parents selected for their high artemisinin content will generate interesting progenies: good genetic progress can be expected through mass selection. However, a dominance variance (σ^2_D) is also present, indicating the crossing of some specific genotypes could result in particularly high artemisinin yielding hybrid lines.

CONCLUSION AND PROSPECTS

Artemisinin content is a very variable trait in *Artemisia annua*. High broad-sense heritabilities have been calculated for this characteristic, indicating that it is governed mainly by genetic factors. Moreover, narrow-sense heritability was also shown to be relatively high. Thus, a good response can be expected by the breeding of lines of *Artemisia annua* rich in artemisinin. Actually, by selecting, from wild populations, plants with high artemisinin content and crossing them, hybrid lines particu-

Table 9   Genetic variances (σ^2) and heritability (H: broad-sense; h^2: narrow-sense) estimated for the artemisinin content among 5 clones of *Artemisia annua* (Delabays, 1997).

Parameters	
σ^2_{gca}	0.127
σ^2_{sca}	0.079
σ^2_A	0.503
σ^2_D	0.310
H (genotypes)	0.998
H (individuals)	0.983
h^2 (genotypes)	0.618
h^2 (individuals)	0.608

larly rich in the molecule have been created (Debrunner *et al.*, 1996). Thus, the high heritability of artemisinin content has been experimentally confirmed. Recently, lines containing up to 1.4% artemisinin in their dry leaves have been developed (Delabays, 1997) and a first cultivar, called "Artemis", resulting from the hybridization between two clones selected for their high artemisinin content, was registered in Switzerland in spring 1999.

Today, artemisinin and its derivatives are important drugs for the treatment of malaria. With the development of these new agents, one of the first problems to solve concerned the production of the artemisinin. Neither its synthesis nor its production by means of cell, tissue or organ cultures are viable (Nair *et al.*, 1986; Woerdenbag *et al.*, 1990 Ferreira and Janick, 1996). At the present time, only its extraction from selected and cultivated strains of *Artemisia annua* is possible. Thanks to the progress achieved in the cultivation of the species as well as in its genetic improvement, the production of sufficient quantities of artemisinin has been facilitated.

REFERENCES

Akhila, A., Thakur, R. and Popli, S.P. (1987) Biosynthesis of artemisinin in *Artemisia annua*. *Phytochemistry*, 26, 1927–1930.

Anonyme. (1979) Qinghaosu Antimalaria Coordinating Research Group. Antimalarial studies on qinghaosu. *Chin. Med. J.*, 92, 811–816.

Anonyme. (1980) Qinghaosu Research Group. *Sci. Sin.* 23, 280.

Benakis, A. (Ed.), (1996) Qinghaosu and its derivatives. *Jap. J. Trop. Med. Hyg.*, 24, suppl. 1, pp. 101

Briggs, D. and Walters, S.M. (1988) Plant variation and evolution. Cambridge University Press, Cambridge, pp. 412

Brown, G.D. (1993) Production of anti-malarial and anti-migraine drugs in tissue culture of *Artemisia annua* and *Tanacetum parthenium*. *Acta Hort.* 330, 269–276.

Charles, D.J., Simon, J.E., Wood, K.V. and Heinstein, P. (1990) Germplasm variation in artemisinin content of *Artemisia annua* using an alternative method of artemisinin analysis from crude plant extracts. *Journal of Natural Products*, 53, 157–160.

Debrunner, N., Dvorak, V., Magalhaes, P. and Delabays, N. (1996) Selection of genotypes of *Artemisia annua* for the agricultural production of artemisinin. *In*: Proceedings of the "International Symposium: Breeding Research on Medicinal and Aromatic Plants", June 30 – July 4, 1996, Quedlinburg, Germany, pp. 222–225.

Delabays, N. (1992a) Notes sur la stratégie de domestication d'une plante sauvage utilisée pour ses métabolites secondaires. *Rev. Suisse Agric.*, 24, 93–98.

Delabays, N. (1992b) The selection of *Artemisia annua* L. and the genetics of its artemisinin (qinghaosu) content. *In*: Proceedings of the second Médiplant Conference, 14–15 may 1992, Conthey, pp. 41–50.

Delabays, N. (1997) Biologie de la reproduction chez l'*Artemisia annua* L. et génétique de la production en artémisinine. Contribution à la domestication et à l'amélioration génétique de l'espèce. Thèse de doctorat, Université de Lausanne, pp. 169

Delabays, N., Benakis, A. and Collet, G. (1993) Selection and breeding for high artemisinin (Qinghaosu) yielding strains of *Artemisia annua* L. *Acta horticulturae*, 330, 203–207.

Delabays, N., Jenelten, U., Paris, M., Pivot, D. et Galland, N. (1994) Aspects agronomiques et génétiques de la production d'artémisinine à partir d'*Artemisia annua. Rev. Suisse Vitic. Arboric. Hort.*, **26**, 291–296.

Duke, M.V., Paul, R.N., Elsohly, H.N., Sturtz, G. and Duke, S.O. (1994) Localization of artemisinin and artemisitene in foliar tissues of glanded and glandless biotypes of *Artemisia annua* L. *Int. J. Plant. Sci.* **155**, 365–372.

El-Feraly, F.S., Al-Meshal, I.A., Al-Yahya, M.A. and Hifnawy, M.S. (1986) On th epossible rôle of qinghao acid in the biosynthesis of artemisinin. *Phytochemistry*, **25**, 2772–2778.

Falconer, (1989) Introduction to quantitative genetics. John Wiley and Sons, New york, pp. 284

Ferreira, J.F.S. (1994) Production and detection of artemisinin in *Artemisia annua* L. Thesis, Purdue University, pp. 125

Ferreira, J.F.S. and Janick, J. (1995) Floral morphology of *Artemisia annua* with special reference to trichomes. *Int. J. Plant Sci.*, **156**, 807–815.

Ferreira, J.F.S. and Janick, J. (1996) Roots as enhancing factor for the production of artemisinin in shoot culure of *Artemisia annua. Plant Cell, Tissue and Organ Culture*, **44**, 211–217.

Gallais, (1990) Théorie de la sélection en amélioration des plantes. Masson, Paris, pp. 588

Gardner, C.O. (1963) Estimates of genetics parameters in cross-fertilizing plants and their implications in plant breeding. *In*: Statistical analysis and plant breeding, Hanson, W.D. and Robinson, H.F., eds, Nat. Ac. Sci., Washington, **982**, 225–252.

Griffing, B. (1956) Concept of general and specific combining ability in relation to diallel crossing systems. *Aust. J. Biol. Sci.*, **9**, 463–493.

Hien, T.T. and White, N.J. (1993) Qinghaosu. *The Lancet*, **341**, 603–608.

Hu, S.L., Xu, Z.L., Pan, J.G. and Hou, Y.M. (1993) The glandular trichomes of *Artemisia annua* L. and their secretions. *J. Res. Educ. Indian Med.* **12**, 9–15.

Jha, J., Jha, T.B. and Mahato, S.B. (1988) Tissue culture of *Artemisia annua* L.: a potential source of an antimalarial drug. *Curr. Sci.*, **57**, 344–346.

Jung, M., Elsohly, H.N. and McChesney, J.D. (1990) Artemisinic acid: a versatile chiral synthon and bioprecursor to natural products. *Planta Med.* **56**, 624.

Khanna, K.R. & Shukla, S. (1991) Genetics of secondary plant products and breeding for their improved content and modified quality. *In: Biochemical aspects of crop improvement*, KHANNA, K.R. ed, CRC Press, Boca Raton, pp. 283–323.

Kim, N.-C. and Kim, S.-U. (1992) Biosynthesis of artemisinin from 11,12-dihydroarteannuic acid. *J. Korean Agr. Chem. Soc.*, **35**, 106–109.

Klayman, D.L. (1985) Qinghaosu (Artemisinin): an antimalarial drug from China. *Science*, **228**, 1049–1054.

Kudalasseril, G.J., Lam, L. and Staba, E.J. (1987) Effect of sterol inhibitors on the incorporation of ^{14}C-isopentenl pyrophosphate into artemisinin by cell-free system from *Artemisia annua* tissue cultures and plants. *Planta Med.*, **28**, 280–284.

Laughlin, J.C. (1993) Effect of agronomic practices on plant yield and antimalarial constituent of *Artemisia annua. Acta Hort.*, **331**, 53–61.

Liersch, R., Soicke, H., Stehr, P. and Tullner, H.U. (1986) Formation of artemisinin in *Artemisia annua* during one vegetation period. *Planta Medica*, **7**, 387–390.

Magalhaes, P.M. (1996) Seleçao, melhoramento e nutriçao da *Artemisia annua* L. para cultivo em regiao intertropical. Tese, Universidade de Campinas, pp. 116

Nair, M.S.R. and Basile, D.V. (1993) Bioconversion of arteannuin B to artemisinin *J. Nat. Prod.*, **456**, 1559–1566.

Nair, M.S.R., Acton, N., Klayman, D.L., Kendrick, K. and Mante, S. (1985) Tissue cultures and cell-free extracts studies on qinghaosu. *Abstracts Int. Cong. Natural Products*, Chapel Hill, July 7–12, 1985, abst. 75.

Nair, M.S.R., Acton, N., Klayman, D.L., Kendrick, K., Basile, D.V. and Mante, S. (1986) Production of artemisinin in tissue cultures of *Artemisia annua*. *J. Nat. Prod.*, 49, 504–507.

Nguyen, H.T. and Sleper, D.A. (1983) Theory and application of half-sib mating in forage grass breeding. *Theor. Appl. Gen.*, 64, 187–196.

Nyquist, W.E. (1991) Estimation of heritability and prediction of selection response in plant populations. *Critical Rev. Plant Sciences*, 10, 235–322.

Raharinaivo, J.N. (1993) Contribution au lancement de la culture du qinghaosu ou *Artemisia annua* L. comme plante antipaludique à Madagascar. Mémoire de fin d'Etudes, Ecole Supérieurs des Sciences Agronomiques, Antananarivo, pp. 52.

Sangwan, R.S., Agarwal, K., Luthra, R., Thakur, R.S. and Singh-Sangwan, N. (1993) Biotransformation of arteannuic acid into artennuin-b and artemisinin in *Artemisia annua*. *Phytochemistry*, 34, 1301–1302.

Simon, J.E., Charles, D.J., Cebert, E, Grant, L., Janick, J. and Whipkey, A. (1990) *Artemisia annua* L.: A promising aromatic and medicinal. *In*: Janick, J. and Simon, J.E. (eds), *Advances in new crops*. Timber Press, Portland, pp. 522–526.

Singh, A., Vishwakarma, R.A. and Husain, A. (1988) Evaluation of *Artemisia annua* strain for higher artemisinin production. *Planta medica*, 7, 475–476.

Sprague, G.F. and Tatum, L.A. (1942) General versus specific combining ability in single crosses of corn. *J. Amer. Soc. Agron.*, 34, 923–932.

Tang, W. and Eisenbrand, G. (1992) Chinese drugs of plants origin. Chemistry, pharmacology and use in traditional and modern medecine. Springer-Verlag, Berlin, pp. 159–174.

Wallart, T.E., Van Uden, W. and Pras, N. (1994) Possible biosynthetic pathway of artemisinin in a Chinese *Artemisia annua* strain. Proceeding of the symposium on "Plant Cell, Tissue and Organ Cultures in Liquid Media, Prague, 8–11 juillet 1994, pp. 171–172.

Whipkey, A., Simon, J.E.; Charles, D.J. and Janick, J. (1992) *In vitro* production of artemisinin from *Artemisia annua* L. *J. Herbs Spices Med. Plants*, 1, 15–25.

White, N. (1994) Artemisinin: current statut. *Trans. Roy. Soc. Trop. Med. Hyg.*, 88, SI, 3–4.

Woerdenbag, H.J., Lugt, C.B. and Pras, N. (1990) *Artemisia annua* L.: a source of novel antimalarial drugs. *Pharm Weekbl*, 12, 169–81.

Woerdenbag, H.J., Pras. N., Chan, N.G., Bang, B.T., Bos, R., Van Uden, W., Boi, N.V., Batterman, S. and Lugt, C.B. (1994) Artemisinin, related sesquiterpenes and essential oil in *Artemisia annua* during one vegetation period in Vietnam. *Planta Med.* 60, 272–275.

12. PHYTOCHEMISTRY OF *ARTEMISIA ANNUA* AND THE DEVELOPMENT OF ARTEMISININ-DERIVED ANTIMALARIAL AGENTS

RAJENDRA S. BHAKUNI, DHARAM C. JAIN AND
RAM P. SHARMA

*Central Institute of Medicinal and Aromatic Plants, P.O. CIMAP,
Lucknow-226015, India.*

INTRODUCTION

The purpose of this chapter is to review the phytochemistry of *Artemisia annua* and to discuss the biosynthesis, extraction and isolation of artemisinin as well as derivatives of artemisinin which have been prepared in order to enhance the therapeutic efficacy of the parent. Semi-synthetic derivatives of artemisinin, products produced by microbial transformation and synthetic 1,2,4-tioxane analogues will be considered.

CHEMICAL CONSTITUENTS

Owing to its antimalarial properties *A. annua* has been the subject of a detailed chemical investigation by scientists throughout the world in order to isolate both major and minor secondary metabolites. Several classes of compounds have been found as detailed below.

Monoterpenes

A large number of monoterpenoid compounds have been characterised from the essential oil of *A. annua* by gas chromatography-mass spectrometry (GC-MS) analysis. The yield of the oil generally varies between 0.3% and 0.4% (v/w). In 1993, Woerdenbag *et al.*, reported 4.0% and 1.4% in plants grown from Chinese and Vietnamese seeds respectively. The principal constituents of Chinese oil were artemisia ketone (63.9%), artemisia alcohol (7.5%), myrcene (5.1%), α-guainene (4.7%) and camphor (3.3%). The Vietnamese oil contained camphor (21.8%), germacrene D (18.3%) and 1,8-cineole (3.1%). The same authors (Woerdenbag *et al.*, 1994), reported that the maximum oil content of plants grown in Vietnam occurred before flowering and contained 55% monoterpenes. Hethelyi *et al.* (1995), analysed the oil content from fresh flowering shoots in Hungary. The oil content varied from 0.48–0.81% and was mainly composed of artemisia ketone and artemisia alcohol varying between 33–75% and 15–56% respectively. *A. annua* growing in Italy was also analysed for its essential oil composition.

The oil of *A. annua* grown at Lucknow in India was found to contain artemisia ketone (58.8%), 1,8-cineole (10.2%) and germacrane D (2.4%) while a Himalayan grown variety contained a smaller percentage of artemisia ketone (52.3%) but an increased amount of 1,8-cineole (13.1%). The highest yield of artemisia ketone (80.9%) was found in *A. annua* grown in Bulgaria followed by a Netherlands variety (63.9%) and a USA variety (63.1%), (Ahmad and Mishra, 1994). The other minor monoterpenoids detected were α- and β-myrcene hydroperoxide (Rucker *et al.*, 1987), camphene (Youan, 1955), geraniol and allo-ocemene (Liu *et al.*, 1988), α-cadinene (Lawerence, 1982; Youan, 1955), trans-α-ocemene, α-selinene, α-thujone, α-farnesene, linalool, menthol and isomenthol (Lawerence, 1982), α- and β-pinene (Libby and Sturtz, 1989; Liu *et al.*, 1988; Tsunemtsu and Tadashi, 1957; Tu *et al.*, 1981a), sabinene and pinocarveol (Libbey and Sturtz, 1989), pinocarvenone, borneol and limonene (Lawerence, 1982), and nerol (Ulubulen and Halfon, 1976).

Sesquiterpenes

Most attention has been focussed on this group of compounds since it is a sesquiterpene lactone, artemisinin **1**, also known as arteannuin or qinghaosu which was found to be responsible for the antimalarial activity of *A. annua* by Chinese investigators in 1972 (Anon. 1979; Liu *et al.*, 1979; Tu *et al.*, 1981). (Note that two alternative numbering systems are used in the literature as shown in **1a** and **1b**; compounds are numbered according to the system used by the authors referred to). Artemisinin contains an endoperoxide group, uncommon in natural products, and the structure of this interesting sesquiterpene lactone endoperoxide has been fully confirmed by variety of methods (Anon., 1977; Blasko *et al.*, 1988; Huang *et al.*, 1988; Jeremic *et al.*, 1973; Leban *et al.*, 1988; Madhusudan *et al.*, 1989; Mi, 1987; Wang *et al.*, 1985, 1986) and by total synthesis (Xu *et al.*, 1986; Avery *et al.*, 1987). An account of the elucidation of the structure of artemisinin is included in a review by Klayman *et al.* (1985). Recently, the absolute configuration of artemisnin has been confirmed by means of X-ray crystallography (Lisgarten *et al.*, 1998).

Artemisinin belongs to the amorphane sub-group of the cadinanes which incorporate a *cis*-decalin skeleton (Bardoloi *et al.*, 1989). It is an unusual compound in the sense that it lacks a nitrogen containing heterocyclic ring system which is found in most other antimalarial compounds. Artemisinin is poorly soluble in water and oil but soluble in most organic solvents (Zeng *et al.*, 1983).

Constituents related to artemisinin reported from the aerial parts of *A. annua* are: arteannuins A to G, **2–8** respectively, (Kasymov *et al.*, 1986; Liu *et al.*, 1979; Mishra, 1986; Mishra *et al.*, 1993; Stefanovic *et al.*, 1977; Tian *et al.*, 1982; Tu *et al.*, 1981a; 1982; Zhu *et al.*, 1984), *epi*-deoxyarteannuin B **9** (El-Feraly *et al.*, 1989; Roth and Acton, 1987a), dihydro, *epi*-deoxyarteannuin B, **10**, (Brown, 1992), compound **11**, and deoxyarteannuin B **12**, (Uskobovic *et al.*, 1974), artemisinic (arteannuic) acid, **13**, (Deng *et al.*, 1981; Kim *et al.*, 1989; Roth and Acton, 1987), artemisinic acid methyl

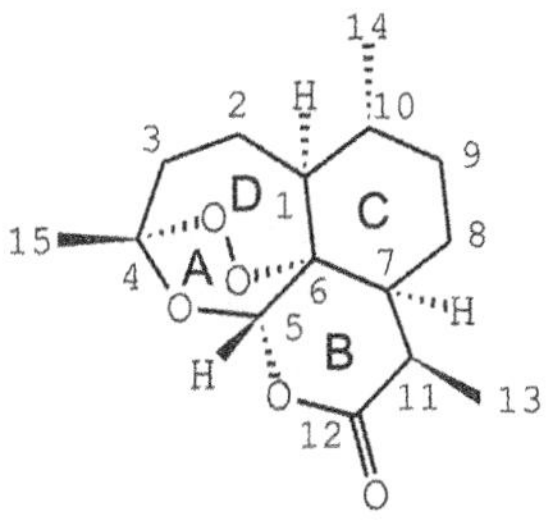

1a
19, 11(13)-ene

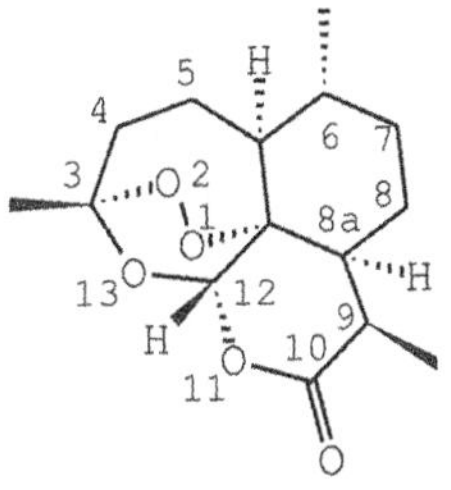

1b

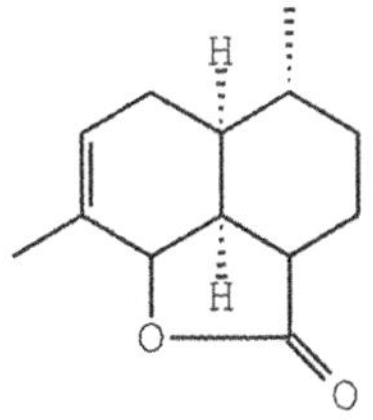

2

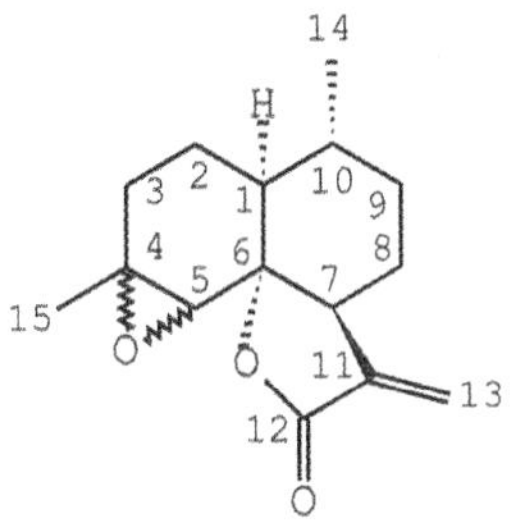

3, α-epoxy
4, β-epoxy
12, 4(5)-ene
26, 4-OH, 5-OH

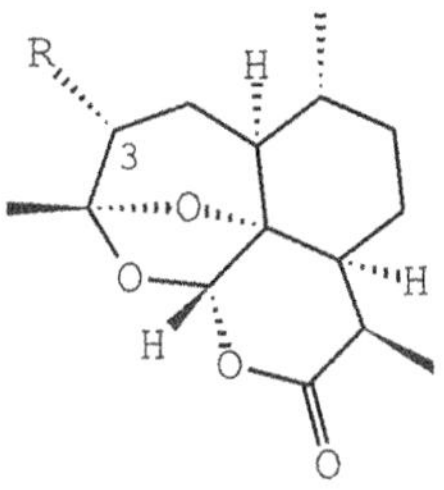

5, R=OH
18, R=H

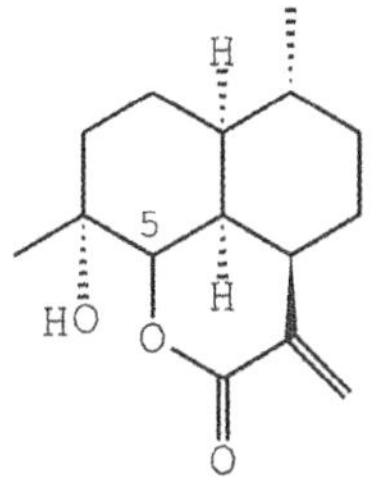

6, 5β-O-
7, 5α-O-

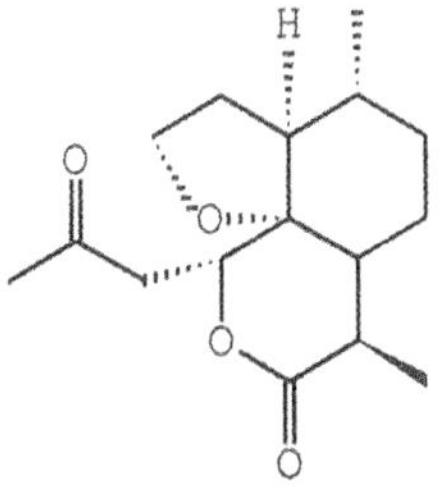

8

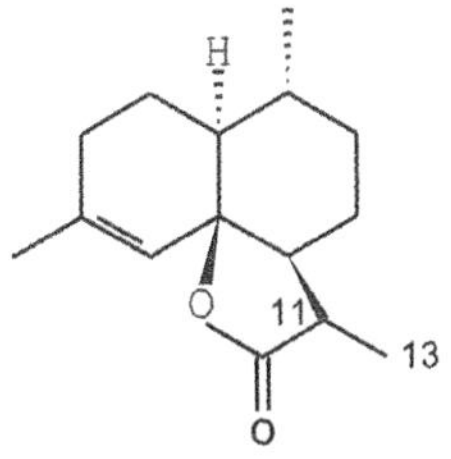

9, 11(13)-ene
10

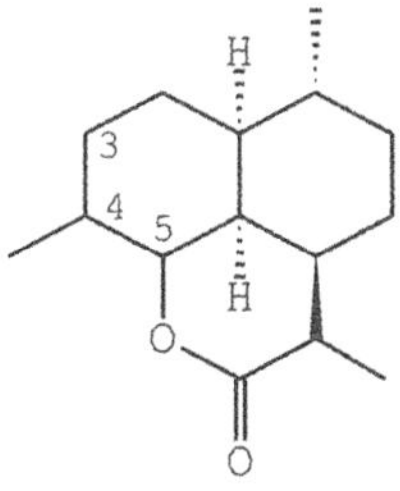

11, 5β-O-, 3(4)-ene

22, 5α-O-, 4(15),
    11(13)-diene

23, 5α-O-, 3(4),
    11(13)-diene

ester, **14**, (Zhu *et al.*, 1982), epoxyartemisinic acid, **15**, (Wu and Wang, 1984), 11,R(−)-dihydroartemisinic acid, **16**, (Huang *et al.*, 1987), 6,7-dehydroartemisinic acid, **17**, (El-Feraly *et al.*, 1989), deoxyartemisinin, **18**, (Tu *et al.*, 1981a; 1982), artemisitene, **19**, (Leppard *et al.*, 1974), artemisin, **20**, Bolt *et al.*, 1963), artemisinol, **21**, (Zhu *et al.*, 1982), annulide, **22**, and isoannulide, **23**, (Brown, 1993), α-hydroxysantonin, **24**, (Pinhey and Sternhell, 1995), 4,5-secocadinane, **25**, and dihydroxycadinanolide, **26**, (Brown, 1994), the bis-norcadinanes norannuic acid, **27**, (Mishra *et al.*, 1993), cadin-4,7(11)-dien-12-al, **28**, cadin-4(15),11-dien-9-one, **29**, and 3-isobutylcadin-4-en-11-ol, **30**, (Ahmad and Mishra, 1994).

The plant growth regulators abscisic acid, **31**, and abscisic acid methyl ester, **32**, have also been reported from *A. annua* (Shukla *et al.*, 1991). Artemisinin, **1**, arteannuin B, **3**, and artemisinic acid, **13**, are the major secondary metabolites of the species. Artemisinin has been isolated in yields of 0.1 to 0.7% from different varieties of *A. annua* grown in China while arteannuin B and artemisinic acid have been reported to be present in amounts 2–4 and 7–8 times more (respectively) than artemisinin in plants which contained 0.1% of the latter.

Triterpenoids and Steroids

Only a few compounds of this class have been reported: α-amyrin, β-amyrin, β-amyrin acetate, α-amyrenone, oleanolic acid, baurenol and taraxasterone belonging to the oleane skeleton were isolated from the aerial parts (Ahmad and Mishra, 1994; Ulubulen and Halfon, 1976). The friedelane triterpenoids, friedelin and friedelan-3-β-ol were isolated from the leaves (Guo, 1994). Friedelin and taraxeryl acetate were reported from the roots of the plant (Agrawal and Bisnoi, 1996). Sterols, β-sitosterol and stigmasterol were isolated from aerial parts (Guo, 1994; Tian *et al.*, 1982; Tu *et al.*, 1985) as well as from the roots of *A. annua* (Agrawal and Bisnoi, 1996).

Flavonoids

A large number of flavonoids have been reported from the plant *A. annua*. Some of them, artemetin, **33**, and casticin, **34**, have been shown to enhance the antimalarial activity of artemisinin (Yang *et al.*, 1989). The non-glycosidic flavonoids isolated from the leaves and stem are: artemetin, **33**, (Baera *et al.*, 1988; Liu *et al.*, 1981), chrysosplenetin, (Bhardwaj *et al.*, 1985), chrysosplenol D, (Yang *et al.*, 1989), casticin, **34**, (Baera *et al.*, 1988), kaempherol, quercetin, luteolin, axillarin and patuletin, (Marco *et al.*, 1990), cirsilineol, eupatorin, penduletin, cirsimatrin, rhamnocitrin, cirsiliol, chrysoeriol, tamarixetin, rhamnetin, quercetin-3-methyl ether, quercetagetin-3,4'-dimethylether, 5,2',4'-trihydroxy-6,7,5'-trimethoxyflavone and 5,7,8,3'-tetrahydroxy, 3,4' dimethoxyflavone, (Yang *et al.*, 1989), 3,5-dihydroxy-6,7,3',4'-tetramethoxyflavone (Liu and Li, 1980), 5,3' dihydroxy, 3,6,7,5'-tetramethoxyflavone, (Jeremic *et al.*, 1980), 3,5,3'-trihydroxy-6,7,4' trimethoxyflavone, (Djermanovic *et al.*, 1975), quercetagetin-6,7,4'-trimethyl ether, quercetagetin-6,7,3'4'-tetramethylether and 5-hydroxy-3,6,7,4'-tetramethylflavone (Tu *et al.*, 1985a).

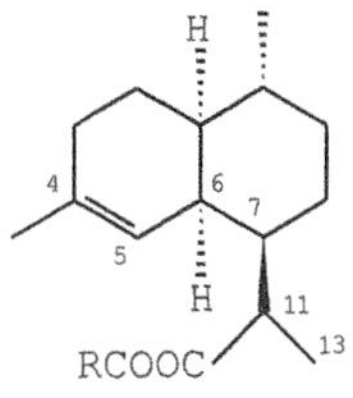

13, R=H, 11(13)-ene
14, R=CH3, 11(13)-ene
15, R=H, 4,5α-epoxy, 11(13)-ene
16, R=H
17, R=H, 4(5), 6(7), 11(13)-triene

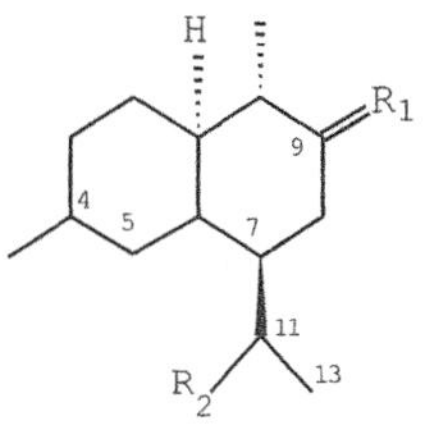

21, R1=H H, R2=CH2OH,
    4(5)-ene

28, R1=H H, R2=CHO,
    4(5), 7(11)-diene

29, R1=O, R2=CH3,
    4(5), 11(13)-diene

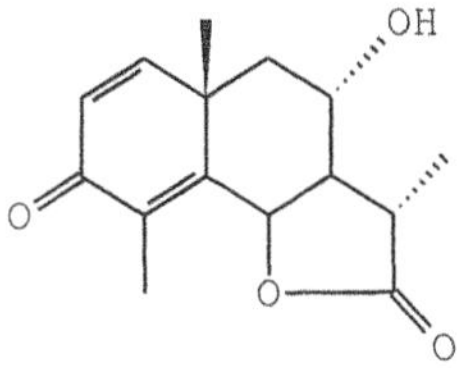

20

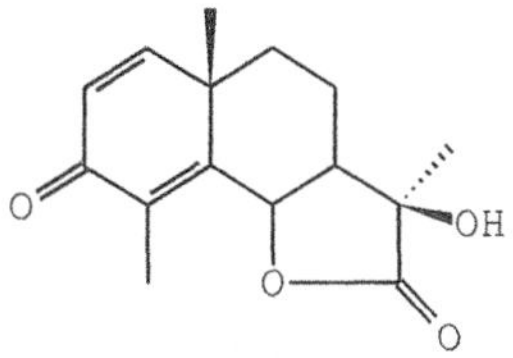

24

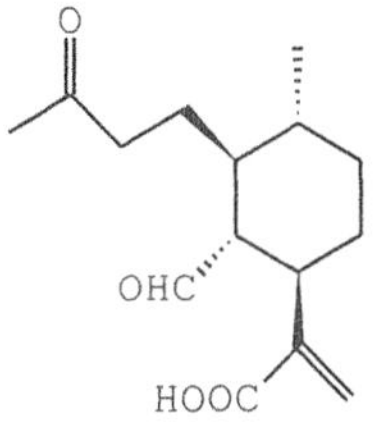

25

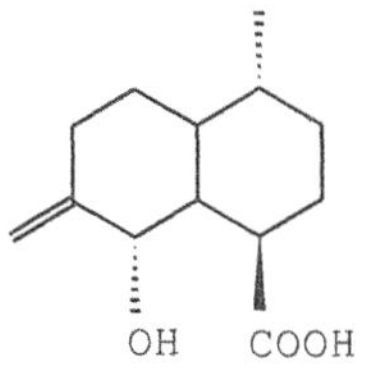

27

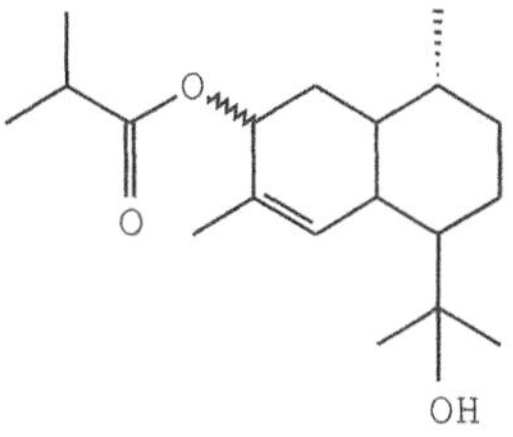

30

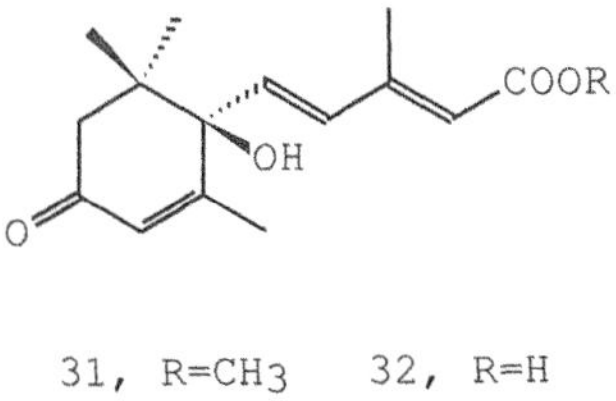

31, R=CH3   32, R=H

33, R=CH₃   34, R=H

The flavonoid glycosides quercetin-3-rutinoside, luteolin-7-O-glucoside and 3-O-glucosides of kaempferol, quercetin, patuletin and 6-methoxy kaempferol were identified from the aerial parts of *A. annua* (Marco *et al.*, 1990).

Coumarins

Methoxylated coumarin derivatives which have been isolated from the aerial parts of *A. annua* are: scopoletin (Saitbaeva *et al.*, 1970), scoparon, scopolin, 7-hydroxy-6,8-dimethoxycoumarin and 2,2-dihydroxy, 6-methoxychromone (Yang *et al.*, 1995). Coumarin (Liu and Li, 1980) and 2,2,6-trihydroxychromone (Yang *et al.*, 1994), are non-methoxylated coumarins.

Phenolic and Aromatic Compounds

5-nonadecylresorcinol-3-O-methyl ether (Brown, 1992), annphenone, a phenolic acetophenone (Singh *et al.*, 1997) and an aromatic plant growth regulating compound, *bis*-(1-hydroxy-2-methylpropyl) phthalate (Shukla *et al.*, 1991), have been reported from aerial parts of the plant.

Lipids and Aliphatic Compounds

n-Pentane and n-nonacosane (Ulubulen and Halfon, 1976), octacosanol (Tu *et al.*, 1985a), 2,29-dimethyltriacontane, *n*-triacontanyl triacontanoate, 2-methyltricosa-8-one-23-ol and nonacosanol (Bhakuni *et al.*, 1990), annuadiepoxide and ponticapoxide (Manns and Hartmann, 1992), are constituents reported from aerial parts of the plant.

BIOSYNTHESIS OF ARTEMISININ

Farnesyl pyrophosphate, the fundamental precursor of sesquiterpenes derived from mevalonic acid, (Akhila *et al.*, 1987), is transformed into artemisinic acid, **13**, one of the major constituents of *A. annua* through cyclisation and oxidation steps (Akhila *et al.*, 1990). Although the biosynthetic pathway of artemisinin, **1**, has not been completely established, there are two proposals:

The first is that arteannuin B, **3**, another cadinolide of *A. annua* and **1** are generated sequentially from artemisinic acid, **13**, (Akhila *et al.*, 1990; Nair and Basile, 1992; 1993). Brown *et al.* (1994), isolated 4,5-secocadinane, **25**, and dihydroxy-cadinanolide, **26**, from *A. annua*. Following the isolation of the above the latter authors suggested a plausible biosynthetic route for the conversion of arteannuin B into artemisinin (Scheme 1). The mechanism proposed is that the epoxide ring of arteannuin B is first cleaved to yield **26** which then undergoes Grob fragmentation to yield the enol form of **25**. The enol tautomer, **25a**, subsequently rearranges to give

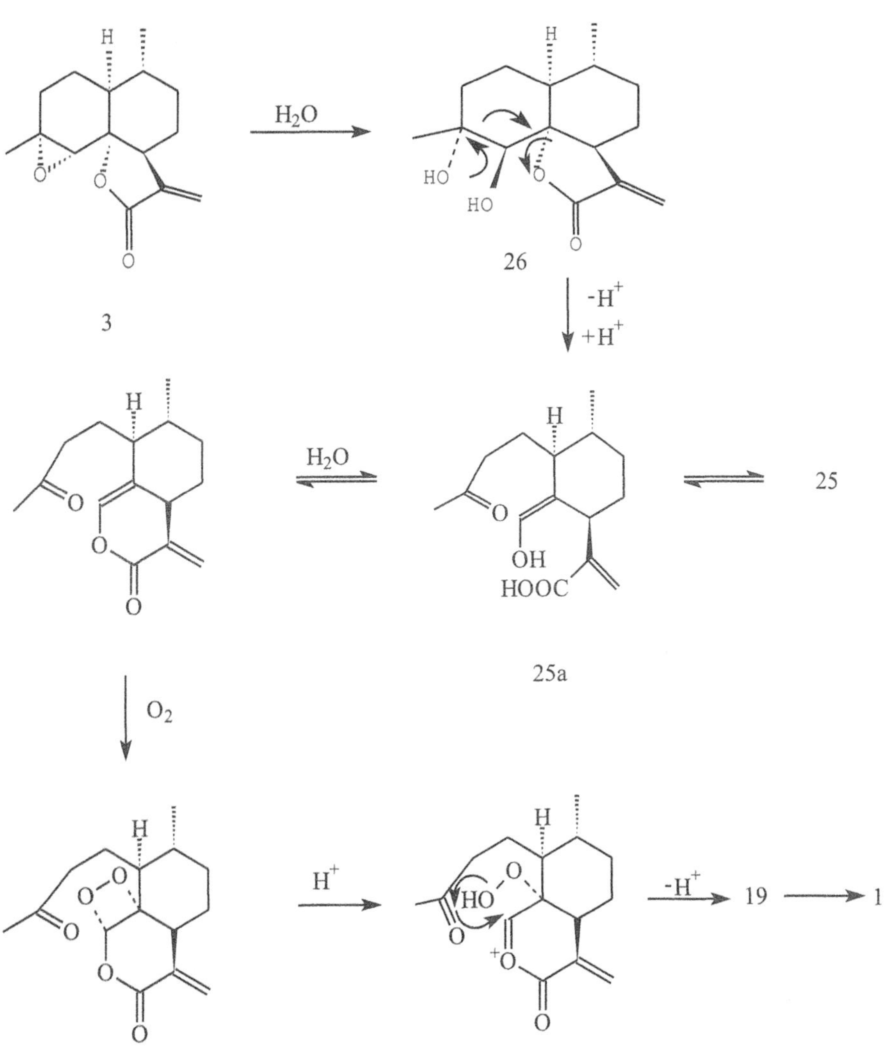

Scheme 1   Postulated biosynthesis of artemisinin from arteannuin B, 3.

the 1,2,4-trioxane system of **19** (artemisitene) *via* enzymic oxygenation at the enolic double bond (Acton and Klayman, 1985). Reduction of the 11–13 double bond completes the transformation to yield artemisinin; Woerdenbag *et al.* (1994), have reported the conversion of **19** into artemisinin. This proposed biosynthetic route is strongly supported by several partial or total chemical syntheses of artemisinin (Avery *et al.*, 1992; Ravindranathan *et al.*, 1990; Schmidt and Holfheinz, 1983; Xu *et al.*, 1986; Ye and Wu, 1990; Zhou and Xu, 1994).

The second view is that **1** and **3** are biosynthesised independently from artemisinic acid, **13**, (Sangwan *et al.*, 1993; Wang *et al.*, 1988, 1993). El-Feraly *et al.* (1986), converted **13** into **3**, whereas Kim and Kim, (1992), showed biotransformation of dihydroartemisinic acid **16**, into **1** enzymatically in tumour homogenate. The latter authors did not observe the enzymic conversion **13** into **16** *in vivo* but as **16** is a constituent of *A. annua* the plant would be expected to have the enzymic system required for this bioconversion. Since artemisinic acid, **13**, has been transformed chemically (non-enzymatically) into **1** *via* two steps, reduction and photo-oxidation by several workers, (Scheme 2) (Acton and Roth, 1992; Haynes and Vonwiller, 1990; Roth and Acton, 1989; 1991; 1991a), it is suggested that artemisinin, **1**, could be generated by this route *in vivo*; the latter authors have not reported the formation of arteannuin B, **3**, following the incorporation of **16** in tumour homogenate.

From the above it is not clear whether **1** is biosynthesised directly from **13** or through intermediate **3**. Thus, a detailed biosynthetic study is required to come to a definite conclusion.

ISOLATION AND EXTRACTION OF ARTEMISININ

Chinese workers have reported that 30 other species of *Artemisia* have been examined but none yield extracts with antimalarial activity (Anon., 1981). American workers have extracted 10 species of *Artemisia* but again, none (except *A. annua*) were found to contain artemisinin (Klayman *et al.*, 1984). In India, screening of 13 *Artemisia* species was carried out but only *A. annua* contained artemisinin (Balachandran *et al.*, 1987).

The Chinese literature does not provide details of isolation methods used to prepare artemisinin from *A. annua* but it does indicate that ethyl ether, petroleum ether and even gasoline have been used as solvents although extraction with hexane for several days at room temperature was also effective (Lou and Shen, 1987).

In the USA, *A. annua* extraction has been done using several low-boiling solvents such as dichloromethane, chloroform, ether and acetone; however, petroleum ether (30–60° C) was found to be the most selective and therefore considered to be the solvent of choice (Klayman *et al.*, 1984). The extract prepared at room temperature using dried, powdered plant material is filtered and submitted to evaporative crystallisation. The crude product is purified by chromatography on silica gel with chloroform-ethyl acetate as the eluate. Recrystallisation of artemisinin may be effected using cyclohexane (Klayman *et al.*, 1984) or 50% ethanol (Anon., 1979). In another report the plant material is extracted with ethanol; the extracts are filtered

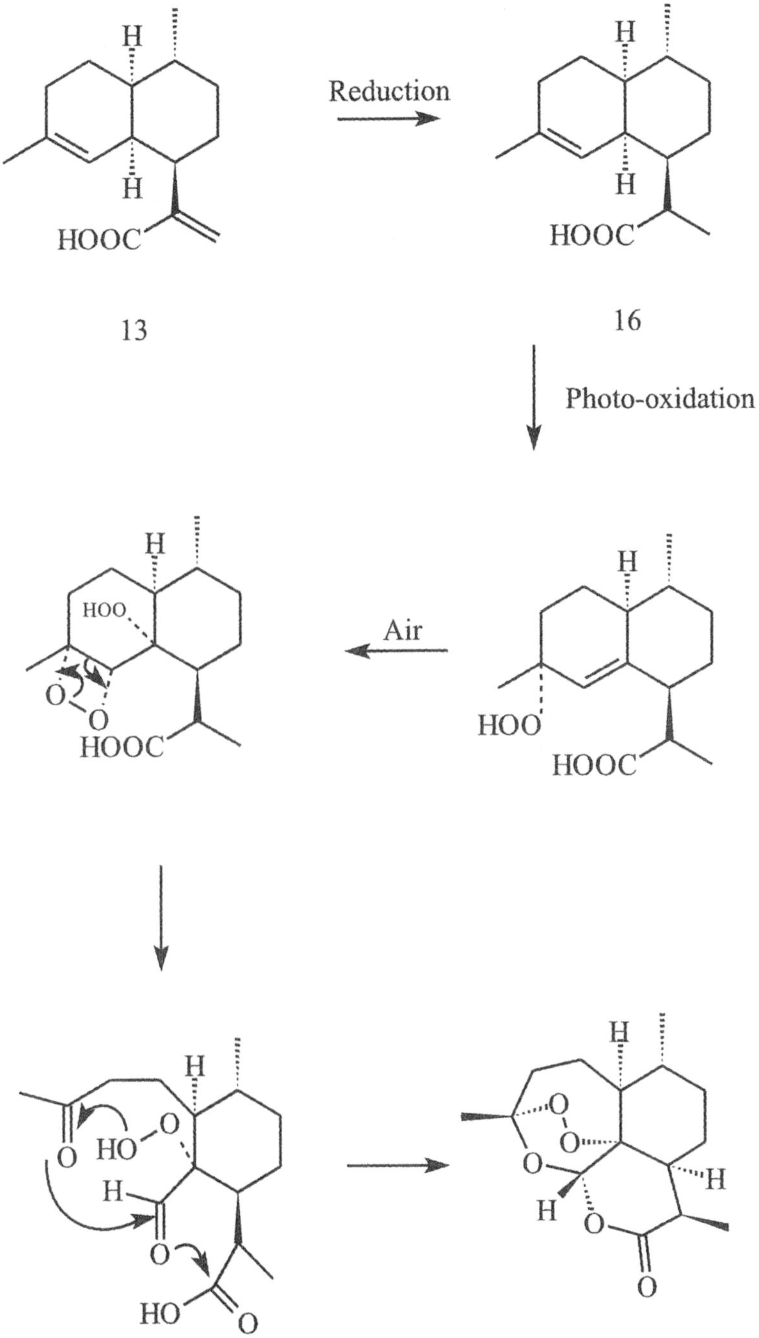

Scheme 2 Proposed mechanism of conversion of artemisinic acid, 13 into artemisinin.

and evaporated and the residue is extracted with hexane or cyclohexane; the extracts are then submitted to evaporative re-crystallisation. The amount of artemisinin recovered depends upon the efficiency of the extraction process and the initial concentration in the plant (Haynes and Vonwiller, 1994).

In an improved method, petroleum ether extracts were treated with a protic solvent to remove much of the accompanying waxes and then further purification of the extract over silica gel gave artemisinin (Klayman et al., 1984).

In India, air dried leaves and flowers of A. annua were extracted with petroleum ether or n-hexane. The concentrated residue was extracted with methanol/ethanol to decrease the bulk of the material which was then loaded onto a column of silica gel. Elution of the column with ethyl acetate/hexane afforded a fraction which on crystallisation from ethyl acetate/hexane gave pure needles of artemisinin (Thakur and Vishwakarma, 1990).

Very little information was found in the literature concerning the large scale isolation of artemisinin from the plant. One procedure reported involves the use of an Ito-multilayer separator extractor. In another large scale extraction procedure the hexane extract of dried leaves of A. annua was partitioned with aqueous acetonitrile; after concentration, the acetonitrile phase was chromatographed on a silica gel column. Artemisinic acid, 13, arteannuin B, 3, and two other sesquiterpenes were isolated (Elsohly et al., 1990).

The limited availability of artemisinin as well as the demand for more potent antimalarial drugs has necessitated the development of synthetic methods for the production of artemisinin. Several total syntheses of artemisinin have been reported Avery et al., 1992; Liu et al., 1993; Ravindranathan, 1994; Ravindranathan et al., 1990; Schmidt and Hofheinz, 1983; Xu et al., 1986; Zhou and Xu, 1994). However, all of these give low yields of the final product and are not an economically viable alternative to plant extraction for commercial scale production. In order to overcome the disadvantage of multistep organic synthetic reactions, the preparation of artemisinin from close biosynthetic precursors has been tested as an alternative. Artemisinin could easily be obtained from artemisinic acid in an overall yield of about 30–35% (Acton and Roth, 1992; Haynes and Vonwiller, 1991). Assuming that the abundance of artemisinic acid in the source is about ten times that of artemisinin as has been reported in several studies on Chinese A. annua material, the relative amount of artemisinin would be increased by four fold. At present, selection and cropping of plant material will be the method of choice for increasing the production of artemisinin. The use of plant cell and tissue cultures has not yielded fruitful results so far.

ARTEMISININ DERIVATIVES AND ANALOGUES

Dihydroartemisinin Derivatives

Since the endoperoxide group is an essential requirement for the antimalarial activity of artemisinin the main emphasis has been to prepare artemisinin derivatives and related compounds without disturbing the peroxy linkage. Reduction of 1 with sodium borohydride yields dihydroartemisinin, 35 in which the lactone function is converted into a lactol (hemiacetal) group (Jeremic *et al.*, 1973; Cao *et al.*, 1982), dihydroartemsinin is more potent than 1 against *P. falciparum in vitro* (Gu *et al.*, 1980). The lactol, 35, has been used to prepare several hundred derivatives such as ethers 35a (Li *et al.*, 1979; 1981), esters 35b, carbonates 35c, etc. The ethers 35a (Li *et al.*, 1979; 1981), have the advantage of being more oil soluble than 1 (Dutta *et al.*, 1987). Artemether, the β-methylether, 36, is the most active compound of all the ether derivatives. Arteethers, the α-isomer 37 and β-isomer 38 were prepared by Brossi and colleagues (Brossi *et al.*, 1988; Buchs and Brossi, 1989), and these are equipotent and 2–3 times more potent than 1. Later, arteethers 37 and 38 were prepared independently by El-Feraly *et al.* (1992) and Vishwakarma *et al.* (1992) respectively. (Note that the names artemether and arteether refer to the β-isomers unless otherwise stated). Not unexpectedly, the deoxy compound, 42, was found to be 100–300 times less potent *in vitro* than its peroxy precursor showing that, as with artemisinin the peroxy linkage is necessary for biological activity.

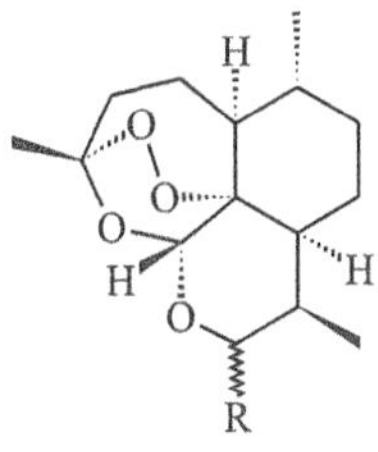

35, R = OH	37, R = α–OC$_2$H$_5$	41, R$_1$ = R$_3$ = H, R$_2$ = OH
35a, R = OAlkyl	38, R = β-OC$_2$H$_5$	42, R$_1$ = R$_3$ = H, R$_2$ = OC$_2$H$_5$
35b, R = OCOR	39, R = β-OCOCH$_2$CH$_2$COOH	43, R$_1$ = R$_2$ = H, R$_3$ = OC$_2$H$_5$
35c, R = OCOOR	40, R = β-OCH$_2$C$_6$H$_4$COOH	72, R$_1$ = OH, R$_2$ = OC$_2$H$_5$, R$_3$ = H
36, R = β–OCH$_3$		

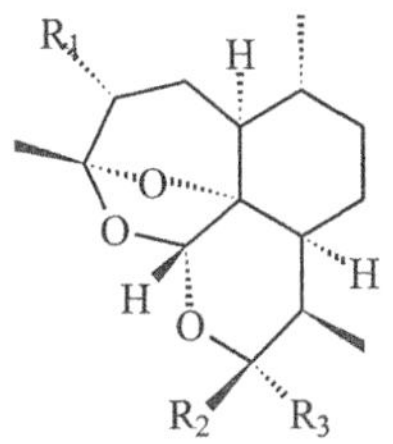

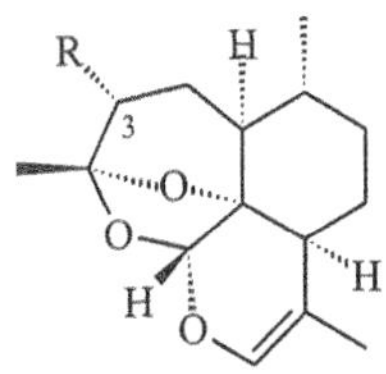

44, R = H
84, R = OH

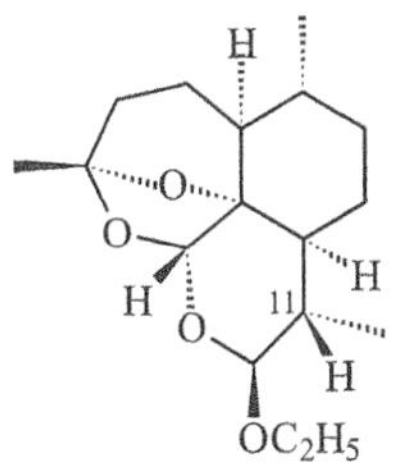

45

Deoxydihydroartemisinin **41**, deoxyethers **42** and **43**, and the enol ether, **44**, were converted into 11-*epi*-deoxyarteether, **45**, which again is much less potent than **1** (Hufford *et al.*, 1990; Xu *et al.*, 1989).

Two interesting derivatives, β-artesunic acid, **39**, the dihydroartemisinin half ester of succinic acid and β-artelinic acid, **40**, the β-*p*-carboxybenzyl ether of **35** have been prepared (Lin *et al.*, 1989; Liu, 1986; Vishwakarma, 1992). Sodium artesunate, the water soluble sodium salt of **39** administered intravenously in saline to mice was about 5.2 times more potent than **1** against both chloroquine-resistant and chloro-quine sensitive strains of *P. berghei* (Yang *et al.*, 1982; Zhao *et al.*, 1986); it is highly effective against cerebral malaria and although **39** is more toxic than **1**, it is less toxic than artemether, **36**. It is stable in an ampoule as a freeze dried powder but in about one week in solution at physiological pH it loses the succinate moiety through hydrolysis.

To solve this stability problem, Lin *et al.* (1987), prepared a series of stable water soluble derivatives of dihydroartemisinin, **35** in which the water solubilising car-boxyl group is on a moiety that is joined by an ether rather than an ester linkage. All the major condensation products are β-isomers. The esters **46** were generally found to be more active against the W-2 (Indochina) than the D-6 (Sierra Leone) strains of

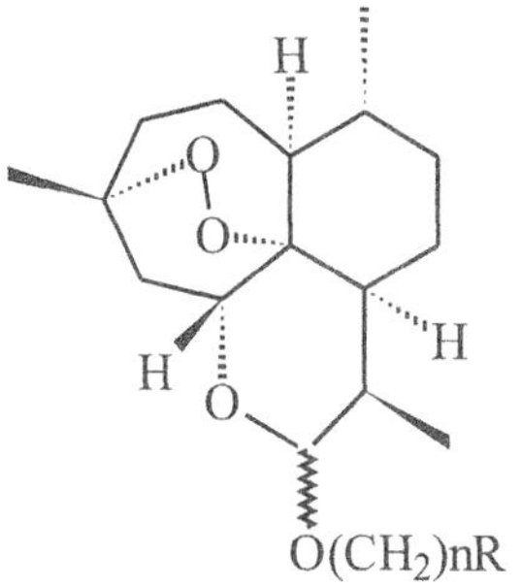

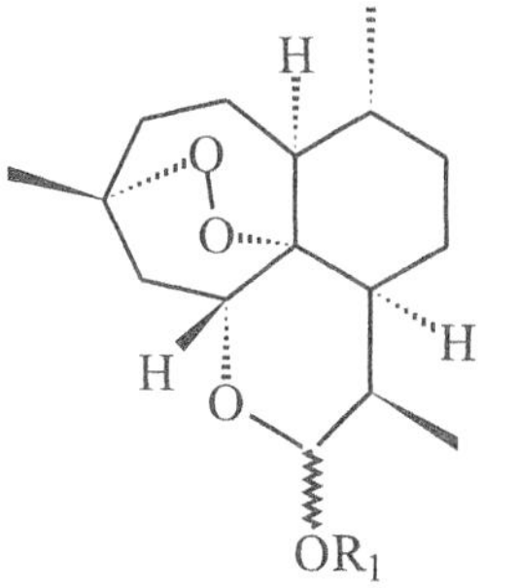

46,   R = COOCH$_3$, COOC$_2$H$_5$,

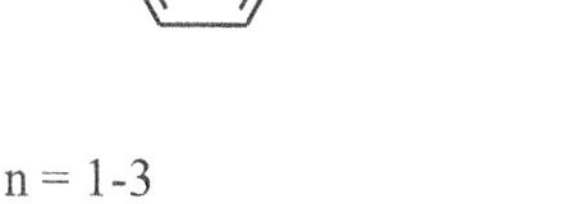

n = 1-3

47

a, R$_1$ = (R) CH$_2$CH(CH$_3$) COOCH$_3$

b, R$_1$ = (S) CH$_2$CH(CH$_3$) COOCH$_3$

c, R$_1$ = (S) CH(CH$_3$)CH$_2$ COOCH$_3$

d, R$_1$ = (R) CH(CH$_3$)CH$_2$ COOCH$_3$

P. falciparum. Esters **46** possessed activity comparable to that of the parent compounds **1** and **35**; however conversion of the esters to their corresponding carboxylates or free acids with the exception of artelinic acid, **40**, drastically decreases the potency in both parasite strains. Artelinic acid, **40**, which is both stable and water soluble exhibited superior *in vivo* activity against *P. berghei* than **1** (Lin *et al.*, 1987).

Lin and co-workers, (1987), also prepared a new series of hydrolytically stable and water soluble derivatives of lactol, **39**, (**47a-d**), with optically active side chains and examined the impact of the stereospecificity of the introduced alkyl side chain on biological properties. The ester derivatives were found to be more active than **1**, artemether, **36**, and arteether, **38**, whereas their corresponding free acids were found to be less active by 10–100 fold. So far the sodium salt of artelinic acid, **40**, remains the most active of the water soluble derivatives of dihydroartemisinin **35**.

Ring-Substituted Derivatives

A series of bromo – and heterocylic or aromatic amino analogues with good water solubility and high antimalarial activity were also prepared (Lin *et al.*, 1990). The lactol **35**, was converted into a key intermediate, 11,12-dehydrodihydroartemisinin, **48**, which is more potent than **1** and as active as **35** (Lin *et al.*, 1989). Compound **48** gave the dibromide, **49**, which led to several amino products; the 3′ fluoroaniline derivative, **50**, was the most active of this series and was more active than **1** against both *P. falciparum* W-2 and D-6 strains *in vitro*.

Lin and Miller, (1995) prepared a series of diastereomeric α-alkylbenzylic ethers from **1**. Artelinic acid, **40**, was used as the model molecule for the design of new analogues. All these compounds possessed at least equal or better *in vitro* activity against *P. falciparum* than **40**. The most active compound, **51**, of this class which has a nitro substituent, showed about 10, 20 and 40 times better antiplasmodial activity than artemether, artemisinin and artelinic acid respectively. Compounds with electron withdrawing groups (e.g. NO_2) have substantially increased activity. The S-diastereoisomers in general are several fold more potent than the corresponding R-isomers. Compounds having a methyl group substituted at the α-methylene group have weaker activity than compounds with a longer carboethoxyalkyl substituent, indicating that the lipophilicity and steric effects of these molecules play important roles in their antiplasmodial activity. This fact is further substantiated by the significantly weaker activity of the carboxylic acids than their corresponding esters.

A further series of stereoisomers of 4-(p-substituted phenyl)-4-(R or S)-[10,(α or β)-dihydroartemisinoxy]-butyric acids were 2–10-fold more potent than artemisinin against *P. falciparum in vitro* (Lin *et al.*, 1997). p-Chlorophenyl and p-bromophenyl derivatives **52** were superior to artelinic acid against *P. berghei* in mice whereas p-fluorophenyl and p-methoxyphenyl analogues were comparable in potency to artelinic acid. It is suggested that an electronic effect besides lipophilicity may play a role in determining the efficacy of this class of compound.

Avery *et al.* (1993), focused their work mainly on tetracyclic 1,2,4-trioxanes derived both from artemisinin (some already discussed), as well purely by synthetic

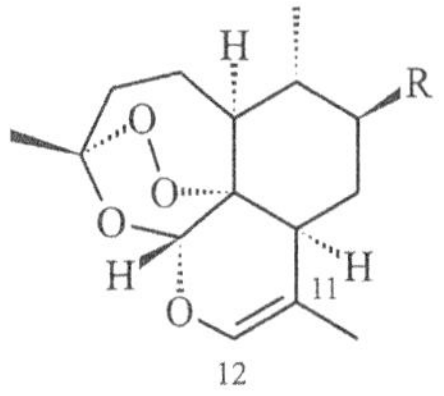

48, R = H
82, R = OH

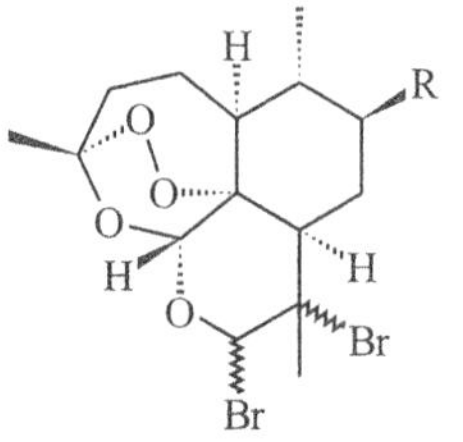

49

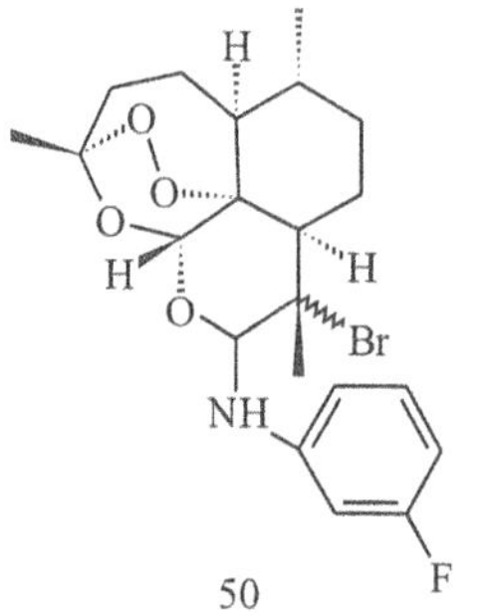

50

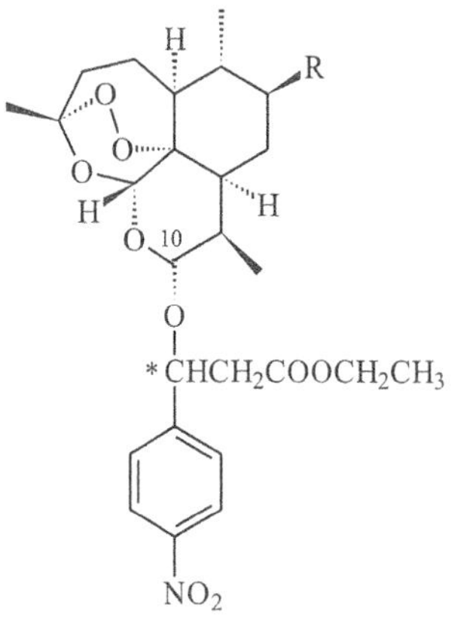

51

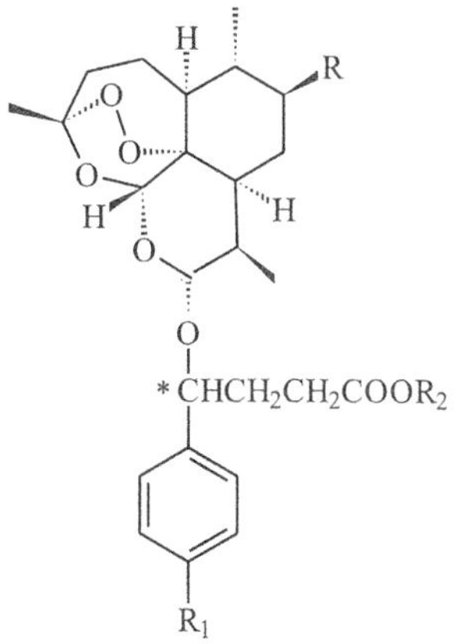

52

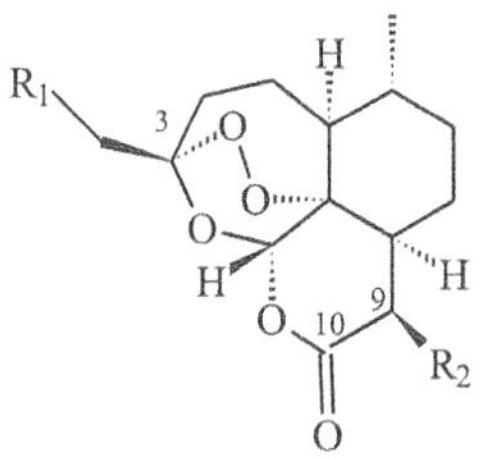

53, 54, R$_1$ = alkyl, arylalkyl
         or carboxyalkyl

53, R$_2$ = H
54, R$_2$ = n C$_4$H$_9$

routes (Avery *et al.*, 1996a). These authors have synthesised a series of active C-9β-substituted artemisinin analogues, 3-substituted (alkyl, arylalkyl- or carboxylalkyl)-9-desmethylartemisinin, **53** and 3-substituted, 9-*n*-butyldesmethylartemisinin derivatives **54** and tested them *in vitro* against the W-2 and D-6 strains of *P. falciparum*. Some of these analogues were much more active than **1**. QSAR and comparative molecular field analysis (CoMFA) of these analogues provided a model with a cross validated r^2=0.793.

Some other ether derivatives like brominated ethers **55**, and cyclisation products (Venogopalan *et al.*, 1993), dimeric ethers (Flippen-Anderson *et al.*, 1989; Galal *et al.*, 1996), a series of ether derivatives with iron chelators attached, (e.g. compound **56**) (Kamchonwongpaisan *et al.*, 1995) and fluorinated ethers (Pu *et al.*, 1995), have been prepared. These derivatives were evaluated for their antimalarial activities but none of them was found to be more active than **1**.

More recently, Nga *et al.* (1998), have prepared a series of fluoroalkyl ether derivatives of dihydroartemisinin. Although these compounds displayed only moderate *in vitro* activities against *P. falciparum*, several derivatives were more active *in vivo* than artemether. It is thought that the fluoroalkyl substituents slow metabolism to dihydroartemisinin which increases their half lives and reduces the incidence of parasite recrudescence. However, further studies are required to demonstrate that reduction of these compounds occurs *in vivo*. The latter authors also prepared 10α-(trifluoromethyl)-dihydroartemisinin which was found to be twice as potent as artemisinin against a multi-drug resistant strain of *P. falciparum* (W-2) *in vitro* and also *in vivo* against *P. berghei* in mice. 12α-(trifluoromethyl)-hydroartemisinin has been made from artemisinin and was found to be more active than the parent against *P. vinckei* in mice (Abouabdellah *et al.*, 1996).

Artemisinin has been transformed into the naturally occurring analogues arteannuin D, **5**, (Haynes and Vonwiller, 1996a; Lee *et al.*, 1989; Lin *et al.*, 1985), artemisitene, **19**, (El-Feraly *et al.*, 1990; Zhang and Zhou, 1988; 1989), and semisynthetic bromo-compounds (Acton and Klayman, 1987; Venogopalan and Bapat, 1993), the epoxide of **48** and several C-11 substituted artemisinins (Acton *et al.*, 1993; Kim *et al.*, 1993; Petrov and Ogyanow, 1991; Pu *et al.*, 1994) and artemisinin related derivatives (Haynes and Vonwiller, 1996a). None of these possessed better antimalarial activity than that of **1**. However, the acid degradation products **57** and **58** of **1** were found to be almost equipotent with **1** *in vitro* against *P. falciparum* (Imakura *et al.*, 1990).

The inactive natural product deoxyartemisinin, **18**, has been prepared from **1**, (Haynes and Vonwiller, 1996; Jeremic *et al.*, 1973). Analogues **5**, **18** and **19** have also been prepared from artemisinic acid, **13** (Haynes and Vonwiller, 1994; Jung *et al.*, 1986; Zhou *et al.*, 1989). Recently, **1** has been converted into **18** and related derivatives (Haynes and Vonwiller, 1996; Jefford *et al.*, 1996).

The antimalarial activities of D-ring enlarged oxides (Pu *et al.*, 1993), C-4 substituted (Rong and Wu, 1993), and silica gel catalysed rearrangement derivatives (Yagen *et al.*, 1994) of artemisinin have not been reported. Similarly, a semi-synthetic derivative of **1** arising from the decomposition of arteether (Idowu *et al.*, 1990), and several related derivatives prepared from **1** (Lin *et al.*, 1986; Shang *et al.*, 1989), carba-analogues (Ye and Wu, 1989), new analogues (Zhang *et al.*, 1988), many related derivatives synthesised from artemisinic acid, **13** (Vonwiller *et al.*, 1995) and a tricyclic trioxane, **59**, obtained from a naturally occurring cadinane (Zaman and Sharma, 1991), have apparently not been tested.

(+)-Deoxoartemisinin, **60**, which is 8-fold more active than **1** against *P. falciparum* *in vitro* and (−)-deoxodeoxyartemisinin which is inactive, have been synthesised from **1** and from artemisinic acid, **13**, (Jung *et al.*, 1989; 1990a; McCheshney and Jung, 1990). Later, the active derivative, **60**, was also prepared by Rong *et al.* (1993), from **1**, as well as by Ye and Wu, (1990), and from **13** by Lansbury and Nowak, (1992). Interestingly, although **60** is more active than **1** *in vitro*, it was only a little more active than **1** *in vivo* when tested in mice infected with *P. berghei*; at 640 mg/Kg both compounds cured 5/5 mice but at 340 mg/Kg none of the mice were cured by **1**, but 2/5 mice were cured by **60**, thus demonstrating that activity *in vivo* does not necessarily parallel that seen *in vitro*.

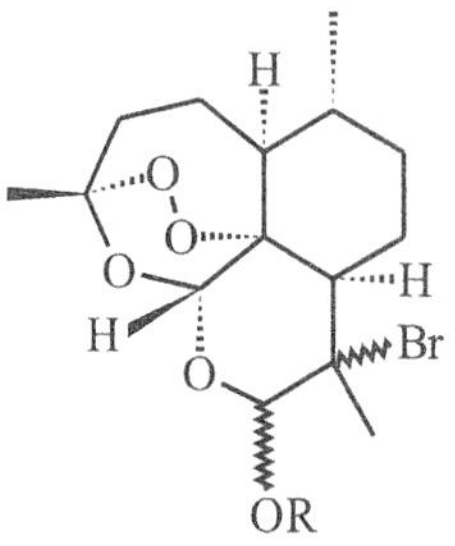

55,   R = C_2H_5; $CH_2\equiv CH$;

CH_2CH=CH_2

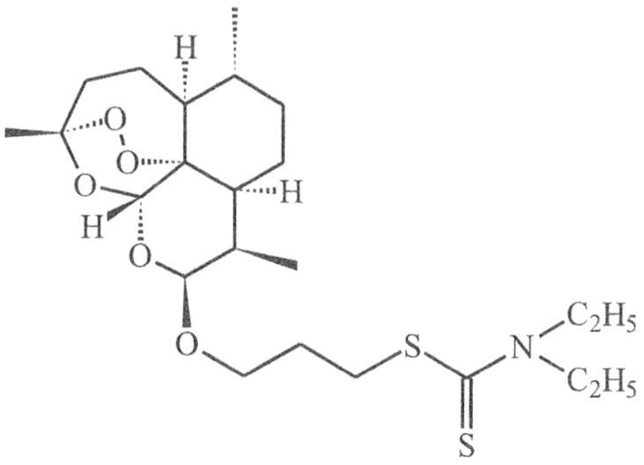

$CH_2CH=CH_2$

56

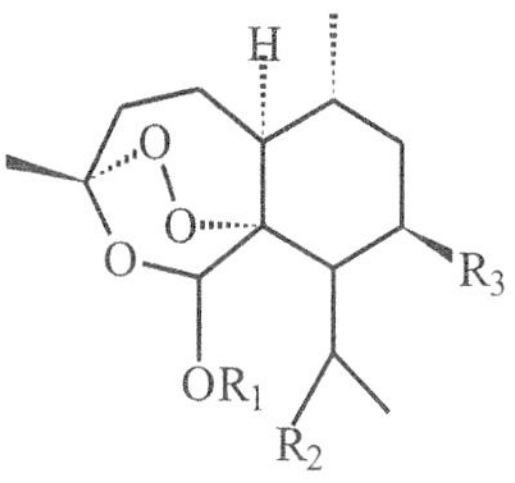

57, $R_1 = CH_3$; $R_2 = COOCH_3$; $R_3 = H$
58, $R_1 = C_2H_5$; $R_2 = COOC_2H_5$; $R_3 = H$
59, $R_1 = H$; $R_2 = CH_3$; $R_3 = OCOCH_3$

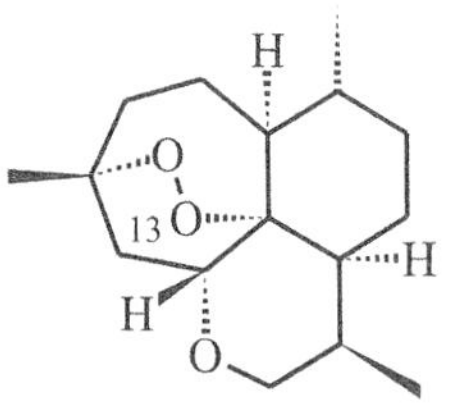

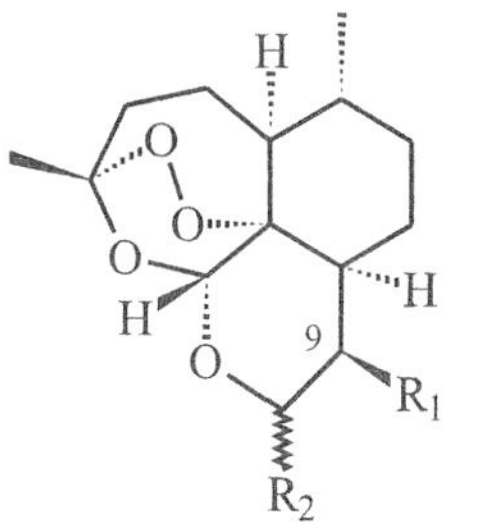

60, $R_1 = CH_3$; $R_2 = H$
62, $R_1 = CH_3$; $R_2 = \alpha\ C_3H_6OH$
63, $R_1 = CH_3$; $R_2 = \beta\ n\ C_3H_7$
64, $R_1 = n\ C_4H_9$; $R_2 = H$
65, $R_1 = n\ C_5H_{11}$; $R_2 = H$

66, $R_1 = CH_3$; $R_2 = \alpha$

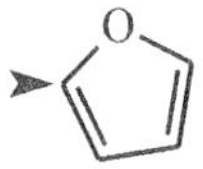

61

Among carba-analogues, some are active; the most active, 10-deoxo-13-carba-artemisinin, **61**, possessed reasonable antimalarial activity, but this was less active than (+)-deoxoartemisinin, **60**. The authors described QSAR of these carba-derivatives (Avery *et al.*, 1995a; 1996). While the peroxidic linkage is essential for antimalarial activity, the lactone carbonyl can be removed (Avery *et al.*, 1990; Jefford *et al.*, 1989; Jung *et al.*, 1990; Posner *et al.*, 1992), or a lactam ring can be substituted for the customary lactone without detriment to antimalarial activity (Avery *et al.*, 1993a; 1995; Torok and Ziffer, 1995). To explain the low activities or inactivities of the carba-analogues they have pointed out that the replacement of the non-peroxidic oxygen with a methylene group does not significantly change the shape of the resultant analogue as the artemisinin ring system is rigid in nature; the above confirms the importance of the 1,2,4-trioxane substructure within the artemisinin tetracycle for potent antimalarial activity. Bustos *et al.* (1989), synthesised (+)-homodeoxyartemisinin, a seven membered D-ring analogue of **60** from **13**. It possessed 20-fold less antimalarial activity than that of **1** suggesting that enlargement of the D ring decreases activity.

Jung *et al.* (1991), prepared the 12-(3′-hydroxy, *n*-propyl) deoxoartemisinins; the α-isomer, **62**, showed 5-fold greater *in vitro* activity against *P. falciparum*. Pu *et al.* (1995), reported the synthesis of 12β-allyldeoxoartemisinin derivatives from dihydroartemisinin. All compounds were tested *in vitro* against *P. falciparum* and the most active compound, 12β-propyldeoxoartemisinin, **63**, exhibited activity and toxicity comparable to that of arteether.

In their search for new antimalarial artemisinin derivatives Avery *et al.* (1996b), developed a new method of preparation for **60** in greater than 95% yield. The active derivative **60** was then converted into a series of novel 3- and 9-substituted 10-deoxoartemisinin analogues which were assayed against W-2 and D-6 clones of *P. falciparum*. Several of the analogues were much more active than **1** and artemether **36**. The butyl analogue, **64**, exhibited about 21-fold more antiplasmodial activity against D-6 than **1**.

The 9β-pentyl analogues of artemisinin, **65** and deoxoartemisinin have been prepared from artemisinic acid **13** by conjugate addition of *n*-butyl trimethylsilyl-methylcuprates followed by photo-oxidation and cyclisation (Vroman *et al.*, 1997). Antiplasmodial activities of the latter are not reported but in a preliminary *in vitro* assay **63** appeared to be less neurotoxic than dihydroartemisinin.

The synthesis and antiplasmodial activities of a series of hydrolysable C-10 carbon-substituted 10-deoxoartemisinin derivatives have been reported by Posner *et al.* (1999). These were prepared from artemisinin in three steps involving lactone reduction, replacement of the anomeric lactol hydrogen by fluorine using diethylaminosulfur trifluoride followed by boron trifluoride-promoted substitution of F by aryl, hetroaryl and acetylide nucleophiles. The most active compound *in vitro* was 10α-(2′-furyl)-10-deoxoartemisinin, **66**, with activity similar to that of artemisinin against *P. falciparum* chloroquine-sensitive strain N54; interestingly, the corresponding C-10β, (C-9α-methyl)-isomer was 35-fold less potent. The above compounds, together with the 2′-(5′-methyl)-furyl- and 2′-*N*-methylpyrrole artemisinin derivatives were found to be more potent than artemisinin *in vivo* against *P. berghei* in mice. The results of

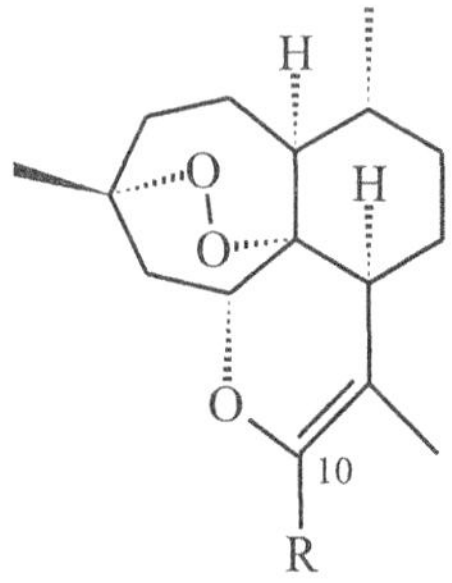

67, R = C(O)nBu
68, R = C(O)Ph
69, R = C(OH)Ph$_2$

preliminary *in vivo* acute toxicity tests indicate that the two furans are comparable to artemether in terms of toxicity in mice.

Recently, O'Dowd *et al.* (1999), have prepared a series of C-9,10 unsaturated C-10 carbon substituted heteroaryl-artemisinin analogues. These were prepared from artemisinin using selective organometallic and phosphorus ylide nucleophiles which added to the carbonyl while leaving the peroxide group untouched. Nine of the 11 derivatives made had *in vitro* antiplasmodial activities comparable to that of artemisinin. The most potent were **67–69** which were ~2–fold more potent than artemisinin.

A new class of novel N–substituted 11-aza-artemisinins have been synthesised from **1** by Torok and Ziffer (1995), and Torok *et al.* (1995). In *in vitro* and *in vivo* assays against drug resistant strains of malaria parasites several derivatives possessed equal or greater activity than **1**. The N-(2′-acetaldehydo-)-11-aza-artemisinin, **70**, the most active of the series was found to be 26-fold more active *in vitro* and 4-fold more active *in vivo* with respect to 1 (Torok *et al.*, 1995).

In order to study structure-activity relationships of artemisinin and the effect of heteroatom substitution at O-11, Avery *et al.* (1995), synthesised a novel class of artemisinin analogues, N-alkyl-11-aza-9-desmethylartemisinins. These amide ana-logues were tested *in vitro* against W-2 and D-6 strains of *P. falciparum*. These derivatives were comparable to **1** and in particular, the N-methyl-11-aza-9-desmethylartemisinin, **71**, was found to be 5-fold more active than **1** against W-2 strains of *P. falciparum*. Log P values determined for most of the analogues did not give an apparent correlation between log P and *in vitro* activity.

Artemisinin Derivatives from Microbial Transformations

Microbial metabolic studies of artemisinin and arteether have shown that a number of micro-organisms are capable of metabolising both of these compounds into

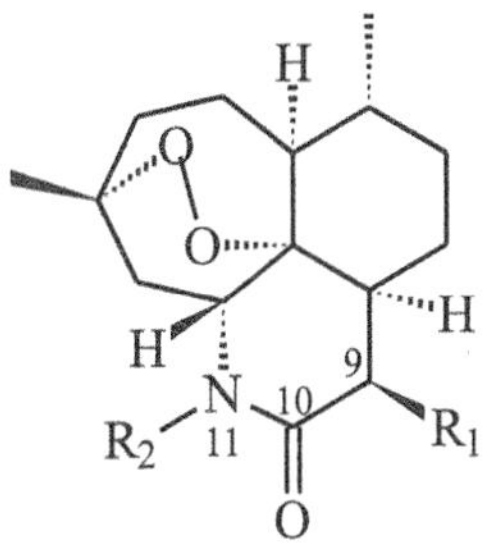

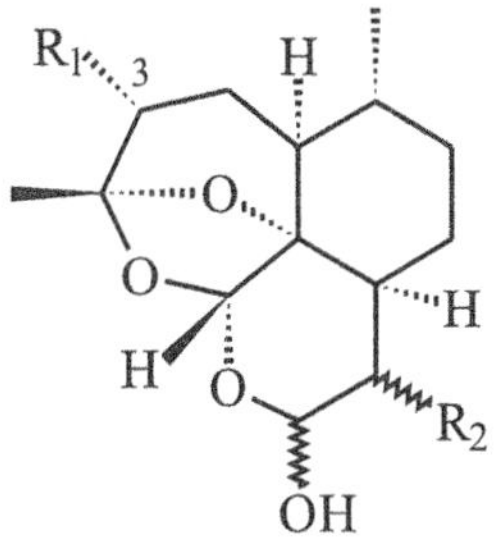

70, $R_1 = CH_3$; $R_2 = CH_2CHO$

71, $R_1 = H$; $R_2 = CH_3$

72, see under 43

73, $R_1 = OH$; $R_2 = \beta\text{–}CH_3$

74, $R_1 = H$; $R_2 = \beta\text{–}CH_3$

83, $R_1 = H$; $R_2 = \alpha\text{–}CH_3$

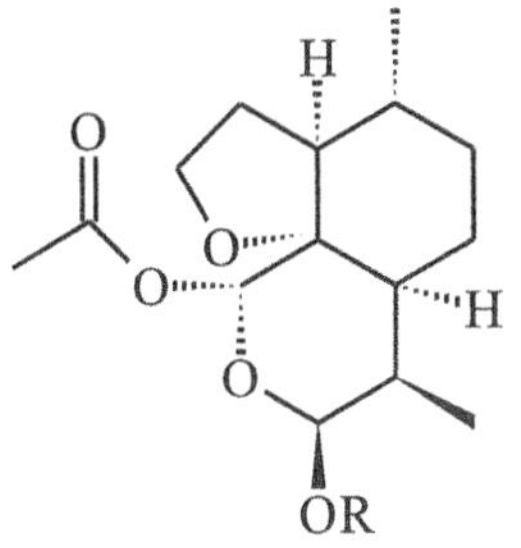

75, $R = C_2H_5$

76, $R = CONHC_6H_5$

various products of interest (Clark and Hufford, 1979; Garcia-Granados *et al.*, 1986; Hikino *et al.*, 1968; Lee *et al.*, 1989). Scale up fermentation of **1** with *Nocardia corallina* (ATCC 19070) and *Penicillium chrysogenum* (ATCC 9480) have resulted in two major microbial metabolites, deoxyartemisinin, **18**, and 3α-hydroxydeoxyartemisinin (arteannuin D), **5**, (Lee *et al.*, 1989).

Aspergillus niger (ATCC 10549) and *Nocardia corallina* (ATCC19070) metabolised arteether, **38**, into four microbial products, 3-α-hydroxyarteether, **72**, 3α-hydroxydeoxydihydroartemisinin, **73**, deoxydihydroartemisinin, **74**, and metabolite **75** (Lee *et al.*, 1990).

Hu and his co-workers, (1991), have reported the conversion of a phenylcarbamoyl derivative of dihydroartemisinin into product **76** and to the 14-hydroxymethyl derivative **77** by employing the fungus *Beauresia sulfurescens*.

Using the same organism, Hu *et al.* (1992), reported three microbial oxidation products, 14-hydroxymethylarteether, **78**, 9β-hydroxyarteether, **79**, and compound

72 from arteether, 38. Compounds 78 and 79 were oxidised chemically to aldehyde 80 and ketone 81, potentially valuable intermediates in their own right.

Khalifa *et al.* (1994), reported large scale fermentation of the semisynthetic active antimalarial derivative anhydrodihydroartemisinin, (11,12-dehydrodihydro-artemisinin), 48, with *Streptomyces lavendulae* L-105 and *Rhizopogon* species (ATCC 36060) and isolated four microbial products, 9β-hydroxyanhydro-dihydroartemisinin, 82, 11-*epi*- deoxydihydroartemisinin, 83, 3α-hydroxydeoxyan-hydrodihydroartemisinin, 84, and a 14-carbon re-arranged product, 85. *In vitro* antiplasmodial assays showed that only metabolite 82 possessed activity but this was lower than that of 48.

Derivatives Based on the 1,2,4-Trioxane Unit

The supply of artemisinin from plant and synthetic sources is currently inadequate to meet that required for the treatment of multi-drug resistant malaria caused by *P. falciparum*. Since the 1,2,4-trioxane unit in artemisinin is required for its antimalarial activity, considerable interest has been shown in the preparation of compounds based on this unit. The rational design and laboratory synthesis of structurally simpler analogues of tetracyclic artemisinin have led to a number of mono-, bi-tri- and tetracyclic 1,2,4-trioxanes, some of which have excellent antimalarial activity (Bunelle *et al.*, 1991; Haynes *et al.*, 1994; Imakura *et al.*, 1988; 1990; 1990a; Jefford *et al.*, 1988, 1988a; 1989; 1993; 1994; 1995; 1995a; 1996; Kepler *et al.*, 1988; Posner *et al.*, 1991; 1992; 1994; 1994a; 1995; 1995a; 1995b; 1995c).

The pioneering work of Posner *et al.* (1991; 1995) was based on synthetic 1,2,4-trioxanes containing the A, B and C portions of 1. Of 13 structurally simple tri-oxanes, 12 possessed considerable *in vitro* antiplasmodial activity (Posner *et al.*, 1991; 1995). The most active compounds, 86 and 87 were much more potent than similar trioxanes prepared by Jefford *et al.* (1989) and were as potent as 1 *in vitro* against the W-2 *P. falciparum* strain. Qualitative structure activity relationship, (QSAR) correlations suggested that structural variations at several different positions of these simple trioxanes can be made without undermining antimalarial activity.

To search for new and active 1,2,4-trioxanes, Posner *et. al.* (1992), prepared a series of 20 different benzylic esters and ethers of tricyclic trioxane primary alcohol 88 and evaluated them *in vitro* against *P. falciparum*. Many of these derivatives were highly active; carboxylate ester, 89, carbamate ester, 90, and sulfonate ester, 91, possessed antiplasmodial potency similar to that of 1; carboxylate esters 92 and 93, carbamate esters 94 and 95, and phosphate esters, 96–98, were found to be about 7-fold more active than 1. Analogues, 90, 92 and 95, which were active, stable, crystalline compounds are potentially important new drug candidates.

Preclinical *in vivo* evaluation of benzylic derivatives of 99 in *Aotus* monkeys infected with multidrug resistant (MDR) *P. falciparum* revealed trioxane 99 to be curative at a dose of 48 mg/kg without recrudescence after 6 months. This result matched that obtained with the clinically used arteether, 38, as a control (Posner *et al.*, 1994). On the basis of these promising findings Posner *et al.* (1995), designed

77, $R_1 = CH_2OH$; $R_2 = R_3 = H$; $R_4 = CONHC_6H_5$
78, CH_2OH; $R_2 = R_3 = H$; $R_4 = C_2H_5$
79, $R_1 = CH_3$; $R_2 = OH$; $R_3 = H$; $R_4 = C_2H_5$
80, $R_1 = CHO$; $R_2 = R_3 = H$; $R_4 = C_2H_5$
81, $R_1 = CH_3$; $R_2 = R_3 = O$; $R_4 = C_2H_5$
82, see under 48
83, see under 74
84 see under 44

85

86, $R = OCH_2C_6H_5$; $R_2 = H$
87, $R_1 = H$; $R_2 = OCH_2C_6H_5$
110, $R_1 = H$; $R_2 = OCH_3$

and synthesised a series of benzylic ethers and some related trioxane analogues as potential next generation antimalarials. The preparation of the potentially chelating derivative, 2-pyridyl system, 100, heteroaromatic carboxylates, 101 and 102, quinolinesulfonate 103, and water soluble quaternary ammonium salts 104 and 105 did not yield trioxanes of especially high *in vitro* antimalarial activity. Among these trioxanes, *p*-fluorobenzyl ether, 106, stands out as the most active and it is more active than 1 *in vitro* against *P. falciparum*, and also has considerable activity in

88,	R = H	94,	R = CON(C$_2$H$_5$)$_2$
89,	R = COCH$_2$NHCOOC$_4$H$_9$ (t)	95,	R = CON(C$_6$H$_5$)$_2$
90,	R = CON(CH$_3$)$_2$	96,	R = PO(OC$_6$H$_5$)$_2$
91,	R = S(O$_2$)C$_6$H$_4$CH$_3$ (p)	97,	R = PO(OC$_2$H$_5$)$_2$
92,	R = COC$_6$H$_4$COOCH$_3$ (p)	98,	R = PS(OC$_2$H$_5$)$_2$
93,	R = COC$_6$H$_4$CON(C$_2$H$_5$)$_4$ (p)	99,	R = CH$_2$C$_6$H$_5$

mice infected with *P. berghei*; in addition, **106** was found to be 10-fold more active than 1 in killing immature *P. falciparum* gametocytes.

To study the molecular mechanisms of action (discussed later in chapter 13) of antimalarial trioxane analogues of **1** with ferrous ion, Posner *et. al.* (1994; 1995a; 1995b; 1996), have designed and synthesised a number of 3- and 4-substituted tricyclic 1,2,4-trioxanes and evaluated their antimalarial activities. Trioxanes **107** and **108** possessed comparable activity to that of **1**.

Jefford *et al.* (1993), have synthesised bicyclic and tricyclic 1,2,4-trioxanes containing the ABC and ACD ring portions of 1. The ABC ring containing trioxane **109** and ACD ring containing derivative **110** were as active as **1** *in vitro* against chloroquine-sensitive and chloroquine-resisitant *P. falciparum* strains.

Several tricyclic seco-artemisinins containing ABC (Avery *et al.*, 1990), and ACD (Avery *et al.*, 1990a; 1993; 1994) rings of 1 and tri- and tetracyclic carba-artemisinin analogues have also been reported. All the seco-analogues retain antimalarial activity and the most active analogue, (−)-5-nor-4, 5-seco-artemisinin, **111**, was as active as **1** *in vitro* against the D-6 strain of *P. falciparum*.

Zouhiri *et al.* (1998) have prepared tricyclic artemisinin analogues bearing a C5α methyl group; these were devoid of antiplasmodial activity which is consistent with the hypothesis that tight binding between haem and the α- face of the artemisinin molecule is necessary for activity.

Among a series of sulfide and sulphone 1,2,4-trioxanes, the sulphones were found to have higher *in vitro* activities against malaria parasites but the most potent was three-fold less active than artemisinin (Posner *et al.*, 1998a).

The preparation of a number of tetracyclic and trioxane dimers has also been reported and several of these compounds have potent *in vitro* antiplasmodial and cytotoxic properties (Posner *et al.*, 1997).

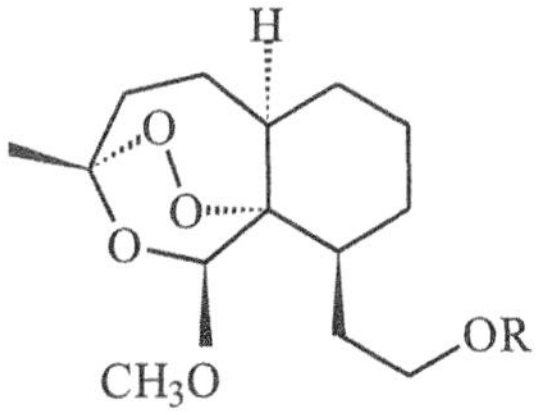

100, R = —CH₂— (2-pyridyl)

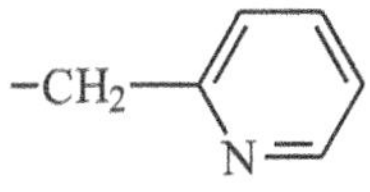

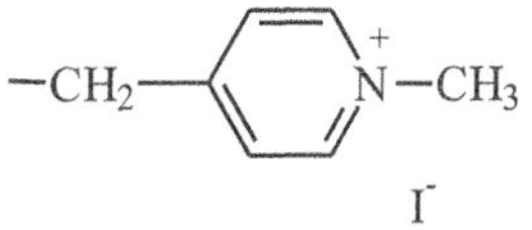

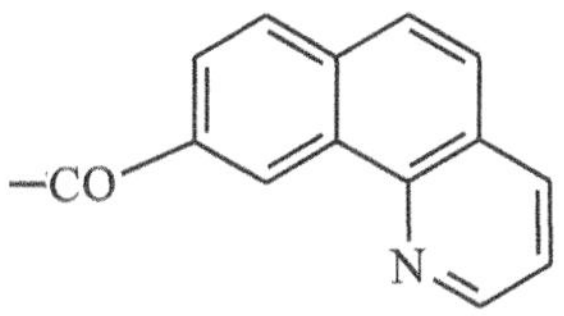

101, R= —CO— (benzo[h]quinolinyl)

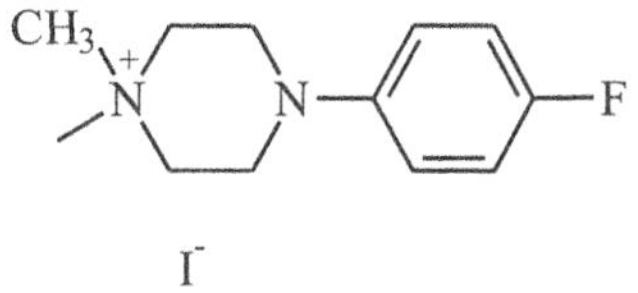

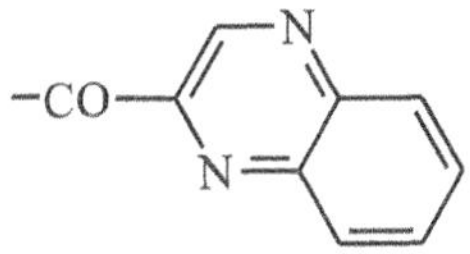

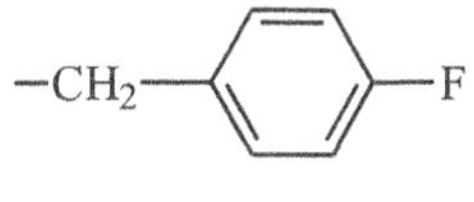

102, R= —CO— (quinoxalinyl)

103, R= —SO₂— (8-quinolinyl)

104, R= —CH₂— (N-methylpyridinium) I⁻

105, R= (N-methyl-piperazinyl)—(4-fluorophenyl) I⁻

106, R= —CH₂— (4-fluorophenyl)

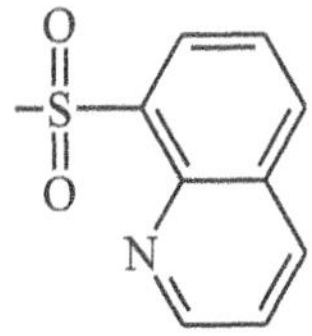

On the basis of a proposed mechanistic understanding of the mode of action of artemisinin Posner *et al.* (1998) synthesised a series of (tricyclic) 3-aryl, 1,2,4-trioxanes. The most active were four 12β-methoxy-3-aryltrioxanes (**112–115**). While the *in vitro* antiplasmodial activities were up to 3-fold less than that of artemisinin, two of the compounds (**113, 114**) were up to twice as potent as artemisinin when given orally to *P. berghei* infected mice. The 3-aryl substitution in these compounds was designed to promote the formation of cytotoxic high-valent iron-oxo species which are postulated to be formed following the reaction of artemisinin derivatives with haem in malaria infected red blood cells. Another mechanism-based strategy involved the synthesis of 4-(hydroxyalkyl)trioxanes which were designed to enhance the formation of C4 radicals

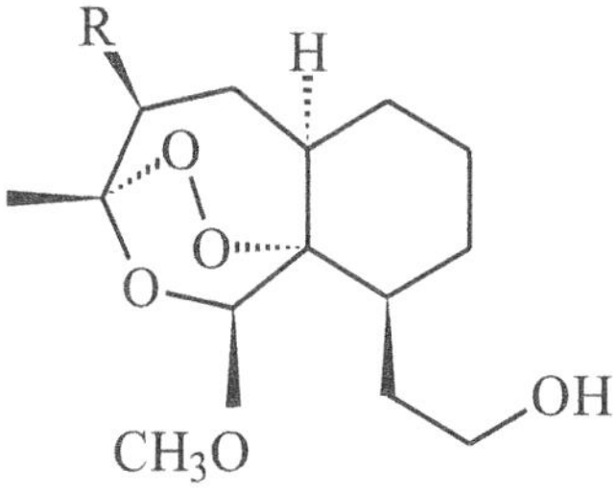

107, R = CH$_3$

108, R = CH$_2$C$_6$H$_5$

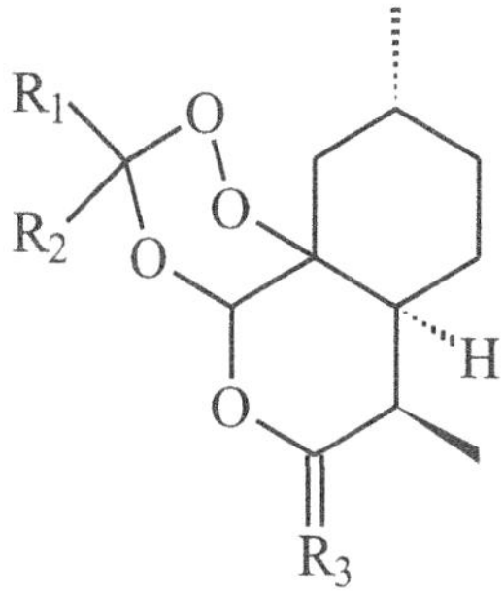

109, R$_1$, R$_2$ = (cyclopentane) R$_3$ = H$_2$

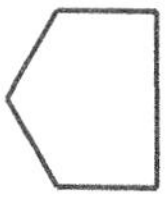

110, see under 87

111, R$_1$ = R$_2$ = CH$_3$; R$_3$ =O

by 1,5 hydrogen atom abstraction and which are postulated to be important in the antimalarial action of artemisinin-like compounds (Cumming *et al.*, 1998). The *in vitro* antiplasmodial activities of these derivatives supported the above hypothesis and the most active compounds (e.g. **116**) were of comparable potency to artemisinin. For a full discussion of the mechanisms involved, see chapter 13.

Singh, (1990), and Singh *et al.* (1995), have reported the synthesis of a series of a new class of active monocyclic 1,2,4-trioxanes **117–120**. The antiplasmodial IC$_{50}$ of trioxanes **117–119** ranges from 2.86 to 222 mg/ml with respect to an IC$_{50}$ of 0.65 mg/ml for 1 against chloroquine resistant strains of *P. falciparum*. However, several of the derivatives **120** possessed promising blood schizontocidal activity against *P. berghei* in mice.

Two synthetically racemic bicyclic *cis*-fused cyclopentene 1,2,4-trioxanes, **121** and **122** displayed high antimalarial activities which are commensurate with those of 1 and arteether, **38** (Peters *et al.*, 1993). Two major *exo*-1,2-diols for each racemate were prepared by Jefford *et al.* (1994). These diols are very useful as new chiral ligands for catalysis. Later, the above authors prepared the enantiomerically pure isomers of both **121** and **122** (Jefford *et al.*, 1995). Jefford, (1996), has also reported a series of 1,2,4-trioxanes, **123**, for the potential treatment of tropical diseases including malaria.

A series of more than 20 cyclic peroxy ketals have also been prepared and assessed for *in vitro* antiplasmodial activities (Posner *et al.*, 1998b). Seven com-

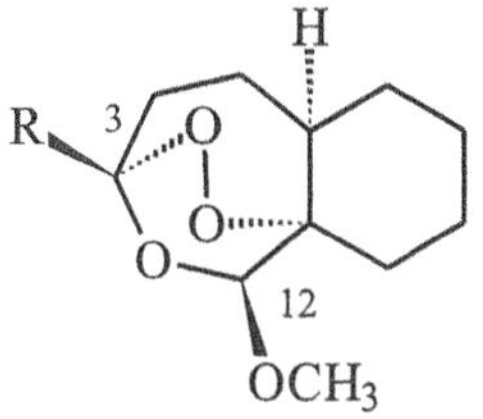

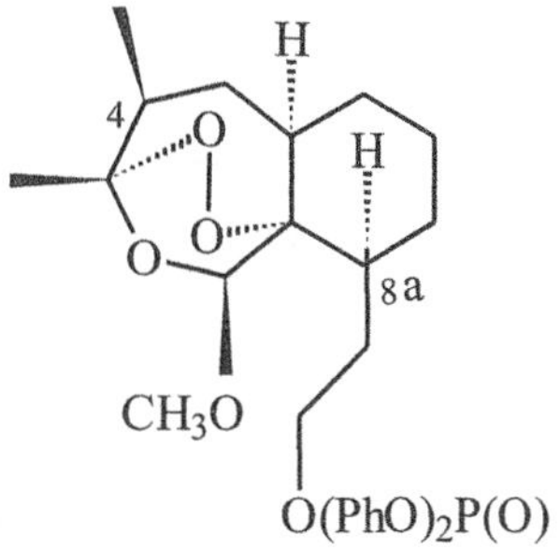

112,   R = p FPh
113,   R = p CH_3OCH_2Ph
114,   R = p (p' $FPhCH_2OCH_2$)Ph
115,   R = p $CH_3C(O)OCH_2Ph$

116

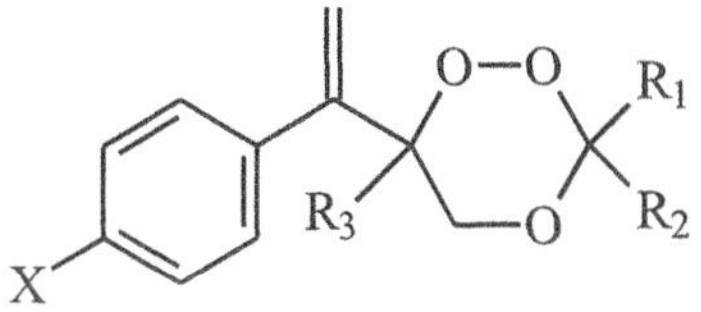

117, $R_1 = R_2 = CH_3$; $R_3 = H$

118, $R_1, R_2 =$ ⬡ ; $R_3 = H$

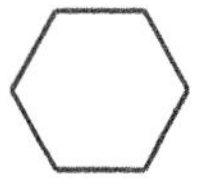

119, $R_1, R_2 =$ ⬡—OCH_3 ; $R_3 = H$

120, $R_1, R_2 =$ ⬡ , ⬡ ; $R_3 = H, OCH_3$

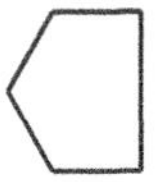

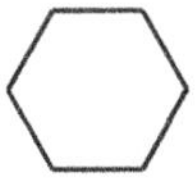

X = H, CH_3, OCH_3, F, Cl

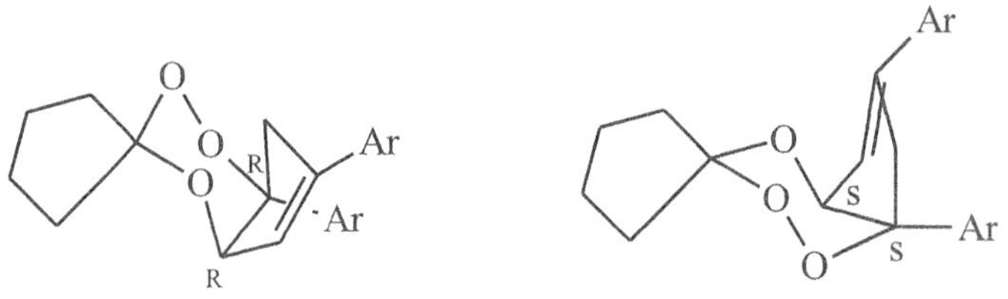

121,   Ar = C_6H_5

122,   Ar = p C_6H_4

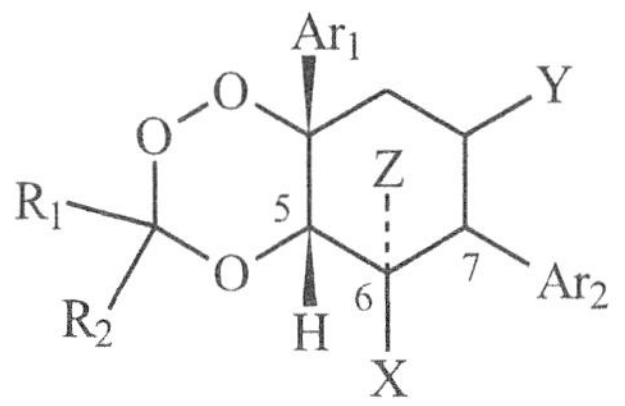

123

R_1, R_2, = linear or branched akyl/alicyclic with
one or more O, S, or N atom

Ar_1, Ar_2, = aromatic

Z = epoxide or Δ 5, 6 or 6, 7

X, Y, = group containing O, N, or S

pounds were found to have activity between one quarter and one tenth of that of artemisinin, the most active being **124**; halogen substituents on the aromatic ring appear to enhance activity.

Vennerstrom *et al.* (1992) reported the synthesis of a dispiro-1,2,4,5-tetraoxane, **125**, which had *in vitro* antiplasmodial activity comparable to that of artemisinin. However, this compound performed poorly *in vivo* when given orally. In an attempt to improve its oral bioavailablity a series of analogues bearing unsaturated and polar functional groups were prepared but although some of these retained *in vitro* antiplasmodial activities their *in vivo* potencies were poor; the reasons for the lack of correlation between *in vitro* and *in vivo* activities are not known (Dong *et al.*, 1999).

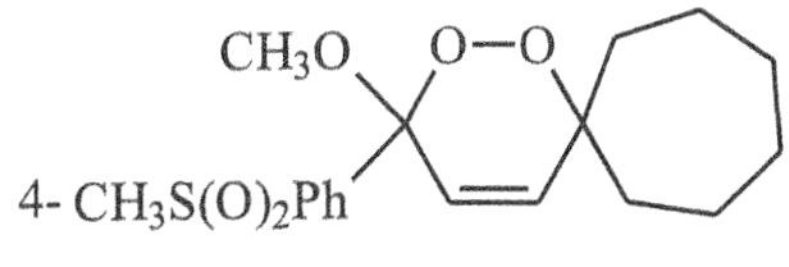

CH₃O corrected: CH_3O O—O

4- $CH_3S(O)_2Ph$

124

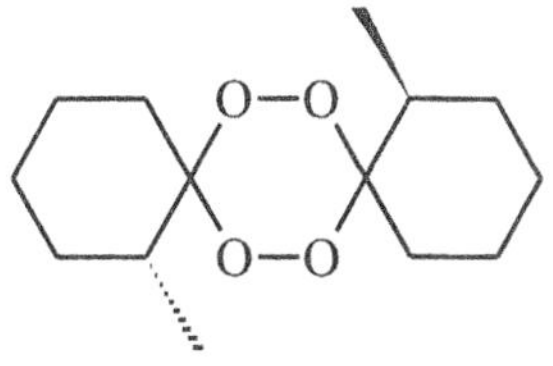

125

CONCLUSION

As a result of the discovery of the highly active antimalarial artemisinin, consider-able effort has been made both to study the phytochemistry of *A. annua* and to syn-thesise compounds based on artemisinin which are more suitable for clinical use. To date, the ether derivative of dihydroartemisinin, artemether and the ester artesunate have become important clinically used antimalarials while arteether and artelinate are currently under development. These derivatives depend upon plant-derived artemisinin for their production and it is to be hoped that it will be possible for them to be produced in countries where they are most needed. However, a number of highly potent, easily synthesised antimalarial compounds which are based on the 1,2,4-trioxane ring have also been prepared and it is anticipated that one or more of these will prove to be clinically useful antimalarial agents.

REFERENCES

Abouabdellah, A., Begue, J.P., Bonnet-Delpon, D., Gantier, J.C., Nga, T.T.T. and Thac, T.D. (1996) Synthesis and *in vivo* antimalarial activity of 12α-trifluoromethyl-dihydroartemisinin. *Bioorg. Med. Chem. Lett.*, 6, 2717–2720.

Acton, N. and Roth, R.J. (1992) On the conversion of dihydroartemisinic acid into artemisinin. *J. Org. Chem.*, 57, 3610–3614.

Acton, N., Karle, J.M. and Miller, R.E. (1993) Synthesis and antimalarial activity of some 9-substituted artemisinin derivatives. *J. Med. Chem.*, 36, 2552–2557.

Acton, N. and Klayman, D.L. (1985) Artemisinin, a new sesquiterpene lactone endoperoxide from *Artemisia annua*. *Planta Med.*, 51, 441–442.

Acton, N. and Klayman, D.L. (1987) Conversion of artemisinin (qinghaosu) to *iso*-artemisitene and to 9-*epi*-artemisinin. *Planta Med.*, 53, 266–268.

Agrawal, P.K. and Bisnoi, V. (1996) Sterols and taraxastane derivatives from *Artemisia annua* and a rational approach based on ¹³C corrected: ^{13}C NMR for the identification of skeletal type of amor-phane sesquiterpene. *Ind. J. Chem.*, 35B, 86–88.

Ahmad, A. and Mishra, L.N. (1994) Terpenoids from *Artemisia annua* and constituents of its essential oil. *Phytochem.*, 37, 183–186.

Akhila, A., Thakur, R.S., and Popli, S.P. (1987) Biosynthesis of artemisinin in *Artemisia annua*. *Phytochem.*, 26, 1927–1930.

Akhila, A., Rani, K. and Thakur, R.S. (1990) Biosynthesis of artemisinic acid in *Artemisia annua*. *Phytochem.*, **29**, 2129–2132.

Anon. (1977) Co-ordinating Group of Research on the structure of Qing Hau Sau. A new type of sesquiterpene lactone-Qing Hau Sau. K'O Hsueh Tung Pao, **22**, 142. (*Chem. Abs.* 1977, **87**, 98788).

Anon. (1979) Qinghaosu Antimalarial Co-ordinating Research Group. Antimalarial studies on Qinghaosu, *Artemisia annua*. *Chin. Med. J.*, **92**, 811–816.

Anon. (1981) Fourth meeting of the Scientific Working Group on the Chemotherapy of Malaria, 1981. Beijing, People's Republic of China; WHO report TDR/CHEMAL SWO QHS/81, **3**, 5.

Avery, M.A., Jennings-White, C. and Chong, W.K.M. (1987) Total synthesis of (+)-artemisinin and (+)-9-desmethylartemisinin. *Tet. Lett.*, **28**, 4629–4632.

Avery, M.A., Chong, W.K.M. and Bupp, J.E. (1990) Tricyclic analogues of artemisinin: Synthesis and antimalarial activity of (+)-4,5-secoartemisinin and (−)-5-nor-4,5-secoartemisinin. *J. Chem. Soc. Chem. Comm.*, 1487–1489.

Avery, M.A., Chong, W.K.M. and Detre, G. (1990a) Synthesis of (+)-8a,9-secoartemisinin and related analogues. *Tet. Lett.*, **31**, 1799–1802.

Avery, M.A., Chong, W.K.M. and Jennings-White, C. (1992) Stereoselective total synthesis of (+)-artemisinin, the antimalarial constituent of *Artemisia annua*. *J. Am. Chem. Soc.*, **114**, 974–979.

Avery, M.A., Gao, F., Chong, W.K.M, Mehrotra, S. and Milhous, W.K. (1993) Structure activity relationships of the antimalarial agent artemisinin. 1. Synthesis and comparative molecular field analysis of C-9 analogues of artemisinin and 10-deoxoartemisinin. *J. Med. Chem.*, **36**, 4264–4275.

Avery, M.A., Gao, F., Mehrotra, S., Chong, W.K.M. and Jennings-White, C. (1993a) The organic and medicinal chemistry of artemisinin and analogues. *Research Trends Trivendrum, India*, 413–468.

Avery, M.A., Gao, F., Chong, W.K.M., Hendrickson, T.F., Inman, W.D. and Crews, P. (1994) Synthesis, conformational analysis, and antimalarial activity of tricyclic analogues of artemisinin. *Tetrahedron*, **50**, 957–972.

Avery, M.A., Bonk, J.D., Chong, W.K.M., Mehrotra, S., Miller, R., Milhous, W.K, Goins, D.K., Venkatesan, S., Wyandt, C., Khan, I. and Avery, B.A. (1995) Structure-activity relationships of the antimalarial agent artemisinin. 2. Effect of heteroatom substitution at O-11: Synthesis and bioassay of N-alkyl-11-aza-9-desmethylartemisinins. *J. Med. Chem.*, **38**, 5038–5044.

Avery, M.A., Fan, P., Karle, J.M., Miller, R. and Goins, K. (1995a) Replacement of the non-peroxide trioxane oxygen atom of artemisinin by carbon: Total synthesis of (+)-13-carba-artemisinin and related structures. *Tet. Lett.*, **36**, 3965–3968.

Avery, M.A., Fan, P., Karle, S.M., Bonk, J.D., Miller, R., and Goins, D.K. (1996) Structure-activity relationships of the antimalarial agent artemisinin. 3. Total synthesis of (+)13-carba-artemisinin and related tetra and tricyclic structures. *J. Med. Chem.*, **39**, 1885–1897.

Avery, M.A., Mehrotra, S., Bonk, J.D., Vroman, S.A., Goins, D.K. and Miller, R. (1996a) Structure activity relationships of the antimalarial agent artemisinin. 4. Effect of substitution at C-3. *J. Med. Chem.*, **39**, 2900–2906.

Avery, M.A., Mehrotra, S., Johnson, T.L., Bonk, J.D., Vroman, J.A. and Miller, R. (1996b) Structure activity relationships of the antimalarial agent artemisinin. 5. Analogues of 10-deoxoartemisinin substituted at C-3 and C-9. *J. Med. Chem.*, **39**, 4149–4155.

Baera, R.T., Nabin Zade, L.I., Zapesos Chinaya, G.C. and Karryer, M.O. (1988) Flavonoids of *Artemisia annua*. *Khim. Prir Soldin*, 289–290.

Balachandran, S., Vishwakarma, R.A. and Popli, S.P. (1987) Chemical investigation of some *Artemisia* species: search for artemisinin or other related sesquiterpene lactones with a peroxide bridge. *Ind. J. Pharm. Sci.*, **49**, 152–154.

Bardoloi, M., Shukla, V.S., Nath, S.C. and Sharma, R.P. (1989) Naturally occurring cadinanes. *Phytochem.*, **28**, 2007–2037.

Bhakuni, R.S., Jain, D.C., Shukla, Y.N. and Thakur, R.S. (1990) Lipid constituents from *Artemisia annua. J. Ind. Chem. Soc.*, **67**, 1004–1006.

Bhardwaj, D.K., Jain, R.K., Jain, S.C. and Manchanda, C.K. (1985) Constituents of *Artemisia annua*: flavones. *Proc. Ind. Nat. Sci. Acad.*, **51**, 741–745.

Blasko, G., Cordell, G.A., Lankin, D.C. (1988) Definitive ¹H and ¹³C NMR assignments of artemisinin (qinghaosu). *J. Nat. Prod.*, **51**, 1273–1276.

Bolt, A.J.N., Cocker, W. and McMurry, T.B.H. (1963) The stereochemistry of artemisin. *J. Chem. Soc.*, 5235–5238.

Brossi, A., Venogopalan, B., Gerpe, L.D., Yeh, H.J.C., Flippen-Anderson, J.L., Buchs, P. *et al.* (1988) Arteether, a new antimalarial drug: Synthesis and medicinal properties. *J. Med. Chem.*, **31**, 645–650.

Brown, G.D. (1992) Two new compounds from *Artemisia annua. J. Nat. Prod.*, **55**, 1756–1760.

Brown, G.D. (1993) Annulide, a sesquiterpene lactone from *Artemisia annua. Phytochem.*, **32**, 391–393.

Brown, G.D. (1994) Cadinanes from *Artemisia annua* that may be intermediates in the biosynthesis of artemisinin. *Phytochem.*, **16**, 637–641.

Bunnelle, W.H., Isbell, T.A., Barnes, C.L. and Qualls, S. (1991) Cationic ring expansion of an ozonide to a 1,2,4-trioxane. *J. Am. Chem. Soc.*, **113**, 8168–8169.

Bustos, D.A., Jung, M., Elsohly, H.N. and McChesney, J.D. (1989) Stereospecific synthesis of (+)-homodeoxoartemisinin. *Heterocyc.*, **29**, 2273–2277.

Buchs, P. and Brossi, A. (1989) Synthesis of artemisinin lactol derivatives as antimalarials. European patent application EP 330520 (Cl.C07D493/18) (Chem Abs, 1990, **112**, 77631V).

Cao, M.Z., Hu, S.C., Li, M.H. and Zhang, S. (1982) Synthesis of carboxylic ester of dihydroartemisinin. *Nanjing Yaoxueyun Xuebao*, **1**, 53–54.

Clark, A.M. and Hufford, G.D. (1979) Microbial transformation of the sesquiterpene lactone – Qing Hao Sau. K'O Hsueh Tung Pao, **22**, 142 (*Chem. Abs.* 1977, **87**, 98788).

Cumming, J.N., Wang, D., Park, S.B., Shapiro, T.A. and Posner, G.H. (1998) Design, synthesis, derivatization and structure-activity relationships of simplified, tricyclic, 1,2,4-trioxane alcohol analogues of the antimalarial artemisinin. *J. Med. Chem.*, **41**, 952–964.

Deng, D.A., Zhu, D.Y., Jao, Y.L., Dai, Z.Y., and Xu, R.S. (1981) Studies on the structure of artemisinic acid. *Kexue Tongbao*, **26**, 1209–1211.

Djermanovic, M., Jokic, A., Mladenooic, S. and Stefanooic, M. (1975) A new flavanol from *Artemisia annua* L. *Phytochem.*, **14**, 1873.

Dong, Y., Matile, H., Chollet, J., Kaminsky, R., Wood, J.K. and Vennerstrom, J.L. (1999) Synthesis and antimalarial activity of 11 dispiro-1,2,4,5-tetraoxane analogues of WR 148999. 7,8,15,16-tetraoxadispiro [5.2.5.2]hexadecanes substituted at the 1 and 10 positions with unsaturated and polar functional groups. *J. Med. Chem.*, **42**, 1477–1480.

Dutta, G.P., Bajpai, R. and Vishwakarma, R.A. (1987) Blood schizontocidal activity of artemisinin (Qinghaosu) and new antimalarial arteether against *Plasmodium berghei. Ind. J. Parasitol.*, **11**, 253–257.

El-Feraly, F.S., Al-Meshal, I.A., Al-Zhaya, M.A. and Hifnaway, M.S. (1986) On the possible role of qinghao acid in the biosynthesis of artemisinin. *Phytochem.*, **25**, 2777–2778.

El-Feraly, F.S., Al-Moshas and Khalifa, S. (1989) *Epi*-deoxyarteannuin B and 6,7-dehydroartemisinic acid from *Artemisia annua*. *J. Nat. Prod.*, **52**, 196–198.

El-Feraly, F.S., Ayalp, A. and Al-Yahaya, M.A. (1990) Conversion of artemisinin to artemisitene. *J. Nat. Prod.*, **53**, 66–71.

El-Feraly, F.S., Al-Yahya, M.A., Orabi, K.Y., McPhail, D.R. and McPhail, A.T. (1992) A new method for the preparation of arteether and its C-9 epimer. *J. Nat. Prod.*, **53**, 66–71.

Elsohly, H.N., Croom, E.M. Jr., El-Feraly, F.S. and El-Sheril, M.M. (1990) A large scale extraction technique of artemisinin from *Artemisia annua*. *J. Nat. Prod.*, **53**, 1560–1564.

Flippen-Anderson, L., George, C., Gilard, R., Yu, Q., Dominguez, L. and Brossi, A. (1989) Structure of an ether dimer of deoxydihydroqinghaosu, a potential metabolite of the anti-malarial arteether. *Acta Cryst.*, **45**, 292–294.

Galal, A.M., Ahmad, M.S., El-Feraly, F.S. and McPhail, A.T. (1996) Preparation and charac-terisation of a new artemisinin derived dimer. *J. Nat. Prod.*, **59**, 917–920.

Garcia-Granados, A., Martinez, A., Onorato, M.E., Arias, and J.M. (1986) Microbial trans-formation of tetracyclic diterpenes; conversion of *ent*-kaurenones by *Aspergillus niger*. *J. Nat. Prod.*, **49**, 126–132.

Gu, H.M., Lu, B.F. and Qu, Z.X. (1980) Antimalarial activity of 25 derivatives of artemisinin against chloroquine-resistant *Plasmodium berghei*. *Chung Kuo Yao Li Hsueh Pao*, **1**, 48–50.

Guo, Q.Z. (1994) Cytotoxic terpenoids and flavonoids from *Artemisia annua*. *Planta Med.*, **60**, 54–57.

Haynes, R.K. and Vonwiller, S.C. (1990) Catalysed oxygenation of allylic hydroperoxides derived from qinghao (artemisinic) acid. Conversion of qinghao acid into dehydroqing-haosu (artemisitene) and qinghaosu (artemisinin). *J. Chem. Soc. Chem. Commun.*, 451–453.

Haynes, R.K. and Vonwiller, S.C. (1991) The development of new peroxidic antimalarials. *Chem. Austral.*, 64–67.

Haynes, R.K. and Vonwiller, S.C. (1994) Extraction of artemisinin and artemisinic acid: preparation of artemether and new analogues. *Trans. Roy. Soc. Trop. Med. Hyg.*, **88**, 23–26.

Haynes, R.K., King, G.R. and Vonwiller, S.C. (1994) Preparation of a bicyclic analogue of qinghao (artemisinic) acid *via* a Lewis acid catalysed ionic Diels-Alder reaction involving a hydroxy diene and cyclic enone and facile conversion into (±)-6,9-Desdimethyl qinghaosu. *J. Org. Chem.*, **59**, 4743–4748.

Haynes, R.K. and Vonwiller, S.C. (1996) The behaviour of qinghaosu (artemisinin) in the presence of heme iron II and III. *Tet. Lett.*, **37**, 253–256.

Haynes, R.K. and Vonwiller, S.C. (1996a) The behaviour of qinghaosu (artemisinin) in the presence of non-heme iron II and III. *Tet. Lett.*, **37**, 257–260.

Hethelyi, E.B., Cseko, I.B., Grosy, M., Mark, G. and Palinkas, J.J. (1995) Chemical composi-tion of *Artemisia annua* essential oils from Hungary. *J. Ess. Oil. Res.*, **7**, 45–48.

Hikino, H., Tokuoka, Y., Hikino, Y. and Takemoto, T. (1968) (–)-Deoxodesoxyartemisinin. *Tetrahedron*, **24**, 3147.

Hu, Y., Highet, R.J., Marison, D. and Ziffer, H. (1991) Microbial hydroxylation of a dihydroartemisinin derivative. *J. Chem. Soc. Chem. Commun.*, 1176–1177.

Hu, Y., Ziffer, H., Li, G. and Yeh, H.J.C. (1992) Microbial oxidation of the antimalarial drug arteether. *Bioorg. Chem.*, **20**, 148–154.

Huang, J., Xia, Z. and Wu, L. (1987) Constituents of *Artemisia annua* L. 1. Isolation and identification of 11R(–)-dihydroarteannuic acid. *Huaxue Xuebao*, **45**, 609–612.

Huang, J., Nicholis, K.M., Cheng, C. and Wang, Y. (1988) Two dimensional NMR studies of arteannuin. *Huaxue Xuebao*, **45**, 305–308.

Hufford, C.D., Lee, I.S., El-Sohly, H.N., Chi, T.H. and Baker, J.K. (1990) Structure elucidation and thermospray high performance liquid chromatography/mass spectroscopy (HPLC/MS) of microbial and mammalian metabolites of the antimalarial arteether. *Pharmaceut. Res.*, **7**, 923–927.

Idowu, O.R., Moggs, J.L., Word, S.A. and Edward, G. (1990) Decomposition reactions of arteether, a semisynthetic derivative of qinghaosu (artemisinin). *Tetrahedron*, **46**, 1871–1884.

Imakura, Y., Yokoi, T., Yamagiashi, T., Koyama, J., Hu, H., McPhail, D.R. *et al.* (1988) Synthesis of desethanoqinghaosu, a novel analogue of the antimalarial qinghaosu. *J. Chem. Soc. Chem. Comm.*, 372–374.

Imakura, Y., Hachiya, K., Ikemoto, T., Yamashita, S., Kiharia, M., Kobayashi, S. *et al.* (1990) Acid degeneration products of qinghaosu and their structure activity relationships. *Heterocycles*, **31**, 1011–1016.

Imakura, Y., Hachiya, K., Ikemoto, T., Kobayash, S. Yamashita, S., Sakakibara, J. *et al.* (1990a) Antimalarial artemisinin analogues: Synthesis of 2,3-desethano-12-deoxoartemisinin related compounds. *Heterocycles*, **31**, 2125–2129.

Jefford, C.W., McGoran, E.C., Boukouvalas, J., Richardson, G., Robinson, B.L. and Peters, W. (1988) Synthesis of new 1,2,4-trioxanes and their antimalarial activity. *Helv. Chim. Acta.*, **71**, 1805–1812.

Jefford, C.W., Wang, Y. and Bernardinelli, G. (1988a) Mechanistic and synthetic studies on the formation of 1,2,4-trioxanes related to arteannuin. Photo-oxygenation of a bicyclic dihydropyran. *Helv. Chim. Acta.*, **71**, 2042–2052.

Jefford, C.W., Velarde, J.A. and Bernardinelli, G. (1989) Synthesis of tricyclic arteannuin like compounds. *Tet. Lett.*, **30**, 4485–4488.

Jefford, C.W., Velarde, J.A. and Bernardinelli, G., Bray, D.H., Warhurst, D.C. and Milhous, W.K. (1993) Synthesis, structure and antimalarial activity of tricyclic 1,2,4-trioxanes related to artemisinin. *Helv. Chim. Acta.*, **76**, 2275–2288.

Jefford, C.W., Misra, D., Dishington, A.P., Timari, G., Rossier, J.C. and Bernardinelli, G. (1994) The osmium catalysed asymmetric dihydroxylation of *cis*-fused cyclopenteno-1,2,4-trioxanes. *Tet. Lett.*, **35**, 6275–6278.

Jefford, C.W., Favarger, F., Vicente, M.G.H. and Jaquier, Y. (1995) The decomposition of *cis*-fused cyclopenteno-1,2,4-trioxanes induced by ferrous salts and some oxophilic reagents. *Helv. Chim. Acta.*, **78**, 452–458.

Jefford, C.W., Kohmoto, S., Jaggi, D., Timari, G., Rossier, J.C., Rudaz, M. *et al.* (1995a) Synthesis, structure, and antimalarial acivity of some enantiomerically pure *cis*-fused cyclopenteno-1,2,4-trioxanes. *Helv. Chim. Acta.*, **78**, 647–662.

Jefford, C.W. (1996) Preparation of 1,2,4-trioxane derivatives for the treatment of tropical diseases incuding malaria. US patent, US5559145 (Cl.514–452, A61K31/335), 24 Sept., 1996; (*Chem. Abs.*, **125**, P317312f).

Jefford, C.W., Vicente, M.G.H., Jaquier, Y., Favarger, F., Mareda, J. *et al.* (1996) The deoxygenation and isomerisation of artemisinin and artemether and their relevance to antimalarial action. *Helv. Chim. Acta.*, **79**, 1475–1487.

Jeremic, D., Jokic, A., Behbud, A. and Stefanovic, M. (1973) New type of sesquiterpene lactone isolated from *Artemisia annua*. *Tet. Lett.*, **32**, 3039–3042.

Jeremic, D., Stefanovic, M., Dokonic, D. and Milosavlyevic, S. (1980) Flavonols from *Artemisia annua*. *Glas Khem Drush, Feofger*, **44**, 615–618.

Jung, M., Elsohly, H.N., McPhail, A.T. and McPhail, D.R. (1986) Practical conversion of artemisinic acid into deoxyartemisinin. *J. Org. Chem.*, **51**, 5417–5419.

Jung, M., Li, X., Bustos, D., Elsohly, H.N. and McChesney, J.D. (1989) A short and stereoselective synthesis of (+)-deoxoartemisinin and (–)-deoxodesoxyartemisinin. *Tet. Lett.*, **30**, 5973–5976.

Jung, M., Bustos, D.A., Elsohly, H.N. and McChesney, J.D. (1990) A concise and stereoselective synthesis of (+)-12-*n*-butyldeoxoartemisinin. *Synlett.*, 743–744.

Jung, M., Li, X., Bustos, D.A., Elsohly, H.N. and McChesney, J.D. Milhous, W.K. (1990a) Synthesis and antimalarial activity of (+)-deoxoartemisinin. *J. Med. Chem.*, **33**, 1516–1518.

Jung, M., Yu, D., Bustos, D., Elsohly, H.N. and McChesney, J.D. (1991) A concise synthesis of 12-(3′-hydroxy-*n*-propyl)-deoxoartemisinin. *Bioorg. Med. Chem. Lett.*, **1**, 741–744.

Kamchonwongpaisan, S., Paityatat, S., Thebtraranonth, Y., Wilairat, P. and Yuthavong (1995) Mechanism based development of new antimalarials: Synthesis of derivatives of artemisinin attached to iron chelators. *J. Med. Chem.*, **38**, 2311–2316.

Kasymov, Sh.Z., Ovezdurdyev, A., Yusupov, M.I., Shamyanov, I.D. and Malikov, V.M. (1986) Lactones of *Artemisia annua*. Khim Prir Soedin 5, 636 (*Chem. Abs.*, **106**, 64344C).

Kepler, J.A., Phillip, A., Lee, Y.M., Morey, M.C. and Corroll, F.I. (1988) 1,2,4-Trioxanes as potential antimalarial agents. *J. Med. Chem.*, **31**, 713–716.

Khalifa, S.I., Baker, J.K., Rogers, R.D., El-Feraly, F.S. and Hufford, C.D. (1994) Microbial and mammalian metabolism studies of the semisynthetic antimalarial anhydrodihydroartemisinin. *Pharmaceut. Res.*, **11**, 990–994.

Kim, S.U. and Lim, H.J. (1989) Isolation of arteannuic acid from *Artemisia annua*. Han'guk Nonghwa Hakhoechi, **32**, 178–179.

Kim, N.C. and Kim, S.U. (1992) Biosynthesis of artemisinin from 11,12-dihydroarteannuic acid. *J. Korean Agric. Chem. Soc.*, **35**, 106–109.

Kim, S.U., Choi, J.H. and Kim, S.S. (1993) Preparation of a cyclopropane containing analogue of artemisinin. *J. Nat. Prod.*, **56**, 857–863.

Klayman, D.L., Lin, A.J., Acton, N., Scovill, J.P., Hoch, J.M., Milhous, W.K. *et al.* (1984) Isolation of artemisinin (qinghaosu) from *Artemisia annua* growing in the United States. *J. Nat. Prod.*, **47**, 715–717.

Klayman, D.L. (1985) Qinghaosu (artemisinin): An antimalarial drug from China. *Science*, **228**, 1049–1055.

Lansbury, P.T. and Nowak, D.M. (1992) An efficient partial synthesis of (+)-artemisinin and (+)-deoxoartemisinin. *Tet. Lett.*, **33**, 1029–1032.

Lawerence, B.M. (1982) Progress in essential oils. *Perfumer and Flavourist*, 7, 43–44.

Leban, I., Gorlic, L. and Japeli, M. (1988) Crystal and molecular structure of qinghaosu, a re-determination. *Acta Pharma. Jugosl.*, **38**, 71–77.

Lee, I.S., Elsohly, H.N., Crown, E.M. and Hufford, C.D. (1989) Microbial metabolism studies of the antimalarial sesquiterpene artemisinin. *J. Nat. Prod.*, **52**, 337–341.

Lee, I.S., Elsohly, H.N. and Hufford, C.D. (1990) Microbial metabolism studies of the antimalarial drug arteether. *Pharmaceut. Res.*, **7**, 199–203.

Leppard, D.G., Rey, M. and Dreiding, A.S. (1974) The structure of arteannuin B and its acid hydrolysis product. *Helv. Chim. Acta.*, **57**, 602–615.

Libby, L.M. and Sturtz, (1989) Unusual essential oils grown in Oregon, II *Artemisia annua*. *J. Ess. Oil Res.*, **1**, 201–202.

Li, Y., Yu, P.I., Chen, Y.X., Li, L.Q., Gai, Y.Z., Wang, D.S. *et al.* (1979) Synthesis of some artemisinin derivatives. *K'O Hsueh T'ung Pao*, **24**, 667–669.

Li, Y., Yu, P.I., Chen, Y.X., Li, L.Q., Gai, Y.Z., Wang, D.S. *et al.* (1981) Studies on artemisinin analogues. Synthesis of ethers, carboxylates and carbonates of dehydroartemisinin. *Acta. Pharmaceut. Sinica*, **16**, 429–439.

Lin, A.J., Klayman, D.L., Hoch, J.M., Silveron, J.V. and George, C.F. (1985) Thermal rearrangement and decomposition products of artemisinin. *J. Org. Chem.*, 50, 4504–4508.

Lin, A.J., Theoharides, A.D. and Klayman, D.L. (1986) Thermal decomposition products of dihydroartemisinin. *Tetrahedron*, 42, 2181–2184.

Lin, A.J., Klayman, D.L. and Milhous, W.K. (1987) Antimalarial activity of new water soluble dihydroartemisinin derivatives. *J. Med. Chem.*, 30, 2147–2150.

Lin, A.J., Lee, M. and Klayman, D.L. (1989) Antimalarial activity of new water soluble dihydroartemisinin derivatives. 2. Stereospecificity of the ether side chain. *J. Med. Chem.*, 32, 1249–1252.

Lin, A.J., Li, L-Q., Klayman, D.L., George, C.F. and Flippen-Anderson, J.L. (1990) Antimalarial activity of new water soluble dihydroartemisinin derivatives: Aromatic amine analogues. *J. Med. Chem.*, 32, 2610–2614.

Lin, A.J. and Miller, R.E. (1995) Antimalarial activity of new dihydroartemisinin derivatives. 6. α-Alkylbenzylic ethers. *J. Med. Chem.*, 38, 764–770.

Lin, A.J., Zikry, A.B. and Kyle, D.E. (1997) Antimalarial activity of new dihydroartemisinin derivatives. 7. 4-(p-substituted phenyl)-4(R or S)-[10(α or β)-dihydroartemisininoxy]butyric acids. *J. Med. Chem.*, 40, 1396–1400.

Lisgarten, J.N., Potter, B.S., Bantuzeko, C. and Palmer, R.A. (1998) Structure, absolute configuration, and conformation of the antimalarial compound, artemisinin. *J. Chem. Crystall.*, 28, 539–543.

Liu, J.M., Ni, M.Y., Fan, J.F., Tu, Y.Y., Wu, Z.H., Wu, Y.L. *et al.* (1979) Structure and reactions of arteannuin. *Acta Chemica Sinica*, 37, 129–143.

Liu, H.M. and Li, K.W. (1980) Studies on the constituents of *Artemisia annua*. Yao Hsuch T'ung Pao 15, 39 (*Chem. Abs.*, 95, 121030).

Liu, H.M., Li, G.L. and Wu, H.Z. (1981) Studies on the chemical constituents of *Artemisia annua*. Yao Huch Pao, 16, 65–67.

Liu, X. (1986) Antimalarial reduced artemisinin succinate. Faming Zhuanli Shenging Gongkai Shuomingshu; CN patent 85100781 (*Chem. Abs.*, 107, 78111).

Liu, Q., Yang, Z.Y., Deng, Z.B., Sa, G.H. and Wang, X.J. (1988) Preliminary analysis on chemical constituents of essential oil from inflorescence of *Artemisia annua*. *Acta Botanica*, 30, 223–225.

Liu, H.J., Yeh, W.H. and Chew, S.Y. (1993) A total synthesis of the antimalarial natural product (+)-qinghaosu. *Tet. Lett.*, 34, 4435–4438.

Lou, X. and Shen, C. (1987) The chemistry, pharmacology and clinical application of qinghaosu (artemisinin) and its derivatives. *Medicinal Res. Rev.*, 7, 29–52.

Madhusudan, K.P., Vishwakarma, R.A., Balachandran, S. and Popli, S.P. (1989) Mass spectral studies on artemisinin, dihydroartemisinin and arteether. *Ind. J. Chem.*, 28B, 751–754.

Manns, D. and Hartmann, R. (1992) Annuadiepoxide, a new polyacetylene from the aerial parts of *Artemisia annua*. *J. Nat Prod.*, 55, 29–32.

Marco, S.A., Sanz, J.F., Bea, J.F. and Barbera, O. (1990) Phenolic constituents of *Artemisia annua*. *Pharmazie*, 45, 382–383.

McChesney, J.D. and Jung, M. (1990) Deoxoartemisinin: new compounds for the treatment of malaria. US patent No. 4920147 (*Chem. Abs.*, 113, 97856).

Mi, J., (1987) Circular dichroism of some synthetic intermediates of qinghaosu. *Yaowu Fenxi Zazhi*, 1, 262–266.

Mishra, L.N. (1986) Arteannuin C, a sesquiterpene lactone from *Artemisia annua*. *Phytochem.*, 25, 2892–2893.

Mishra, L.N., Ahmad, A., Thakur, R.S. and Jakupovic, J. (1993) Bisnor cadinanes from *Artemisia annua* and definitive ^{13}C-NMR assignments of β-arteether. *Phytochem.*, 33, 1461–1464.

Nair, M.S.R. and Basile, D.V. (1992) Use of cell free systems in the production of the potent antimalarial, artemisinin. *Ind. J. Chem.*, **31B**, 880–882.

Nair, M.S.R. and Basile, D.V. (1993) Bioconversion of arteannuin B to artemisinin. *J. Nat Prod.*, **56**, 1559–1566.

Nga, T.T.T., Menage, C., Begue, J-P., Bonnet-Delpon, D. and Gantier, J-G. (1998) Synthesis and antimalarial activities of fluoroalkyl derivatives of dihydroartemisinin. *J. Med. Chem.*, **41**, 4101–4108.

O'Dowd, H., Ploypradith, P., Xie, S., Shapiro, T.A. and Posner, G.H. (1999) Antimalarial artemisinin analogs. Synthesis *via* chemoselective C-C bond formation and preliminary biological evaluation. *Tetrahedron*, **55**, 3625–3636.

Peters, W., Robinson, B.L., Rossier, J.C., Mishra and Jefford, C.W. (1993) The chemotherapy of rodent malaria XLIX. The activity of some synthetic 1,2,4-trioxanes. 2. Structure-activity studies on *cis*-fused cyclopenteno-1,2,4-trioxanes against drug sensitive and drug resistant lines of *P. berghei* and *P. yoelli* spp. *Annal. Trop. Med. Parasiotol.*, **87**, 9–16.

Petrov, O. and Ogyanov, I. (1991) An approach to the synthesis of novel 11-hydroxyartemisinin derivatives. *Collection Czech. Chem. Comm.*, **56**, 1037–1041.

Pinhey, J.T. and Sternhell, S. (1995) Structure of α-hydroxysantonin and some aspects of the stereochemistry of related eudesmanolides and guanolides. *Aust. J. Chem.*, **18**, 543–557.

Posner, G.H., Oh, C.H., and Milhous, W.K. (1991) Olefin oxidative cleavage and dioxetane formation using triethylsilyl hydroperoxide: Application to preparation of potent antimalarial 1,2,4-trioxanes. *Tet. Lett.*, **32**, 4235–4238.

Posner, G.H., Oh, C.H., Gerena, L. and Milhous, W.K. (1992) Extraordinarily potent antimalarial compounds: New structurally simple easily synthesised tricyclic-1,2,4-trioxanes. *J. Med. Chem.*, **35**, 2459–2467.

Posner, G.H., Oh, C.H., Wang, D., Gerena, L. and Milhous, W.K., Meshick, S.R. and Asawamahasakda, W. (1994) Mechanism based design, synthesis and *in vitro* antimalarial testing of new 4-methylated trioxanes structurally related to artemisinin: The importance of a carbon-centered radical for antimalarial activity. *J. Med. Chem.*, **37**, 1256–1258.

Posner, G.H., Oh, C.H., Webster, H.K., Ager, A.L. Jr. and Rossan, R.N. (1994a) New antimalarial tricyclic 1,2,4-trioxanes: Evaluations in mice and monkeys. *Am. J. Trop. Med. Hyg.*, **50**, 522–526.

Posner, G.H., Cumming, J.N., Polypradith, P. and Oh, C.H. (1995) Evidence for Fe(IV)=O in the molecular mechanism of action of the trioxane antimalarial artemisinin. *J. Am. Chem. Soc.*, **117**, 5885–5886.

Posner, G.H., McGarvey, D.S., Oh, C.H., Kumar, N., Mehsnick, S.R. and Asawamahasakda, W. (1995a) Structure-activity relationships of lactone ring-opened analogues of the antimalarial 1,2,4-trioxane artemisinin. *J. Med. Chem.*, **38**, 607–612.

Posner, G.H., Oh, C.H., Gerena, L. and Milhous, W.K. (1995b) Synthesis and antimalarial activities of structurally simplified 1,2,4-trioxanes related to artemisinin. *Heteratom Chem.*, **6**, 105–116.

Posner, G.H., Wang, D., Cumming, J.N., Oh, G.H., French, A.N., Bodley, A.L. *et al.* (1995c) Further evidence supporting the importance of, and the restrictions on, a carbon centered radical for high antimalarial activity of 1,2,4-trioxanes like artemisinin. *J. Med. Chem.*, **38**, 2273–2275.

Posner, G.H., Park, S.B., Gonzaley, L., Wang, D., Cumming, J.N., Klinedins, D. (1996) Molecular mechanism of action of antimalarial trioxane analogues of artemisinin. *J. Am. Chem. Soc.*, **118b**, 3537–3538.

Posner, G.H., Wang, D., Gonzaley, L., Tao, X., Cumming, J.N., Klinedins, D. *et al.* (1996a) Mechanism based design of simple symmetrical, easily prepared, potent antimalarial endoperoxides. *Tet. Lett.*, **37**, 815–818.

Posner, G.H., Ploypradith, P., Hapangama, W., Wang, D.S., Cumming, J.N., Dolan, P. *et al.* (1997) Trioxane dimers have potent antimalarial, antiproliferative and antitumour activities *in vitro. Bioorg. Med. Chem. Lett.*, **5**, 1257–1265.

Posner, G.H., Cumming, J.N., Woo, S-H., Ploypradith, P., Xie, S. and Shapiro, T.A. (1998) Orally active antimalarial 3-substituted trioxanes: New synthetic methodology and biological evaluation. *J. Med. Chem.*, **41**, 940–951.

Posner, G.H., O'Dowd, H., Caferro, T., Cumming, J.N., Ploypradith, P., Xie, S.J. *et al.* (1998a) Antimalarial sulfone trioxanes. *Tet. Lett.*, **39**, 2273–2276.

Posner, G.H., O'Dowd, H., Ploypradith, P., Cumming, J.N., Xie, S. and Shapiro, T.A. (1998b) Antimalarial cyclic peroxy ketals. *J.Med. Chem.*, **41**, 2164–2167.

Posner, G.H., Parker, M.H., Northrop, J., Elias, J.S., Ploypradith, P., Xie, S. *et al.* (1999) Orally active, hydrolytically stable, semisynthetic, antimalarial trioxanes in the artemisinin family. *J. Med. Chem.*, **42**, 300–304.

Pu, Y.M., Yeh, H. and Ziffer, H. (1993) An unusual acid catalysed rearrangement of 1,2,4-trioxanes. *Heterocycles*, **36**, 2099–2107.

Pu, Y.M., Yogen, B. and Ziffer, H. (1994) Stereoselective oxidations of a β-methylglycal, anhydroartemisinin. *Tet. Lett.*, **35**, 2129–2132.

Pu, Y.M. and Ziffer, H. (1995) Synthesis and antimalarial activities of 12β-allyldeoxo-artemisinin and its derivatives. *J. Med. Chem.*, **38**, 613–616.

Pu, Y.M., Torok, D.S., Ziffer, H., Pan, X.Q. and Meshnick, S.R. (1995) Synthesis and anti-malarial activities of several fluorinated artemisinin derivatives. *J. Med. Chem.*, **38**, 4120–4124.

Ravindranathan, T., Kumar, M.A., Menon, R. and Hiremath, S.V. (1990) Stereoselective synthesis of artemisinin. *Tet. Lett.*, **31**, 755–758.

Ravindranathan, T. (1994) Artemisinin (qinghaosu). *Curr. Sci.*, **66**, 35–41.

Rong, Y.J. and Wu, Y.L. (1993) Synthesis of C-4 substituted qinghaosu analogues. *J. Chem. Soc. Perkin Trans.*, **1**, 2147–2148.

Rong, Y.J., Ye, B. and Wu, Y.L. (1993) An effective synthesis of deoxoqinghaosu from dihy-droqinghaosu. *Chin. Chem. Lett.*, **4**, 859–860.

Roth, R.J. and Acton, N. (1987) Isolation of artemisinic acid from *Artemisia annua. Planta Med.*, **53**, 502–510.

Roth, R.J. and Acton, N. (1987a) Isolation of *epi*-deoxyarteannuin B from *Artemisia annua. Planta Med.*, **53**, 576.

Roth, R.J. and Acton, N. (1989) A simple conversion of artemisinic acid into artemisinin. *J. Nat. Prod.*, **52**, 1183–1185.

Roth, R.J. and Acton, N. (1991) US Patent 4992,5611.

Roth, R.J. and Acton, N. (1991a) A facile semisynthesis of the antimalarial drug qinghaosu. *J. Chem. Edu.*, **68**, 612–613.

Rucker, G., Mayer, R. and Manns, D. (1987) α- and β-myrcene hydroperoxide from *Artemisia annua. J. Nat. Prod.*, **50**, 287–289.

Saitbaeva, I.M. and Sidyakin, G.P. (1970) Coumarins from *Artemisia annua. Khim Prir Soedin.*, **6**, 758.

Sangwan, R.S., Agrawal, K., Luthra, R., Thakur, R.S. and Sangwan, N.S. (1993) Biotransformation of arteannuic acid into arteannuin B and artemisinin in *Artemisia annua. Phytochem.*, **34**, 1301–1302.

Schmidt, G. and Hofheinz, W. (1983) Synthesis of qinghaosu. *J. Am. Chem. Soc.*, **105**, 624–625.

Shang, X., He, C.H., Zheng, Q.T., Yang, J.J. and Liang, X.T. (1989) Chemical transformations of qinghaosu, a peroxidic antimalarial. II. *Heterocycles*, **28**, 421–424.

Shukla, A., Farooqui, A.H.A. and Shukla, Y.N. (1991) Growth inhibitors from *Artemisia annua*. *Indian Drugs*, **28**, 376–377.

Singh, C. (1990) Preparation of β-hydroperoxides by photooxygenation of allylic alcohols and their elaboration into 1,2,4-trioxanes. *Tet. Lett.*, **33**, 6901–6902.

Singh, C., Mishra, D., Saxena, G. and Chandra, S. (1995) *In vivo* potent antimalarial 1,2,4-trioxanes: Synthesis and activity of 8-(α-arylvinyl)-6,7,10-trioxspiro(4,5)decanes and 3-(α-arylvinyl)-1,2,5-trioxspiro(5,5)undecanes against *P. berghei* in mice. *Bioorg. Med. Chem. Lett.*, **5**, 1913–1916.

Singh, A.K., Pathak, V. and Agrawal, P.K. (1997) Annphenone, a phenolic acetophenone from *Artemisia annua*. *Phytochem.*, **44**, 555–557.

Stefanovic, M., Miljovic, D and Velmirovic, S. (1977) Chemical transformations of arteannuin B-a cadinane sesquiterpene isolated from *Artemisia annua* L. *Glas. Hem. Drus. Beograd.*, **42**, 227–236.

Thakur, R.S. and Vishwakarma, R.A. (1990) *Artemisia annua* L. A valuable source of the antimalarial artemisinin. *CSIR News Bulletin* (*India*), **40**, 26–29.

Tian, Y., Wei, Z. and Wu, Z. (1982) Studies on the chemical constituents of qing hao (*Artemisia annua*), a traditional Chinese herb. *Zhongcaoyao*, **13**, 9–11.

Torok D.S. and Ziffer, H. (1995) Synthesis and reactions of 11-azaartemisinin and derivatives. *Tet. Lett.*, **36**, 829–832.

Torok, D.S., Ziffer, H., Meshnick, S.R., Pan, X.Q. and Ager, A. (1995) Synthesis and antimalarial activities of N-substituted 11-azaartemisinins. *J. Med. Chem.*, **38**, 5045–5050.

Tsunemtsu, T. and Tadashi, N. (1957) Essential oil of *Artemisia annua*. I. Isolation of a new ester compound. *Yakugaku Zasshi*, **77**, 1307–1309.

Tu, Y.Y., Ni, M.Y., Chung, Y.Y. and Li, L.N. (1981) Chemical constituents in *Artemisia annua* L. and the derivatives of artemisinin. *Chung Yao Tsung Pao*, **6**, 31. (*Chem. Abs.*, **95**, 1756164).

Tu, Y.Y., Ni, M.Y., Zhong, Y.R., Li, L.N., Cui, S.L., Zhang, M.Q. *et al.* (1981a) Studies on the chemical constituents of *Artemisia annua*. Acta. Pharmaceut. *Sinica*, **16**, 366–370.

Tu, Y.Y., Ni, M.Y., Zhong, Y.R., Li, L.N., Cui, S.L., Zhang, M.Q. *et al.* (1982) Studies on the constituents of *Artemisia annua*. *Planta Medica*, **44**, 143–144.

Tu, Y.Y., Yiu, J.P., Z1, Li, Huang, M.M. and Liang, X.T. (1985) Chemical constituents of sweet wormwood (*A. annua*). *Zhongcaoyao*, **16**, 200. (*Chem. Abs.*, **107**, 36599).

Tu, Y.Y., Zhu, Q. and Shen, X. (1985a) Constituents of young *Artemisia annua*. *Zhongyao Tongbao*, **10**, 419–420. (*Chem. Abs.*, **104**, 31768b).

Ulubulen, A. and Halfon, B. (1976) Phytochemical investigation of the herb of *Artemisia annua*. *Planta Medica*, **29**, 258–260.

Uskobovic, N.R., Williams, T.H. and Blount, J.F. (1974) The structure and absolute configuration of arteannuin B. *Helv. Chemica Acta.*, **57**, 600–602.

Vennerstrom, J.L., Fu, H-N., Ellis, W.Y., Ager, A.L., Jr., Wood, J.K., Andersen, S.L. *et al.* (1992) Dispiro-1,2,4,5-tetraoxanes: A new class of antimalarial peroxides. *J. Med. Chem.*, **35**, 3023–3027.

Venogopalan, B. and Bapat. C.P. (1993) Functionalisation of iso-artemisitene. *Tet. lett.*, **34**, 5787–5790.

Venogopalan, B., Shinde, S.L. and Karnik, P.J. (1993) Role of radical initiated cyclisation reactions in the synthesis of artemisinin based on novel ring skeletons. *Tet. Lett.*, **34**, 6305–6308.

Vishwakarma, R.A., Mehrotra, R., Tripathi, R., and Dutta, G.P. (1992) Stereoselective synthesis and antimalarial activity of α-artelinic acid from artemisinin. *J. Nat. Prod.*, 55, 1142–1144.

Vonwiller, S.C., Warner, J.A., Mann, S.T. and Haynes, R.K. (1995) Copper (II) trifluoromethanesulfonate-induced cleavage oxygenation of allylic hydroperoxide derived from qinghao acid in the synthesis of qinghaosu derivatives. Evidence for the intermediacy of enols. *J. Am. Chem. Soc.*, 117, 11098–11105.

Vroman, J.A., Khan, I. and Avery, M.A. (1997) Conjugate addition of trimethylsilylmethyl cuprates to artemisinic acid: Homologation and subsequent conversion to 9β-modified analogs. *Synlett.*, 1438–1440.

Wang, Z., Nakahima, T.T., Kopesky, K.R. and Molina, J. (1985) Qinghaosu: proton and carbon-13 nuclear magnetic resonance spectral assignments and luminescence. *Canad. J. Chem.*, 63, 3070–3074.

Wang, B., Yin, M., Shong, G., Chan, Z., Zhao, T, Wang, M. *et al.* (1986) Carbon-13 NMR spectroscopy of arteannuin analogues. *Huazue Xuebao*, 44, 834–838. (*Chem. Abs.*, 107, 40113).

Wang, Y., Xia, Z., Zhou, F., Wu, Y. and Huang, J. (1988) Studies on the biosynthesis of arteannuin. Artemisinic acid as a key intermediate in the biosynthesis of arteannuin and arteannuin B. *Xuaxue Xuebao*, 46, 1152–1153.

Wang, Y., Xia, Z., Zhou, F., Wu, Y. and Huang, J. (1993) Studies on the biosynthesis of arteannuin. IV. The biosynthesis of arteannuin and arteannuin B by the leaf homogenate of *Artemisia annua* l. *Chin. J. Chem.*, 11, 457–463.

Woerdenbag, H.J., Bos, R., Salomons, M.C., Hendriks, H., Pras, N. and Malingre, T. (1993) Volatile constituents of *Artemisia annua* L. (Asteraceae). *Flav. Frag. J.*, 8, 131–137.

Woerdenbag, H.J., Pras, N., Noguyen, G.C., Bui, T.B., Bos, R., and Van-Uden, W. (1994) Artemisinin related sesquiterpenes and essential oil in *Aremisia annua* during a vegetation period in Vietnam., 60, 272–275.

Wu, Z. and Wang, Y. (1984) Structure and synthesis of arteannuin and related compounds XI. Identification of epoxyarteannuic acid. *Huaxue Xuebao*, 42, 596–598.

Xu, X.X., Zhu, J., Huang, D.Z. and Zhou, W.S. (1986) Total synthesis of arteannuin and deoxyarteannuin. *Tetrahedron*, 42, 819–825.

Xu, X.X., Zhu, J., Huang, D.Z. and Zhou, W.S. (1989) Structure and synthesis of arteannuin and related compounds. XXIII. Lactone configuration of the deoxyarteannuin degradation product. *Huazue Xuebao*, 47, 771–774.

Yagen, B., Pu, Y.M., Yeh, H.J.C. and Ziffer, H. (1994) Tandem silica gel catalysed rearrangements and subsequent Baeyer-Villiger reactions on artemisinin derivatives. *J. Chem. Soc. Perkin. Trans.*, 843–846.

Yang, Q., Shiweizi, Li, R., Gan, J. (1982) The antimalarial and toxic effect of artesunate on animal models. *J. Trad. Chin. Med.*, 2, 99–103.

Yang, S., Roberts, M.F. and Phillipson, J.D. (1989) Methoxylated flavones and coumarins from *Artemisia annua*. *Phytochem.*, 28, 1509–1511.

Yang, S., Roberts, M.F., O'Neill, M.J., Bucar, F. and Phillipson, J.D. (1995) Flavonoids and chromones from *Artemisia annua*. *Phytochem.*, 38, 255–257.

Yang, Y.Z., Little, B. and Meshnick, S.R. (1994) Alkylation of proteins by artemisinin, effects of heme, pH and drug structure. *Biochem. Pharm.*, 48, 569–573.

Ye, B. and Wu, Y.L. (1989) Synthesis of carba-analogues of qinghaosu. *Tetrahedron*, 45, 7287–7290.

Ye, B. and Wu, Y.L. (1990) An effective synthesis of qinghaosu and deoxoqinghaosu from arteannuic acid. *J. Chem. Soc. Chem. Comm.*, 726–727.

Youan, T. (1955) Essential oil of *Artemisia annua*. *Perfum. Essent. Oil. Rec.*, **46**, 75–78.

Zaman, S.S. and Sharma, R.P. (1991) Some aspects of the chemistry and biological activity of artemisinin and related antimalarials. *Heterocycles*, **32**, 1593–1638.

Zeng, M., Li, L., Chen, S., Li, C., Liang, X., Chen, M. *et al.* (1983) Chemical transformations of qinghaosu: A peroxidic antimalarial. *Tetrahedron*, **39**, 2941–2946.

Zhang, L. and Zhou, W.S. (1988) Structure and synthesis of arteannuin and conversion into artemisitene. *Youji Huaxue*, **8**, 329–330 (*Chem. Abs.*, **111**, 78408).

Zhang, L. and Zhou, W.S. (1989) Structure and synthesis of arteannuin and related compound XXIV. Conversion of arteannuin to natural Δ11–13 arteannuin. *Huaxue Xuebao*, **47**, 117–1119. (*Chem. Abs.*, **112**, 217297).

Zhang, J.L., Li, J.C. and Wu, Y.L. (1988) Synthesis of qinghaosu analogs *via* ozonisation. *Yaoxue Xuebao*, **23**, 452–455.

Zhao, Y., Hanton, W.K. and Lee, K.H. (1986) Antimalarial agents: Artesunate, an inhibitor of cytochrome oxidase activity in *Plasmodium berghei*. *J. Nat. Prod.*, **49**, 139–142.

Zhou, W., Xu, S. and Zhang, L. (1989) Studies on the structure and synthesis of arteannuin and related compounds. XXII. The regioselective synthesis of arteannuin D. *Huaxue Xuebao*, **47**, 340–344. (*Chem. Abs.*, **112**, 36202).

Zhou, W.S. and Xu, X.X. (1994) Total synthesis of the antimalarial sesquiterpene peroxides qinghaosu and yingzhaosu A. *Acc. Chem. Res.*, **27**, 211–216.

Zhu, D., Zhang, S., Liu, B., Fan, G., Liu, J and Xu, R. (1982) Study on antibacterial constituents of qing hao (*Artemisia annua* L.) *Zhongcaoyao*, **13**, 6. (*Chem. Abs.*, **97**, 107028v).

Zhu, D., Deng, D., Zhang, S. and Ku, R. (1984) Stucture of artemisilactone. *Hauoexue Xuebao*, **42**, 937–939. (*Chem. Abs.*, **102**, 21192).

Zouhiri, F., Desmaele, D., d'Angelo, J., Riche, C., Gay, F. and Ciceron, L. (1998) Artemisinin tricyclic analogues: Role of a methyl group at C-5α. *Tet. Lett.*, **39**, 2969–2972.

13. THE MODE OF ACTION OF ARTEMISININ AND ITS DERIVATIVES

COLIN W. WRIGHT[1] AND DAVID C. WARHURST[2]

[1]*School of Pharmacy, University of Bradford, West Yorkshire, BD7 1DP, UK.*

[2]*Department of Infectious and Tropical Diseases, London School of Hygiene and Tropical Medicine, University of London, Keppel Street, London, WC1E 7HT, UK.*

INTRODUCTION

Malaria is responsible for the deaths of 1–2 million people each year, most of them children. Four species of malaria parasite may infect man, (*Plasmodium falciparum, P. vivax, P. malariae,* and *P. ovale*) but almost all of the deaths are caused by *P. falciparum* as a result of cerebral malaria. Fever, characteristic of infection with all four species of malaria parasite is mediated by cytokines such as tumour necrosis factor (TNF). Specific pathological features of falciparum malaria relate to the accumulation of erythrocytes infected with the later developmental stages of the blood cycle in the capillaries of vital organs such as the brain, kidney, intestine, lungs etc. The widespread development of *P. falciparum* resistant to chloroquine and the emergence of strains resistant to other antimalarials as well (multi-drug resistance), especially in south-east Asia has made the need for new antimalarials very urgent (Bradley, 1995).

A brief description of the life cycle of *P. falciparum* follows: Parasites in the form of sporozoites are transmitted *via* the saliva of female *Anopheles* mosquitoes when they feed on human blood. The sporozoites travel in the bloodstream to the liver where they penetrate liver cells and multiply before breaking out in the form of merozoites which invade red blood cells. Each merozoite develops into a trophozoite which then divides giving rise to 16 or more merozoites, and at this stage the mature trophozoite is referred to as a schizont. The red cells rupture releasing the merozoites which then invade fresh red cells thus establishing a continuous cycle (erythrocytic cycle).

The cyclical release of parasites from the red blood cells stimulates cytokine production by the immune system and explains the intermittent fever characteristic of malaria (on every third day with *P. falciparum* infection), though it must be noted that the "classical" symptoms of malaria are not always seen in malaria patients. Some merozoites develop into male or female gametocytes which, if ingested by a mosquito during a blood meal, give rise to sporozoites following sexual reproduction and development.

In the red blood cells, malaria parasites digest the host cell haemoglobin, releasing the amino acids and leaving the haem ring intact. (Note: In order to avoid confusion, haem containing ferrous iron, (ferroprotoporphyrin IX), will be designated haem (FeII) while haem containing ferric iron, (haemin, ferriprotoporphyrin IX) will be described as haem (FeIII). In cases where the oxidation state of the iron is not clear, the term haem will be used). As free haem is toxic to the parasite, it is polymerised as haem (FeIII) into a black, non-toxic, crystalline product known as haemozoin or malaria pigment. Initially, it was thought that this took place by the action of an enzyme, haem polymerase, (Slater and Cerami, 1992), but it now appears that the formation of haemozoin takes place spontaneously (Dorn et al., 1995). (Note that in man haem is broken down by haem oxygenase, but this enzyme is absent in malaria parasites).

Quinine and 4-aminoquinoline antimalarials such as chloroquine are thought to act by binding to haem forming a toxic complex, thus preventing haemozoin formation and inhibiting parasite growth (e.g. Egan et al., 1994). As will be shown in the following discussion, artemisinin-like compounds also depend upon an interaction with haem for their antimalarial properties, but the mechanism by which parasite growth is inhibited appears to be different from that of quinine.

THE ANTIMALARIAL MODE OF ACTION OF ARTEMISININ AND DERIVATIVES

There is now an increasing amount of experimental evidence which suggests that the antimalarial mode of action of artemisinin, 1 and related compounds, 2–6 is a two-step process. In the first, artemisinin is "activated" by interacting with the haem residue which remains following the digestion of haemoglobin by the malaria parasite; the result of this process is the production of highly reactive free radicals. In the second step, the reaction of artemisinin-derived free radicals with various molecular components of the parasite leads to the disruption of normal metabolic processes and parasite death. The evidence supporting the above will now be critically discussed in some detail.

The Requirement of Haem Iron for the Antimalarial Action of Artemisinin

The discovery that deoxyartemisinin, 7 which lacks the endoperoxide group is devoid of antimalarial activity clearly showed that the presence of this group is essential for activity (Anon. 1982). The well established observation that iron causes the catalytic decomposition of peroxides to yield free radicals, (reviewed by Halliwell and Gutteridge, 1989), stimulated investigations into the interaction of iron with artemisinin and its relevance to the antimalarial mode of action of artemisinin-like drugs (Meshnick, 1994). The iron-rich nature of the red blood cell contents was consistent with this mechanism. However, several discrete "pools" of iron exist in the red cell i.e. haemoglobin, methaemoglobin (haemoglobin FeIII), haem (FeII and FeIII), haemozoin, free iron and iron present in haemoproteins (cytochromes).

[^{14}C]-artemisinin did not react with haemoglobin in uninfected red cells, (Meshnick et al., 1991), although using isolated haemoproteins a proportion, (5–18%) of added labelled artemisinin bound to catalase, cytochrome C and haemo-

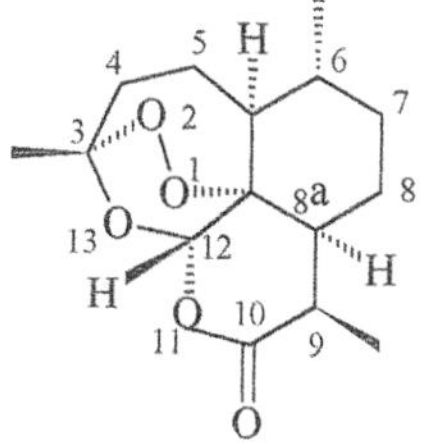

1   artemisinin

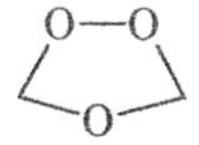

1a   1,2,4-trioxane ring

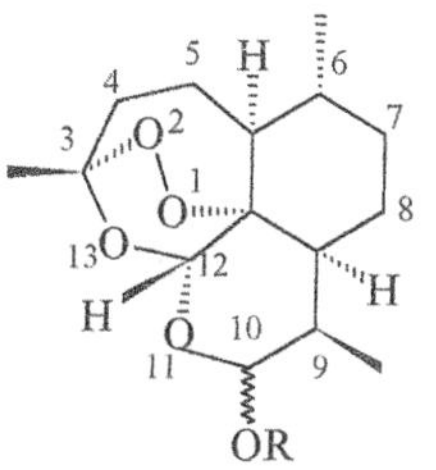

2	R=H	dihydroartemisinin
3	R=CH$_3$	artemether
4	R=C$_2$H$_5$	arteether
5	R=COCH$_2$COONa	sodium artesunate

6    R=CH$_2$—⟨benzene⟩—COONa   sodium artelinate

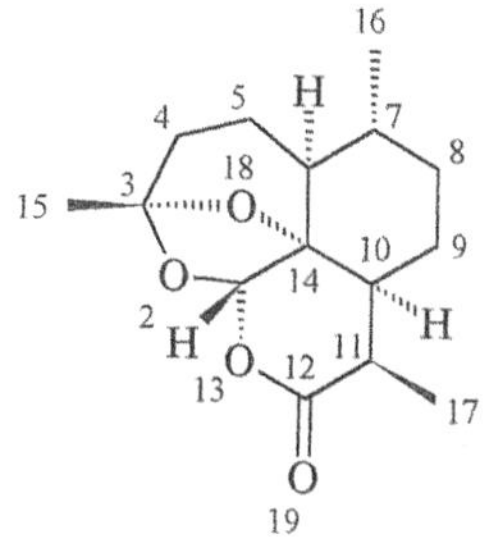

7   deoxyartemisinin

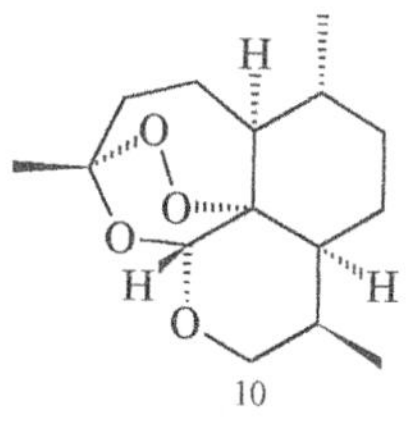

8   deoxoartemisinin

Figure 1

globin and, in the case of the latter two most of the drug was bound to protein rather than to the haem moiety; there was no binding to free globin (Yang *et al.*, 1994). Haem did react to form covalent adducts (see below under parasite molecules targetted by artemisinin), and artemisinin also reacted with isolated haemozoin although the incorporation of label was about 5-fold less than that found with intact parasites (Hong *et al.*, 1994); it is suggested that this is because the drug acts more rapidly with haem (FeII) than with haem (FeIII), (for evidence see below), as haem is maintained as haem (FeII) in intact red cells.

The action of artemisinin against *P. falciparum in vitro* was antagonised by iron chelators suggesting that free iron or haem was responsible (since some chelators e.g. desferrioxamine bind haem as well as iron). Desferrioxamine also antagonised the action of artemether, **3**, and the action of this and other chelators was prevented by pre-saturation with iron (Meshnick *et al.*, 1993). The binding of chelators to haem would prevent the interaction of haem-iron with the endoperoxide moiety. It has also been suggested that free iron could be released by the alkylation of haem with artemisinin. Iron chelators would then prevent this free iron from reacting with the endoperoxide group to produce more free radicals (Cumming *et al.*, 1997 and references therein).

The necessity for haem-iron, (rather than free-iron) involvement in the mode of action of artemisinin is supported by the high degree of selectivity seen in the action of this drug against malaria parasites present in red blood cells while leaving other cells, with the possible exception of neuronal cells unaffected (see page 268). In addition chloroquine, which binds to haem (FeIII) preventing its polymerisation into haemozoin antagonises the action of artemisinin (Chou *et al.*, 1980). Interestingly, chloroquine resistant *P. berghei* which lacks visible haemozoin is resistant to artemisinin suggesting that haemoglobin breakdown is essential for the antimalarial action of artemisinin (Peters *et al.*, 1986). Further support for this conclusion has been provided by electron micrographs which reveal that artemisinin, (Anon. 1979), and artesunate, **5**, (Li *et al.*, 1981), initially affect parasite food vacuoles. In addition, autoradiography studies have confirmed that artemisinin is located in the parasite food vacuole (Maeno *et al.*, 1993).

Following the observation that artemisinin causes changes to membranous structures including the mitochondria, Zhao *et al.* (1986), investigated the possibility that the drug might act by inhibiting the mitochondrial haem-containing enzyme, cytochrome oxidase. The cytochrome oxidase of *P. berghei* trophozoites was completely inhibited by sodium artesunate at 1 mM *in vitro* and *in vivo* at 100 mg/kg when given intravenously to infected mice; however, these doses are much greater than are required to inhibit the growth of malaria parasites so that inhibition of this enzyme is unlikely to be the primary mode of action of artemisinin-like compounds.

Taken together, the above suggests that artemisinin-like compounds are likely to react most readily with haem (FeII), released from haemoglobin as a result of parasite digestion before its conversion into haemozoin.

The Nature of the Interaction between Haemin and Artemisinin

A number of research groups have investigated the ability of iron containing compounds to cleave the peroxide group of artemisinin and related compounds including the trioxanes which are compounds based on the 1,2,4-trioxane ring **1a** of artemisinin **1** (Wu *et al.*, 1998 and references therein). Ferrous (FeII) compounds readily react but with ferric (FeIII) compounds the reaction is much slower under the same conditions; in a study of the reduction of artemisinin using cyclic voltammetry it was found that artemisinin cannot be reduced by FeIII (Zhang *et al.*, 1992). The rate of the reaction is also pH dependent, being much slower at pH 6–7 than at pH 4 (Haynes and Vonwiller, 1996). It is noteworthy, in this context that the pH of the malaria parasite food vacuole where haemoglobin is digested is estimated to be about 5.3 (Krogstad, *et al.*, 1985).

Whereas ferric iron reacts only slowly with artemisinin, Meshnick *et al.* (1993), reported that artemisinin reacts strongly with haem in both FeII and FeIII forms. However, in the case of artesunate, the reaction is reported to be much slower with haem (FeIII) than with haem (FeII), (Adams and Berman 1996). Similar slowing of the reaction was found when artemisinin was incubated with haem (FeII) without a reducing agent. When haemoglobin is digested by the malaria parasite, haem (FeII) is released but this may readily oxidise to haem (FeIII), although it has been suggested that the high level of glutathione present in red blood cells may maintain haem in its reduced (FeII) form (Chen *et al.*, 1998). Questions remain concerning the the oxidation states of haem in parasitised erythrocytes and its reactivity towards artemisinin derivatives.

The interaction of artesunate, 5 with haem has been studied using cyclic voltammetry (Chen *et al.*, 1998). Note that in these experiments, haem (FeIII) was initially solubilised in sodium hydroxide solution resulting in the formation of haematin in which a hydroxyl group is co-ordinated onto the iron; however, spectroscopic studies indicate that the molecules actually exist as dimers linked by an Fe-O-Fe bridge (Brown *et al.*, 1980). In the presence of haem (FeIII) at concentrations as low as 2×10^{-8} M the reduction of artesunate, (1 mM), was facilitated (reduction potential reduced by 680 mV) indicating that the reduction was catalytic; this was supported by UV spectra which showed that no complex was formed between haem and artesunate. No reduction was observed when deoxydihydroartemisinin and succinic acid were used in place of artesunate, thus confirming that reduction of the peroxy bridge had occurred. During the course of the reaction haem (FeIII) is electrolytically reduced to haem (FeII) and the latter then reduces artesunate with regeneration of haem (FeIII).

The above results are consistent with those reported for artemisinin, (Zhang *et al.*, 1992), suggesting that the modes of action of the two drugs are similar. For the above reductive process to occur in the parasite-infected red blood cell, a mechanism for the reduction of haem (FeIII) would be needed; as suggested above, glutathione, may fulfill this role.

Paitayatat *et al.*, 1997, examined the ability of a number of artemisinin derivatives to bind with haem (FeII); actually this was FeII haematin since it was prepared by dissolving haem (FeIII) in sodium hydroxide and then adding sodium dithionite as a reducing agent. The absorption peak of haem (FeII) at 415 nm was immediately reduced by artemisinin and other derivatives which possess antimalarial activity; deoxyartemisinin, 7, as expected did not affect absorbance. With the exception of artesunate, there was a correlation between the dissociation constant and the log IC_{50} antiplasmodial activity.

Using molecular modelling techniques, Shukla *et al.* (1995), explored the mode of binding of artemisinin and of deoxyartemisinin, 7, with haem (FeII) and haem (FeIII).

With haem (FeIII) and artemisinin, the lowest energy docking configuration was found to be when the peroxide bridge oxygens were in close proximity to the haem iron (this projects slightly above the plane of the porphyrin ring and artemisinin binds on the same side); the oxygen at C-10 was also involved as well. However, binding could also occur with the 3 non-peroxide oxygens of artemisinin locating onto the iron since the energy of this configuration is only 1.6 kcal/mol above that for the peroxide-oxygen binding mode. A third possibility involving binding with one peroxide oxygen O-1 and the non-peroxidic oxygen O-13 was found to be of much higher energy. In the case of haem (FeII), three modes of binding are also possible but in this case the energy difference between them was small, only 0.2 kcal. The configuration in which the peroxide oxygens bound to the FeII iron had the lowest energy at 233.5 kcal/mol but this is significantly higher than that for peroxide binding with haem (FeIII), (220 kcal/mol). It is interesting that the reaction of Fe II compounds with artemisinin is much faster than those of Fe III since the latter would appear to provide the most favourable complex. However, the binding of artemisinin to Fe II or Fe III haemin is only the initial step in a sequence of events leading to radical generation and it is likely that the subsequent step(s) are more favourable with FeII. Further studies are needed in order to explain the above observations.

When the above modelling was repeated with deoxyartemisinin, 7, the most stable arrangement was found to be one in which the oxygens O-2, O-13 and O-19 interacted with haem (FeIII); the lack of involvement of O-18 indicates that the interaction of the latter with haem is different from that of artemisinin (Shukla et al., 1995), but in any case 7 is inactive as an antimalarial because it lacks the endoperoxide group.

In another study, Grigorov et al. (1997), examined the quantitative structure-activity relations (QSAR), of a series of synthetic trioxanes and showed that two hydrophobic features and hydrogen bonding ability are essential for antimalarial activity in these compounds. Molecular modelling of an active trioxane-haem (FeII) complex indicated that the peroxide bond of the trioxane lies close to the FeII atom of haem, thus supporting the above hypothesis.

In order to test the hypothesis that the action of artemisinin-like antimalarials requires binding of the peroxide group with haem iron, Zouhiri et al. (1998), synthesised tricyclic 1,2,4 trioxanes possessing an α-methyl group at C5 in place of the hydrogen normally found in this position in antimalarial trioxanes. As expected, antiplasmodial activity in the latter compounds was markedly reduced, presumably as a result of steric hindrance preventing close contact between the iron and the peroxide moiety.

Mechanistic Scheme for the Reaction of Artemisinin with Iron

The reaction of artemisinin with ferrous (FeII) compounds has been used as a model by several research groups in order to postulate the identity of products which may have a significant role in the parasite killing action of artemisinin. The use of various reactants and reaction conditions in different laboratories has led to some confusion as divergent results and a variety of reaction products have been reported. However, many of these results have been rationalised by Wu *et al.* (1998) who have proposed a "United mechanistic framework for the Fe(II) induced cleavage of qinghaosu (artemisinin) and its derivatives". Scheme 1 illustrates the pathways proposed in this scheme which explains the formation of the major products found in laboratory experiments (adapted from Wu *et al.*, 1998). The main features of the scheme are discussed below, but the reader should consult the original work for further information.

It is generally agreed that the initial event is the transfer of a single electron from an Fe(II) ion to the peroxy bond of artemisinin **1**. Two possible radicals (**1A**, **1B**, scheme 1), result according to whether the iron combines with O-1 or O-2, although for reasons which will be explained later, it is assumed that **1A** and **1B** are rapidly interchangeable. Each of the two radicals may then undergo further reactions.

Radical **1A** undergoes a 1,5-H shift to give **9** which leads to the formation of deoxyartemisinin, **11** *via* two possible pathways; firstly, C3-O2 scission leads to the enol **10** which could easily form **11** by addition of the OH to the enol double bond. During this process the loss of Fe(IV)=O, (Fe^{2+}=O) occurs which may be significant as this species has been suggested to be responsible for the parasiticidal affect of artemisinin (see page 260). In the second pathway leading to deoxyartemisinin formation, the intermediate **9** acquires a hydrogen atom to yield **12** and then loses Fe(III)=O, (Fe$^+$=O) to form the anion, **13** which then deprotonates to form **11**. Alternatively, intermediate **9** may undergo radical substitution at O-2 with the loss of Fe^{2+} to give **14**, which then rearranges to **15**.

Unlike radical **1A**, radical **1B** cannot abstract a hydrogen from C-4, (and no α-hydrogen is present on C-3), so that reaction takes place by β-scission. C3-C4 scission yields the C-4 primary radical **16**, which may undergo radical substitution at O-1 with the loss of Fe^{2+} to give **17** (arteannuin G). It is also possible that C3-O13 scission in **1B** may occur leading to **19** and/or **20**; this will be discussed below (see page 258).

Evidence which supports the above routes of product formation (*via* **1A** and **1B** respectively), was obtained by examining the reaction of simple 1,2,4-trioxanes (based on **1a** as models of artemisinin) with iron (Posner and Oh, 1992 and reviewed in Cumming *et al.*, 1997). Using ^{18}O labelled compounds the former authors found that the major products formed were analogous to **10** and **11**, (deoxyartemisinin), and proposed that these products could arise by mechanisms analogous to those shown in scheme 1. Interestingly, the formation of different products according to which oxygen atom is attached to the iron is paralleled in nature by the haem-induced prostaglandin-endoperoxide rearrangement (Ullrich and Brugger, 1994). In addition, **11** and **17** are also known microbial degradation products of artemisinin (Lee and Hufford, 1990) and analogous metabolites are similarly produced from artemether by microbial (Hu *et al.*, 1992) and mammalian (Chi *et al.*, 1991), metabolism as well as from deoxoartemisinin (Khalifa *et al.*, 1995).

Application of the Scheme to Artemisinin

We will now consider how the above scheme is able to explain the apparently disparate published results of experiments in which the reaction of various iron compounds with artemisinin has been investigated. It is clear that the nature and the

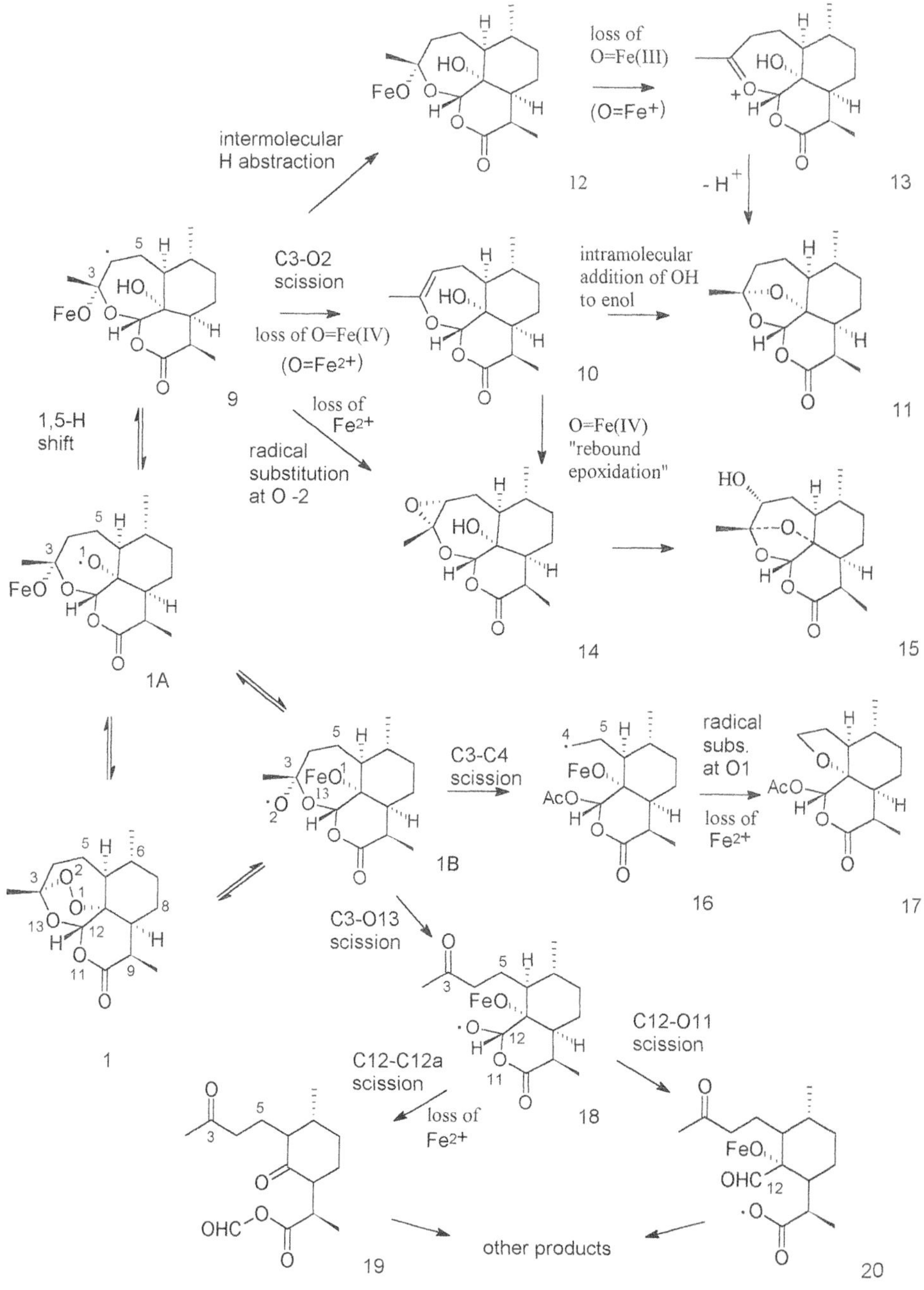

Scheme 1 Proposed mechanistic framework for the Fe(II) induced cleavage of artemisinin. (Adapted from Wu *et al.*, 1998).

relative proportions of the products arising in a particular reaction are very much dependent upon both the nature of the solvent used and on the anion of the iron salt.

For example, when tetrahydrofuran, (THF) is used as the solvent with ferrous bromide and artemisinin the major product is deoxyartemisinin, 11, whereas acetonitrile with ferrous chloride gives 17 as the major product while 11 is not formed at all. In the less polar THF the Fe-O bond is stronger than it is in the more polar acetonitrile so that the radical substitution reactions which lead to products 14,15, and 17 are not favoured since the Fe-O bond must be broken. Instead, β-scission (C3-O2), is more likely and hence the pathway to deoxyartemisinin, 11 is encouraged. The reversibility of many of the reactions, especially the interchange between radicals 1A and 1B is a key feature of the scheme. Hence, the formation of 11 will result in the removal of 9, which in turn will drive the reaction *via* radical 1A.

When acetonitrile is used with ferrous chloride, β-scission (C3-O2), which may be considered to be the elimination of an O atom radical, is not favoured because in more polar solvents the oxygen, (as Fe-OR) is more like a solvated alkoxide anion making this process very difficult. Conversely, the favouring of radical substitution results in the production of 17 along with a smaller proportion of 15. When aqueous acetonitrile with ferrous sulphate is used, the products are similar but 15 is the major product together with a smaller amount of 17 i.e. in this case the reaction proceeds mainly *via* radical 1A rather than *via* 1B as in the former case where acetonitrile and ferrous chloride were used. Again, it is suggested that this finding can be explained in terms of the effects of the solvents and anions and provided that the conversions of 1A to 9 and of 16 to 17 are considered to be the rate limiting steps in the 1A and 1B pathways repectively. With acetonitrile/ferrous chloride, the solvent will favour and therefore accelerate reactions 9 to 14 and 16 to 17. Since only the rate limiting step of the 1B pathway is affected, the result is 17 as the major product.

When aqueous acetonitrile/ferrous sulphate is used, the solvent will have a greater accelerating effect than acetonitrile alone but this will be counteracted by the sulphate anion because reactions of artemisinin with ferrous sulphate have been shown to be much slower than those with ferrous chloride which is the most reactive salt. The sulphate anion reduces the ease with which Fe^{2+} can deliver an electron to the peroxy bridge to form 1A or 1B, and that with which Fe^{3+} can receive an electron in the radical substitution reactions. In the presence of sulphate anion, reaction 16 to 17 will therefore be slowed as compared with when the chloride anion is used, while the conversion of 1A to 9 will not be affected; hence the reaction is able to proceed *via* 1A to give 15 as the major product.

Application of the Scheme to Artemisinin Derivatives

With dihydroartemisinin, 2, artemether, 3, and artesunate, 5, the major products formed by reaction of the above with ferrous sulphate in aqueous acetonitrile are analogous to those formed with artemisinin i.e. identical to 15 and 17 except for the substituent at C-10. In the case of artesunate, a significant amount of the dihydro-analogue of **15** was also formed while artemether gave rise to only a very small amount of this derivative.

In contrast, when deoxoartemisinin, **8** (Figure 1), was reacted with ferrous bromide in THF, the major product was the C-10 deoxo-analogue of **19** together with a small quantity of deoxo-deoxyartemisinin i.e. **11** without the C-10 carbonyl (Avery *et al.*, 1996). The pathway followed *via* **1A** leading to **11** is therefore similar to that which occurs with artemisinin **1** under the same conditions (see above). However, with deoxoartemisinin the route *via* **1B** is favoured and the major product is the deoxo-analogue of **19**. The formation of **17** will be disfavoured since radical substitution at oxygen is relatively slow in THF but, alternatively, **18** may arise by C3-O13 scission. There are two possibilities for **18**; C12-C12a scission with the loss of Fe^{2+} to give **19** or C12-O11 scission to form **20**, but only the latter appears to take place with artemisinin. With deoxoartemisinin, **8**, the reverse is found, hence the deoxo-analogue of **19** forms rather than the deoxo-analogue of **20**. A likely explanation for this is that a carbonyl at C-10 (as found in artemisinin), favours C12-O11 scission because in the resulting radical, **20**, the unpaired electron can be delocalised by the carbonyl and hence it more readily gives rise to other products than does **19**.

It can now be seen that this scheme as proposed by Wu *et al.* (1998), provides an explanation for the experimentally observed finding that the nature of the substituent on C-10 affects the nature of the products even though it is some distance from the peroxy bridge of the molecule (although it is important to note that the replacement of the carbonyl at C-10 with a methylene group will affect the molecule as a whole as the ring system is strained). By analogy, other artemisinin derivatives which do not possess a C-10 carbonyl such as those mentioned above would be expected to behave similarly to deoxoartemisinin under the same reaction conditions. The reason why the above reactions follow the **1B** route may be that the C3-C13, C12-C12a and Fe-O1 scissions are able to occur rapidly at a rate significantly higher than the reactions of the **1A** route leading to **11**.

While the above experiments have provided a valuable insight into the interactions of iron compounds with artemisinin, the relationship between the latter and the reaction which takes place with haem is not well understood. Experiments in which artemisinin is reacted with haem (FeII) are complicated by the requirement to prepare haem FeII *in situ* from haem (FeIII), e.g. by using a thiol. The latter, being a good hydrogen atom donor (i.e. radical scavenger), may enhance intermolecular H-abstraction thus increasing the formation of **11** *via* **12**, hence changing the product ratio; in addition, the thiol as well as the haem may form adducts with the products, thus interfering with their recovery. The use of imidazole to complex the iron and thus mimic haem has been reported by Haynes and Vonwiller, (1996a); with ferrous chloride, imidazole and acetonitrile, **17** was the major product but a significant amount of **15** and a small quantity of **11** was also formed; Jefford *et al.* (1996), obtained a similar result without the imidazole. Haemin/benzylmercaptan in THF also gave rise to **17** as the major product with a little 15 and only a trace of **11**, (Posner *et al.*, 1995), whereas ferrous bromide in THF yields **11** as the major product (see above).

Species Responsible for Parasite Death

While the above evidence suggests that the reaction of artemisinin-like antimalarials with haem generates one or more cytotoxic species which are responsible for the antiplasmodial action of these agents, the identity of the "killer" molecule(s) remains speculative. Species postulated to fulfill this role are carbon-centred radicals, the high valence iron species Fe(IV)=O and the epoxide **14**.

Evidence for carbon-centred radicals

The involvement of free radicals in the antimalarial action of artemisinin is suggested by the finding that free-radical scavengers such as ascorbic acid and vitamin E (α-tocopherol) antagonise its action both *in vitro* (Krungkrai and Yuthavong, 1987, Meshnick *et al.*, 1989) and *in vivo* (Levander *et al.*, 1989), while oxidant drugs such as miconazole and doxorubicin have been shown to potentiate its action (Krungkrai and Yuthavong, 1987). In addition, α-tocopherol inhibited the oxidation of free thiol groups in isolated erythrocyte membrane proteins which occurs during the reaction of artemisinin with haem (FeIII) (Meshnick *et al.*, 1993). Further, lipid peroxidation end products have been detected in artemisinin treated parasites (Meshnick *et al.*, 1989). Recently, Wu *et al.* (1998), have shown that a secondary radical is formed when artemisinin is reacted with ferrous sulphate in aqueous acetonitrile (see below). The radical was trapped by adding the spin trapping reagent 2-methyl-2-nitropropane to the reaction mixture and recording the electron spin resonance spectrum. This finding is consistent with the formation of a C-4 radical although there is no evidence to show that the trapped radical contained iron.

As discussed previously and shown in scheme 1, the nature and relative proportions of the degradation products arising from the reaction of iron with artemisinin or its derivatives is dependent upon which of the two main pathways is followed. Both of the postulated pathways involve the formation of carbon-centred radicals, but these are secondary in the case of the route *via* **1A**, (compound **9**) and primary in one branch only of the route *via* **1B** (compound **16**). From the experiments described above, it is clear that the route taken is greatly influenced by the nature of the iron salt and the conditions of the reaction.

In order to determine which of the two routes is likely to be more important *in vivo*, Posner *et al.* (1994), have examined the structure activity relationships of a number of derivatives of the trioxane alcohol, **21**, which were designed so that the 1,5-H shift necessary for the route *via* **1A** is either blocked or enhanced.

In artemisinin-like compounds the 1,5-H shift can only take place if there is an α-hydrogen at C-4, i.e. on the same side of the ring as the (cleaved) peroxide group. Inverting the configuration of C-4 in **21** to give **22** which has a C-4α methyl group, resulted in more than 100-fold loss of *in vitro* activity against *P. falciparum* suggesting that the route *via* **1A** is far more important in the mode of action of trioxane alcohols than the route *via* **1B**. Further, iron degradation experiments showed that an analogue in which both C-4 hydrogens were replaced by methyl groups, **23**, (and which also had low activity against *P. falciparum*), gave rise only to the ring contracted ester analogous to **17**, i.e. degradation occured only *via* the **1B** route (Posner *et al.*, 1994), whereas the potent trioxane alcohol, **21** gave rise to analogues of both **15** and **17** as expected (Posner and Oh, 1992). The loss of the *in vitro* activities of **22** and **23** were not, therefore due to steric effects preventing degradation *via* the 1B pathway.

The synthesis and structure-activity relationships of a further series of C-4 substituted 1,2,4-trioxanes, (Posner *et al.*, 1995b) and reviewed in Cumming *et al.* (1997), confirmed the above; these compounds were C-4β substituted so that the 1,5-H shift results in the formation of tertiary C-4 radicals which are more stable than the corresponding secondary C-4 radicals. Some of these compounds, e.g. **24**, showed enhanced antiplasmodial activity compared to the parent (unsubstituted at C-4), but surprisingly, those predicted to yield the most stable C-4 radicals, e.g. **25**, were less active and appeared to give rise to degradation products *via* **1B**. Taken together however, the above results suggest that the pathway *via* **1A** is important for the activity of artemisinin-like compounds in malaria parasites and therefore that the 1,5-H shift which leads to secondary carbon centred radicals may be essential for the generation of highly active derivatives.

In the route *via* 1B, the formation of a primary C-4 radical **16** resulting from the cleavage of C3–C4 in radical **1B** is proposed. However, no evidence of primary radical formation was found in the artemisinin/ferrous sulphate/aqueous acetonitrile reaction, but in this system the **1A** route predominates so that **9** rather than **16** would be expected to be the main radical produced. However, Robert and Meunier, (1997), have shown that a primary C-4 radical is likely to be formed when artemisinin reacts with *meso*-tetraphenylporphyrin co-ordinated to manganese (II). (This haem model was chosen since, like artemisinin it is non-polar, and the manganese ion can easily be removed so that characterisation of adducts formed is less difficult). The isolation of a covalent adduct in which the C-4 of artemisinin was bonded to a porphyrin ring provides evidence for the existence of a primary C-4 radical analogous to **16** in scheme 1.

Evidence for the involvement of high-valent Fe(IV)=O species

Experimental results which support the formation of high-valent Fe(IV)=O species have been reviewed by Cumming *et al.* (1997). As shown in scheme 1, the formation of epoxide **14** from **9** may occur *via* two pathways. Radical substitution at O-2 with the release of Fe^{2+} leads directly to **14** while C3-O2 scission gives **10** with the release of Fe(IV)=O, (Fe^{2+}=O). Epoxide **14** may then arise by "rebound epoxidation" by the high valent iron species which is itself reduced to Fe^{2+}. Support for the existence of high valent iron species has been obtained by the use of "reporter" reactions. In the presence of iron and artemisinin (but not one or the other), hexamethyl Dewar benzene was rearranged to hexamethylbenzene, (Traylor and Miksztal, 1987), oxidation of methylphenylsulphide to the corresponding sulphoxide and oxidation of tetralin to 1-hydroxytetralin were observed (Groves and Viski, 1990). The oxidation reactions were not affected by the removal of oxygen from the solvents, eliminating the possibility that molecular oxygen was responsible for the oxidations.

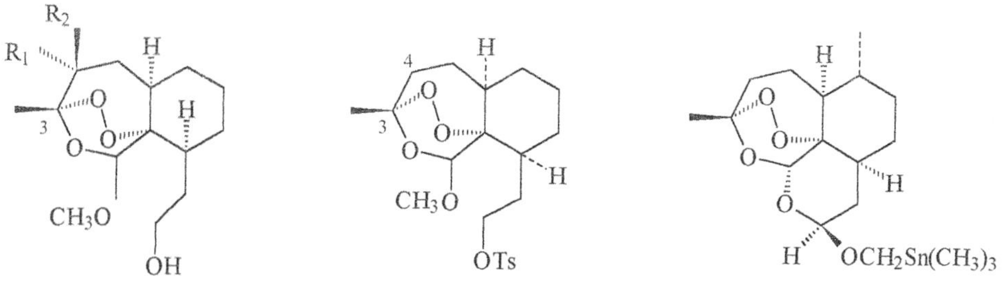

21 R_1 = H, R_2 = CH_3
22 R_1 = CH_3, R_2 = H
23 R_1 = CH_3, R_2 = CH_3
24 R_1 = H, R_2 = $PhCH_2$
25 R_1 = H, R_2 = $(CH_3)_3SiCH_2$
26 R_1 = H, R_2 = $(CH_3)_3Sn$

Figure 2

When artemisinin is degraded with iron, 4-hydroxydeoxyartemisinin, **15**, is formed as a single 4α-stereoisomer by the action of the α-hydroxyl group on the α-epoxide, **14** which could arise by one or both of the above routes (Posner et al., 1995). In contrast, Posner and Oh, (1992), found that the degradation products of trioxane tosylate, **27** contained both C-4α and C-4β hydroxytrioxanes. This mixture could only have arisen *via* a mixture of α- and β-epoxides which in turn must have been produced by an intermolecular reaction since the 3α-iron oxy bond cannot access the β-face of the molecule i.e. the high valent iron species can react on either face. Cummings et al. (1997), consider that, "taken as a whole the above strongly implicate involvement of a high-valent iron-oxo intermediate in the iron-induced degradation of artemisinin and its analogs". However, by analogy with trioxane **27**, we would expect artemisinin to give rise to *both* C-4α and C-4β hydroxy isomers if a high valent iron-oxo intermediate is involved, but, as stated above only the 4α isomer has been reported.

In support of the involvement of high valent Fe(IV)=O, haem iron-oxo species are known to be involved in the action of horse radish peroxidase and cytochrome P-450 enzymes; also, a number of non-haem iron containing mono-oxygenases have been shown to effect the epoxidation of olefins which suggests that high valent iron-oxo species could be functioning as intermediates (Cumming et al., 1997, and references therein).

In order to provide further evidence in support of high-valent Fe=O as a reaction intermediate Posner et al. (1996), synthesised a number of 1,2,4-trioxane analogues with various substituents at C-3 and C-4 which were designed either to facilitate or inhibit the release of Fe(IV)=O following the reaction of the trioxane with Fe(II). The substitution into the molecule of a (trimethyltin)methyl group at C-4 (compound **26**), as a better radical leaving group than Fe(IV)=O resulted in a ten-fold reduction of antiplasmodial activity compared to an analogue with a C-4β-(trimethylsilyl)methyl group, **25**. When the former was degraded in the presence of iron the product contained 15% of a 4-methylene deoxytrioxane derivative showing that trimethyltin was lost by at least a proportion of the compound and no rearrangement of hexamethyl Dewar benzene was observed. However, it is possible that the presence of the tin in the molecule may have reduced the antiplasmodial activity due to other reasons; to assess this possibility, Posner *et al.* (1996), prepared a tin-containing ether of dihydroartemisinin, **28**, and reported that, as this compound has measurable antiplasmodial activity this indicates that the presence of tin does not *necessarily* destroy a trioxanes activity; however, it was more than 100-fold less active than artemisinin so that, on the contrary, it is quite possible that the presence of tin *per se* may well reduce antiplasmodial activity, especially as dihydroartemisinin, **2**, is more active than artemisinin itself.

Better evidence for the formation of Fe(IV)=O was obtained with two C-3 substituted analogues, **29, 30**, which were designed so that the postulated C-4 radical intermediate (analogous to **9** in scheme 1) could undergo a second 1,5-H shift leading to a tertiary radical, (from **29**), or an even more stable benzylic radical, (from **30**) so that the release of Fe(IV)=O would be less favoured (scheme 2). Unexpectedly, these compounds were found to be about 6-fold more active than their 3β-methyl and 3β-ethyl analogues against *P. falciparum in vitro* but, significantly, on reaction with ferrous bromide in tetrahydrofuran rearrangement of hexamethyl Dewar benzene was seen suggesting the presence of a high-valent iron oxo intermediate. The potential of the above compounds to undergo a second 1,5-H shift appears to have failed to prevent β-scission of Fe(III)-O· (↔ Fe(IV)=O) perhaps because the former is too slow to compete with the latter; the enhanced antiplasmodial activity is suggested to be due to another effect e.g. increased transport across membranes.

Derivatives with either vinyl or phenyl substituents at C-3 designed to conjugate with, and hence stabilise the double bond formed when Fe(IV)=O leaves were found to have similar and markedly increased activity (respectively) compared to an analogue possessing a C-3β ethyl group, but on reaction with iron only the vinyl trioxane produced rearrangement of hexamethyl Dewar benzene to hexamethylbenzene (Posner *et al.*, 1996).

Wu *et al.* (1998), have argued against the above evidence for the involvement of high-valent Fe=O, by suggesting that the "reporter" reactions do not necessarily require Fe(IV)=O as an oxidant. For example, the rearrangement of hexamethyl Dewar benzene may be initiated by the loss of one electron which could be received by radical **1A**; this would lead to an increased proportion of **12** and hence **11** as is actually observed (Posner *et al.*, 1995). Wu *et al.* (1998), also propose that since, in aqueous media, product **11** is not formed, this is evidence that the C3–O2 β-scission required for the release of Fe(IV)=O does not occur; however, as discussed above, and shown in scheme 1, C3–O2 β-scission may lead to **15** so that absence of **11** does not preclude C3–O2 scission. In addition, the latter authors point out that nothing is known about the oxidative ability of Fe(IV)=O which itself is a speculated species; in the parasite, by analogy, haem Fe(IV)=O would be present and it cannot be assumed that this would behave in the same way as the iron in non-haem containing sytems.

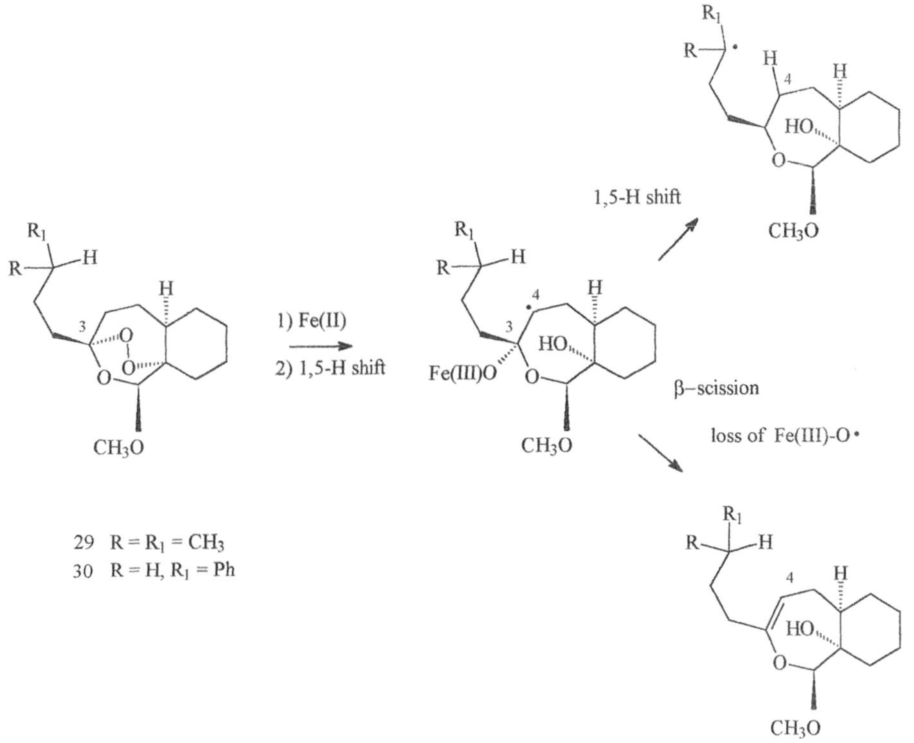

Scheme 2   Proposed products resulting from the Fe(II) induced cleavage of compounds **29** and **30**. (Adapted from Posner *et al.*, 1996).

Taken together, the above indicates that while there is some experimental evidence in support of the role of high valent iron-species in the antiplasmodial action of artemsinin-like compounds, this is by no means proven and there are some anomalous results which have yet to be explained.

Evidence for the formation of alkylating agents

Posner *et al.* (1995), proposed that **14** may act as an alkylating agent but Wu *et al.*, (1998) consider that rearrangement e.g. to **15** is more likely as O-1 is well placed to compete with other nucleophiles which may attack C-3. If the non-peroxidic O-13 of artemisinin is replaced by a methylene group, the resulting carba-analogue is 25-fold less active against *P. falciparum in vitro* (Avery *et al.*, 1995). A possible explanation for this finding is that the carba-analogue of **10** would be less susceptible to epoxidation by Fe(IV)=O since it is not as electron-rich as **10** itself (Cumming *et al.*, 1997); thus the reduced activity of the carba-analogue of **10** may be consistent with a reduced ability to form the epoxide (i.e. the carba-analogue of **14**). On the other hand, Avery *et al.* (1996), have shown that the presence of an epoxide does not necessarily enhance antimalarial activity since the epoxide, **31**, was devoid of anti-

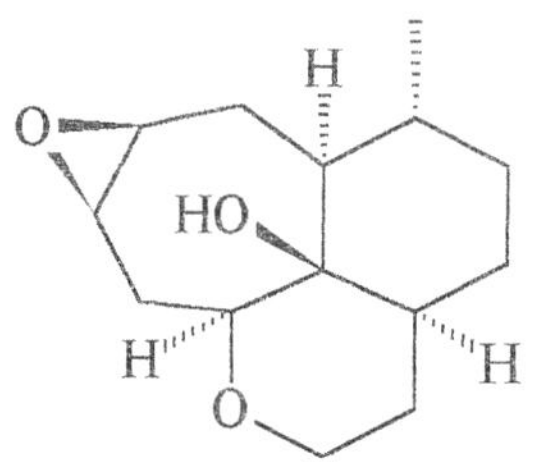

Figure 3

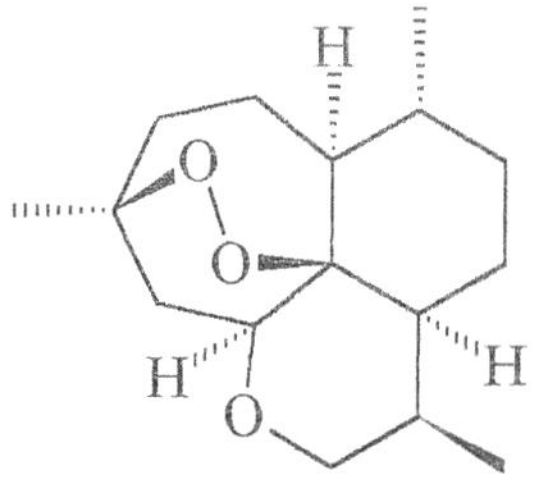

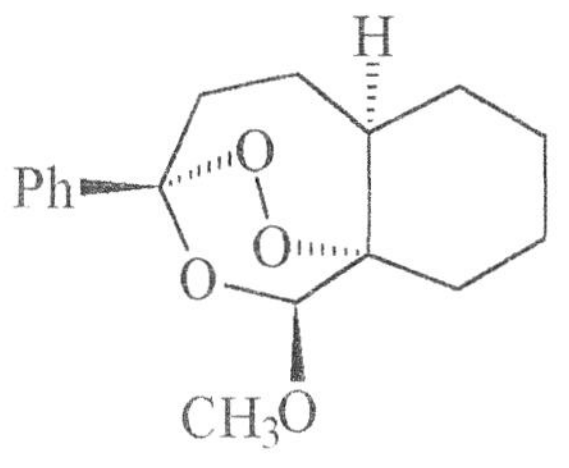

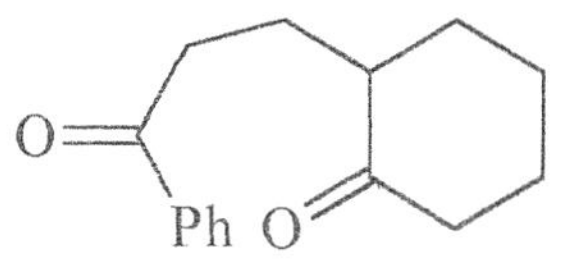

malarial activity while the artemisinin analogue, **32**, was about 2/3 of the potency of artemisinin itself.

One other candidate which has been proposed as an alkylating agent is one of the iron degradation products of the 3-phenyltrioxane, **33**. Reaction of the latter with ferrous bromide yielded a ring-contracted acetal analogous to **17** plus a smaller amount of a diketone, **34** which was tested for antiplasmodial activity because dicarbonyl compounds are known to alkylate proteins (Posner *et al.*, 1996). Although the diketone was more than 100-fold less active than artemisinin, the latter authors suggested that 3-phenyl trioxane may be a prodrug, the active diketone being formed within the parasite; the low antiplasmodial activity of the former is thought to be due to difficulty in reaching the malaria parasite. However, as no analogous diketones have been reported to be formed by iron degradation of artemisinin and its derivatives, it appears unlikely that **34** will prove to be an important intermediate.

Parasite Molecules Targetted by Artemisinin

Although, as discussed previously, haem iron is likely to be responsible for the activation of artemisinin and the subsequent formation of free radicals, there is also evidence which suggests that artemisinin binds to haem forming a toxic complex and/or interferes with haem metabolism.

Hong *et al.*, 1994, reported evidence for the formation of artemisinin-haem adducts in *P. falciparum* infected erythrocytes incubated with labelled artemisinin. When artemisinin was incubated with isolated haemozoin the former disappeared in a time-dependant manner and two products of m/z 856 and 871 were purified using high performance liquid chromatography. On thin-layer chromatography most of the labelled haemozoin from the infected erythrocytes co-migrated with the two adducts formed with isolated haemozoin thus suggesting that similar covalent artemisinin-haemozoin adducts may be formed *in vitro* and *in situ*. However, these adducts were found to be inactive against malarial parasites *in vitro*, and pre-incubation of artemisinin with haem (FeIII) reduced its activity (Mehsnick *et al.*, 1991).

As mentioned above, Robert and Meunier, (1997), reported the formation of a covalent adduct resulting from the reaction between artemisinin and a manganese containing porphyrin. If a similar reaction occurs with artemisinin and haem (rather than with haemozoin) in the malaria parasite, inhibition of haemozoin formation might result; Orjih (1996), reported that haemozoin formation in parasitized erythrocytes was reduced by 95% when artemisinin (250ng/ml) was added to the culture medium.

In contrast, Asawamahasakda *et al.* (1994) found that, in their experiments, artemisinin did not inhibit haemozoin formation, nor did it cause its degradation; however, in another paper Asawamahasakda *et al.* (1994a) suggested that artemisinin may interfere with haemozoin formation by alkylation of "histidine rich protein" which is thought to be involved in the polymerisation of haem to haemo-zoin (Sullivan *et al.*, 1996).

A possible mechanism by which the interaction of artemisinin with haem (FeIII) may lead to parasite death has been proposed by Berman and Adams, (1997), who showed that the redox activity of haem (FeIII) as measured by its peroxidase activity and by its ability to oxidise membrane lipids has been shown to be enhanced by artemisinin. Increased redox activity paralleled major changes in the absorption spectrum of haemin with the loss of the Soret peak. The latter authors propose a model in which artemisinin binds irreversibly to haem in the parasite food vacuole preventing its polymerisation to haemozoin and promoting haem-catalysed oxida-tion of the vacuolar membrane by molecular oxygen which leads ultimately to vacuole rupture and parasite autodigestion. Adams and Berman, (1996), have also examined the kinetics of the reaction between artesunate and haem (FeIII); these are consistent with a three-step two-intermediate mechanism with the final product pos-sessing a degraded tetrapyrrole ring system. In addition, the redox activity of the haem (FeIII)-artesunate complex was shown to be about four fold that of haem (FeIII) alone.

The alkylation of parasite proteins by artemisinin-derived free radicals may be important for the lethal action of these compounds, but there have also been a few reports suggesting that artemisinin-like drugs may interact with some non-parasite proteins. The latter will be discussed first as they are important in the context of understanding the basis of the selective toxicity of artemisinin against malaria parasites.

Using [3H]-dihydroartemisinin and [14C]-artemisinin Yang *et al.* (1993), reported that covalently bound drug-protein adducts were formed when these compounds were incubated with human serum albumin. Mass spectrometry indicated that the drug-albumin adducts had a molecular weight about 478 above that of the protein, consistent with the binding of more than one drug molecule to each molecule of protein. Examination of the effects of iodoacetamide, N-ethylmaleimide, haem (FeIII) and Fe^{2+} on the binding showed that thiol and amine groups were involved and that both iron-dependent and iron-independent reactions had taken place. In contrast, Muhia *et al.* (1994), reported that no loss of artemisinin occurred when the drug was added to haemoglobin-free plasma and incubated at 37° C for 24 hours, but when artemisinin or artemether was added to whole blood the drugs were decomposed to products which were undetectable by HPLC but there was no reduction in antiplasmodial activity. The decomposition was greatly enhanced by haem (FeIII) and to a lesser extent by haemoglobin (as red cell lysate). Experiments with [14C]-artemisinin have provided evidence that this compound is able to alkylate the haemoproteins cytochrome C, catalase and haemoglobin with most of the drug being bound to the protein in the latter two molecules (Yang *et al.*, 1994), but haemoglobin in intact erythrocytes was not alkylated. Although only 5–18% of added drug bound, and Vattanaviboon *et al.* (1997), have calculated (from the former authors' data), that only about 0.003 molecule of drug was covalently bound per haemoglobin molecule, these results are consistent with the above evidence for haemoglobin-mediated decomposition of artemisinin.

Bakshi *et al.* (1997), confirmed that artemisinin concentrations decrease in the presence of whole blood or haemoglobin, and this was enhanced in an oxygen-free atmosphere; it is suggested that oxygen may compete with the peroxide moiety of the drug for the haem in haemoglobin and that this may contribute to the selective toxicity in malaria parasites as the oxygen tension in parasitised cells is low. No reduction was seen when artemisinin was incubated with erythrocyte ghosts suggesting that little binding to red cell membranes occurred; however this finding is in contrast to that of Asawamahasakda *et al.* (1994), who reported that labelled dihydroartemisinin binds to proteins located on the cytoplasmic face of the red cell membrane but it was not taken up by intact erythrocytes. However, the drug was not weakly bound as it was not removed by phospholipase digestion followed by solvent extraction, but it was removed by mercaptoethanol treatment which suggests that sulphydryl groups are involved. The particular proteins bound included the structural proteins spectrin and actin which help to maintain the deformability of the red cell; interestingly high concentrations of artemisinin have been shown to reduce red cell deformability (Scott *et al.*, 1989).

Clearly, further work will be needed to clarify the inconsistencies in the above results; it is possible that traces of haemoglobin-breakdown products could have been responsible for the decomposition of the drugs in some of the experiments. These findings have important implications with respect to the measurement of blood levels of artemisinin-like drugs, especially in patients with malaria in whom haemoglobin degradation products are released into the blood plasma.

When *P. falciparum* infected erythrocytes are incubated with [10–^{3}H]-dihydro-artemisinin at physiological drug concentrations a number of non-abundant parasite proteins are selectively alkylated; the same proteins were alkylated by several different artemisinin derivatives but no alkylation occurred in non-infected erythrocytes (Asawamahasakda *et al.*, 1994b). Four major protein bands with relative molecular masses of 25, 50, 65 and > 200 kDa and two minor bands, (32 and 42 kDa) became labelled. Bhisutthibhan *et al.* (1998), have identified one of these proteins, (the 25kDa band), as the *P. falciparum* translationally controlled tumour protein homologue (TCTP). *In vitro*, the latter authors found that dihydroartemisinin reacts with recombinant TCTP but only in the presence of haem (FeIII). Using Scatchard analysis TCTP was shown to bind 2 molecules of haem (FeIII)/molecule of protein but only one molecule of dihydroartemisinin/molecule of protein. TCPT contains a single cysteine residue which is important for artemisinin binding since pretreatment of TCTP with iodoacetamide reduced drug binding by two-thirds but did not affect haem (FeIII) binding. At the present time the function of TCTP in malaria parasites is unknown and although the above studies suggest that the mode of action of artemisinin is associated with the alkylation of certain parasite proteins, this does not prove that this is the mechanism by which parasite death occurs.

Damage to DNA has also been proposed to be involved in the antimalarial action of artemisinin (Wu *et al.*, 1996). When pUC18 supercoiled DNA, calf thymus DNA or salmon DNA were incubated with artemisinin and ferrous sulphate in aqueous acetonitrile buffer at pH 6.5 at 37° C for 24 hours cleavage of DNA occurred; artemether also caused lesions in supercoiled DNA but with less efficacy than artemisinin.

Other Mechanisms Relevant to Antimalarial Activity

Using flow cytometry, Wenisch *et al.* (1997), examined the effect of artemisinin on phagocytic function of neutrophils by measuring the uptake of fluorescein isothio-cyanate-labelled *Escherichia coli*; in addition, the intracellular release of reactive oxygen intermediates following phagocytosis by neutrophils was also assessed by estimating the amount of dihydrorhodamine 123 converted to rhodamine 123 intra-cellularly. Incubation of cells with artemisinin, artemether and dihydroartemisinin decreased phagocytosing capacity, but in contrast, the intracellular generation of reactive oxygen intermediates was enhanced. Phagocytosis by neutrophil granulo-cytes has been shown to have a role in severe malaria and human granulocyte-stimulating factor has been shown to suppress parasitaemia in rodent malaria (Waki *et al.*, 1993). It is suggested that the increase in reactive oxygen intermediates could be advantageous in malaria and explain the rapid parasite clearance seen with artemisinin derivatives.

As previously discussed in chapter 12, many artemisinin analogues have been made in the search for new antimalarial agents, and these studies have shown that the pres-ence of the trioxane ring is a requirement for activity. Since the interaction of artemisinin with haem iron is required for its activity, Kamchonwongpaisan *et al.* (1995), prepared a number of artemisinin derivatives linked to iron chelators in order to test their hypothesis that if iron can be made available in close proximity to the artemisinin moiety, then enhanced activity should result, provided that the iron is not bound too strongly, since the chelator must "deliver" the iron to the "active site" of the drug. In addition, iron chelating agents themselves have intrinsic antimalarial activity which might be expected to act synergistically with artemisinin. A variety of chelating agents were used, coupled to dihydroartemisinin via the C-10 oxygen.

None of the linked compounds was superior to the parent compound although several retained potency comparable to that of artemisinin (about 5-fold lower than dihydroartemisinin). The activity of the compounds was not affected by the addition of Fe^{3+} to the medium showing that reaction with exogenous iron was not responsible for parasite killing. It is possible that the lack of enhanced activity observed was due to reduced concentration in the parasite as it cannot be assumed that their penetration into the parasite food vacuole is as good as that of the parent molecule. The above authors did not determine whether there is a difference between Fe (II) and Fe (III) chelators, but this may be important if, as suggested above, the reaction of artemisinin is much faster with Fe (II) than with Fe (III). One of the chelating agents linked with dihydroartemisinin, diethyl dithiocarbamate is a potent inhibitor of the antioxidant enzyme superoxide dismutase which is present in the erythrocyte and the parasite; although no enhancement of activity was seen, this kind of approach may yield highly active antimalarials in the future. However, in another study, Posner et al. (1995a), found that trioxane analogues with potential chelating ability did not have increased activities.

TOXICITY OF ARTEMISININ AND ITS DERIVATIVES

Introduction

In addition to their rapid onset of action and high activity against drug-resistant malaria parasites, artemisinin and its clinically used derivatives are remarkable for their lack of reported toxic effects, (Meshnick et al., 1996), although an early Chinese study showed that artemisinin caused the death and resorption of rat and mouse foetuses when given after day 6 of gestation (Anon. 1982a, cited in Klayman, 1985). However, for some time there has been concern that artemisinin-like agents might be neurotoxic and, recently, it has been reported that acute cerebellar dysfunction may be associated with artesunate therapy in man (Miller and Panosian, 1997).

In the first animal studies, occasional deaths were seen in animals given high doses of artemisinin-like drugs; these were presumed to be due to cardiotoxicity since abnormalities in the electrocardiogram (ECG) had been observed in animals as well as in humans (Brewer et al., 1994). Further studies in dogs and rats given moderate daily doses of I.M. arteether and artemether showed that these drugs cause delayed, dose-dependent neurotoxicity with symptoms of gait disturbance, loss of spinal and pain response, brain stem and eye reflexes as well as ECG abnormalities. Neuropathological lesions were characterised by swelling and rounding of nerve cell bodies, increased eosinophilia, vacuolisation of cytoplasm with a loss of Nissl substance (central chromatolysis), swelling and fading of nuclei, and separation and clumping of fibrillar and granular components of nucleoli. Strikingly, lesions were found almost exclusively in the pons and the medulla and particularly affected various nuclei present in these areas, suggesting a receptor-mediated neurotoxicity. Although most prominent in animals receiving 15 mg/kg/day or more, scattered lesions were seen with doses of 5 mg/kg/day given for 28 days. This may be very significant considering that the clinically used dose of artemether in man is 10 mg/Kg/day and that drug toxicity in the mouse is generally lower than that in man due to the higher rate of metabolism associated with a greater surface area to weight ratio.

The delay in toxicity, (after 9 days in dogs), suggested that either prolonged exposure to the drug was required to produce the effects or that the accumulation of toxic metabolites was responsible. The latter is supported by the lack of reported neuropathic changes following high acute doses (although seizures and cardiovascular effects have been noted), and by the poor correlation of parent drug levels to lesion severity; artemether and arteether appear to be rapidly metabolised after absorption, but their major metabolite, dihydroartemisinin appears to have a longer half-life and accumulate with repetitive daily dosing (Brewer *et al.*, 1994). Selective dose-dependent toxicity of arteether to the brain stem was also observed in rhesus monkeys (*Macaca mulatta*) and, significantly, lesions were found in animals which did not exhibit clinical signs of neurotoxicity (Petras *et al.*, 1997).

Further studies were carried out by Genovese *et al.* (1998), in rats in order to investigate the effects of lower doses of artemether than had been used in previous studies and to examine its effects on auditory structures which had been shown in previous studies (Petras *et al.*, 1997), to be damaged by arteether. At low doses (3.125 and 6.25 mg/kg/day × 7 days) one rat in each group exhibited damage to the nucleus trapezoideus showing that auditory systems can be affected at low doses. At 12.5 mg/kg/day all the rats appeared healthy but damaged neurons were observed in the trapezoideus and superior olive nuclei and to a lesser extent in the nucleus ruber and mild damage was present in several other nuclei; the results indicate that at least in rats, auditory dysfunction could be an early sign of arteether-induced neurotoxicity (Genovese *et al.*, 1998a).

As the above experiments were carried out with parenteral oil-based formulations the possibility that the route and vehicle might influence the toxicity of artemether and arteether should be borne in mind; orally active artemisinin derivatives are less hazardous in experimental animals (Nontprasert *et al.*, 1998).

Mechanism of Neurotoxicity

Artemisinin analogues are cytotoxic to neuronal and glial cells in culture (Fishwick *et al.*, 1995). Since the brain contains a high level of iron, Smith *et al.* (1997), examined the effects of haem (FeIII) on the neurotoxicity of artemisinin derivatives *in vitro* in order to determine whether the mechanism of toxicity was analogous to that which is responsible for parasite death.

Artemether and arteether did not inhibit neurite outgrowths of cultured mouse neuroblastoma NB2a cells at concentrations of up to 300 nM while dihydroartemisinin inhibited growth by 35% at 300 nM. In the presence of haem (FeIII), (2 μM), growth inhibition was markedly enhanced but desoxyarteether (deoxyarteether), which does not have the endoperoxide moiety, was non-toxic at 300nM and no enhancement was seen in the presence of haem (FeIII). Similarly, the metabolism of the tetrazolium salt MTT by NB2a cells was significantly inhibited by artemether and dihydroartemisinin in the presence of haem (FeIII) but not by desoxyartemether. In addition, haem (FeIII), (2 μM), significantly increased the binding of ^{14}C-dihydroartemisinin to both rat brain homogenate and to NB2a cells. These results suggest that the neurotoxicity produced by artemisinin analogues may be due to haem catalysed production of free radicals. This is supported by the increase in drug binding induced by haemin which is consistent with the alkylation of proteins and by the finding that the effect of the drug/haemin combination on neurite outgrowth-a measure of toxicity to axonal/neurite mainte-nance-was greater than the effect on MTT metabolism—a measure of general neuronal survival. The presence of haemozoin in significant quantities in the brains of patients with cerebral malaria may possibly support the feasibility of the above mechanism (Smith *et al.*, 1997).

Evidence for the alkylation of neuronal proteins has been obtained by incubating NB2a mouse neuroblastoma cells with ^{3}H-dihydroartemisinin (Kamchonwongpaisan *et al.*, 1997). Gel electrophoresis showed that neuronal proteins with molecular weights of 27, 32, 40 and 81 kD were alkylated but this was much slower and weaker than found with *P. falciparum* proteins from cells incubated under identical condi-tions. Because of the above, Fishwick *et al.* (1998), postulated that the neurotoxicity of artemisinin derivatives might be explained by the binding of the compounds to com-ponents of the cytoskeleton that are necessary for axonal maintenance. The finding that diketone 34, (see above), may be a degradation product of trioxanes suports this proposal since neurotoxic diketones such as 2,5-hexanedione act by forming pyrrole cross-links between cytoskeletal proteins.

In order to test this hypothesis, Fishwick *et al.* (1998), used various techniques to investigate the mechanism of the toxicity of dihydroartemisinin to NB2a neuroblastoma cells. However, no evidence of specific changes in the cell cytoskeleton were detected using Western blotting and monoclonal antibodies and immunocytochemistry failed to detect any changes. Transmission electron microscopy revealed that in dihy-droartemisinin-treated NB2a cells the cristae of the mitochondria were absent or deformed and there was little rough endoplasmic reticulum visible either due to a reduc-tion in the amount or possibly due to ribosome dissociation. Scanning electron microscopy showed that dihydroartemisinin depleted the filopodia-like processes pro-jecting from the surface of the cell body and neurites. None of the above effects were seen in cells treated with deoxyartemisinin. These data suggest that dihydroartemisinin exerts its effects by disruption of the mitochondria and other membranous organelles but there is no evidence to support the suggestion that artemisinin derivatives cause damage to the cytoskeleton (Fishwick *et al.*, 1998).

However, in another study, Fishwick *et al.* (1998a), reported that the increased toxicity to NB2a cells seen when cells are incubated with dihydroartemisinin in the presence of haem (FeIII) is accompanied by a corresponding increase in binding of dihydroartemisinin with protein. Haem (FeIII) increased binding in rat cortex homogenate, NB2a and C6 cells although there were differences in the low and high affinity binding capacities and K_D values between the three preparations. Iodoacetamide and sodium cyanate pre-treatment reduced binding to cortex proteins by approximately 70%, and the majority of the decrease was due to the blocking of thiol groups by iodoacetamide with only a small effect due to the blocking of amine groups by cyanate. Binding was not disrupted by acetone and detergent extraction and boiling indicating that covalent bonds had been formed. Again, deoxyartemisinin had no effect on the cells and the above data are consistent with the iron-mediated degradation of dihydroartemisinin leading to free radical production and the subsequent alkylation of parasite proteins (Fishwick *et al.*, 1998a).

Although the toxic effects of artemisinin derivatives on neuronal cells are most important, these compounds have also been shown to be toxic to some tumour cells. Using the MTT assay and the clonogenic assay Beekman *et al.* (1996), showed that artemisinin and a dimer of dihydroartemisinin were cytotoxic to murine Ehrlich ascites (EN19) cells and to human HeLa S3 cancer cells, but the main effect was growth inhibition rather than cell killing; in contrast, artemisitene, which possesses an exocyclic methylene group at C-9 and which has intrinsic alkylating properties, acted by killing cells.

The inhibition of cell growth observed with artemisinin occurred in the absence of iron, and HPLC analysis showed that recovery of unchanged drug was complete after 24 hours incubation with cells, (Beekman *et al.*, 1997a), although thin-layer chromatography revealed a trace amount of an unidentified compound with a mass 16 units higher than that of artemisinin. When an excess of ferrous ammonium sulphate was added, artemisinin G, 17, (Scheme 1), was formed as the major product. Not unexpectedly, the products of the reaction between artemisinin and the ferrous salt were not toxic to Ehrlich ascites (EN2) cells since it is likely that it is the free-radical intermediates which are toxic. In the presence of ferrous iron up to 10 μM the cytotoxicity of artemisinin was not enhanced, apparently because the culture medium interacts with Fe^{2+}. This is consistent with the observation that the toxicity of artemisinin to neuroblastoma cells was not enhanced by Fe^{2+} (Parker *et al.*, 1994), which again, may possibly be due to interference of the iron by medium constituents. However, Moore *et al.* (1995), have reported that a combination of dihydroartemisinin and ferrous sulphate given orally to rats significantly reduced the growth of implanted fibrosarcoma tumours, while no significant effect was seen with either compound alone, suggesting that in some circumstances ferrous iron can enhance the toxicity of dihydroartemisinin. In another study, Lai and Singh, (1995), showed that toxicity of dihydroartemisinin to a human leukaemia cell line, (molt-4 lymphoblastoid cells), was enhanced in the presence of holotransferrin; much less effect was seen with normal human lymphocytes, perhaps because most cancer cells have high rates of iron intake.

From the above, it appears that artemisinin derivatives are more likely to exhibit cytotoxocity to cells enriched with iron, particularly bound iron such as that found in haem. Since the peroxy group is essential for both iron enhanced cytotoxicity and antimalarial activity of artemisinin derivatives it is unlikely that analogues will be found which are devoid of the cytotoxic effect while retaining high antiplasmodial activity. Beekman *et al.*, 1997, have shown that the cytotoxicities of ether-linked dimers of dihydroartemisinin and of epoxides of artemisitene were stereochemistry dependent, but in these compounds the peroxide moiety was not crucial for the cytotoxic effects although it contributed to them.

METABOLISM AND PHARMACOKINETICS

Studies on the absorption, distribution and metabolism of artemisinin and its derivatives have been hampered by the poor UV absorption of most of these compounds, and White, (1994), has advised that the early data should be interpreted with caution. In addition, the conversion of artemisinin to non-detectable but active metabolites in whole blood, (Muhia *et al.*, 1994), requires that care must be taken when attempting to relate blood levels to therapeutic effects.

Artemisinin, arteether and artemether are rapidly metabolised *in vivo* to dihydroartemisinin (Lee and Hufford, 1990). Because artemisinin is insoluble in both water and oil, derivatives which are more easily formulated are preferred; however, the pharmacokinetics of oral, intramuscular (IM) and rectal artemisinin have been studied in man (Titulaer *et al.*, 1990). Artemisinin was rapidly and well absorbed orally (peak levels 260 ng/ml after 1 hour, elimination half-time 1.9 hours) with a bioavailability 32% of that found with an oil suspension administered by intramuscular injection (peak levels 209 ng/ml after 3.4 hours, elimination half-time 7.4 hours), and animal studies have shown that extensive first-pass metabolism occurs; IM absorption from an aqueous suspension is poor. Rectal absorption of artemisinin was much slower with peak levels (c. 170 ng/ml) achieved after 11.3 hours while the elimination half-time was 4.1 hours (Shen, 1989, cited in White, 1994). With I.M. artemether, peak levels (c. 200 ng/ml) occurred after about 6 hours then declined with an elimination half time of 4–11 hours (Zhao *et al.*, 1988). Metabolism to dihydroartemisinin is slower than with artemisinin so that the parent compound is probably responsible for the majority of the antimalarial effect *in vivo* (White, 1994). Kager *et al.* (1994), reported that arteether was found to have a longer elimination half-life (23 hours after a single dose), but this is surprising considering that the pharmacokinetics of the two esters are very similar in animals and it has been suggested that this difference is due to assay sensitivity rather than a genuine pharmacokinetic difference (White, 1994). It is important to note, however, that the absorption of arteether and artemether following intramuscular injection is erratic, (White, 1998), particularly in severe malaria where rapid absorption may be critical for the survival of the patient (Murphy *et al.*, 1997); poor absorption in severe malaria is thought to be due to metabolic acidosis.

More recently, Teja-isavadharm *et al.* (1996) compared the bioavailability of oral, rectal and IM artemether (5 mg/kg), in healthy volunteers. Artemether and dihydroartemisinin plasma levels were measured and plasma antiplasmodial activity was also assessed. Oral absorption of artemether and its conversion into dihydroartemisinin was rapid (artemether peak levels 406 nmol/litre after 1.7 hours; dihydroartemisinin peak levels 1009 nmol/litre after 1.8 hours). A comparison of plasma concentrations and bioassay results suggested that other unidentified metabolites contributed little to antimalarial activity *in vivo*. IM and rectal absorption were slower and more variable and dihydroartemisinin concentrations were lower than with oral dosing; relative bioavailabilities were 25% (I.M.) and 35% (rectal) in the six hours following administration. It is suggested that further studies should be carried out to determine whether rectal artemether would be effective for the treatment of severe malaria where oral or parenteral administration is not possible.

In contrast to artemether, artesunate given intravenously or orally is very rapidly hydrolysed to dihydroartemisinin which has an elimination half-life of approximately 45 minutes (Yang *et al.*, 1986), so that this compound may be considered to be a pro-drug for dihydroartemisinin. The main problem experienced with the clinical use of artemisinin derivatives is that of recrudescence where parasites are apparently cleared from the blood with drug-treatment but re-emerge some time later, (usually about two weeks) after treatment has been completed. This is not due to drug-resistance since the parasites remain drug-sensitive, but to a failure of treatment to completely eradicate parasites. The reasons for this phenomenon are not well understood but as it is more commonly seen with oral therapy, limited oral bioavailablity has been suggested as a possible cause but other factors such as variation in sensitivity of different blood stages and rapid elimination of the drug may also be involved.

Using Caco-2 intestinal epithelial cells, Augustijns *et al.* (1996), investigated the transepithelial permeability of artemisinin and sodium artesunate. Artemisinin crossed the epithilium more easily than the highly ionised artesunate and transport appeared to take place by passive diffusion; artesunate permeability was pH dependent and was 4-fold higher at pH 6 than at pH 7.4, consistent with a pKa value of 4.6; it was not reduced during transport. The authors suggest that passage across the intestinal epithelium is probably not a limiting factor in the oral absorption of these compounds.

Recrudescence occurs even in some patients who have had seven days treatment with an artemisinin derivative which according to theoretical models should be sufficient to eradicate all parasites (White, 1998). One possible explanation is that release of merozoites from the liver stages is not continuous and may be delayed until drug treatment has ceased (Murphy *et al.*, 1990); in highly endemic areas additional liver stage parasites may be present (due to repeated mosquito bites) which reach the blood after the drug has disappeared. Another explanation could be that some circulating parasites are not killed because they have not entered the growth phase or because they are damaged but not killed. At a very early stage of parasite development it is suggested that there may be insufficient free haem to induce destruction but enough to damage the parasite and temporarily delay its development (White, 1998). When artemisinin plasma levels in orally treated malaria patients were measured on day 1 and day 6 of therapy it was found that the maximum levels attained were about 6-fold higher on day 1 than on day 6 suggesting that the rate of metabolism had increased due to autoinduction (Alin *et al.*, 1996). Further studies will be needed to determine whether this is related to recrudescence.

Artelinic acid, 6, (Fig. 1), is a new water soluble aromatic ether derivative of dihydroartemisinin which is being developed by the Walter Reed Army Institute of Research in the USA. While this compound has similar antimalarial activity to arteether and artemether, it is much more stable to hydrolysis in aqueous solutions than artesunate, and studies suggest that artelinate has more favourable pharmacokinetics than artesunate (Li *et al.*, 1998). In dogs, absorption from the oral and intramuscular routes was rapid and high (80–90%), and the ratio of radioactive artelinate in urine to that in faeces was lower (< 0.7) after oral than after intravenous, (1.2) or intramuscular, (1.3) administration, suggesting presystemic excretion of artelinate metabolites in the bile (Li *et al.*, 1998). Oral bioavailablity in dogs (80%), was much higher than that in rats (33%), and that reported in rabbits (4.6%) (Titulaer *et al.*, 1993). Fortunately, the dog is considered to be a better model than the other species for making predictions in man.

Compared with the above artemisinin derivatives, artelinate produced the highest maximum plasma levels with the largest area under the concentration-time curve while the volume of distribution and clearance values were the lowest; the mean residence time and elimination half-life were the longest. Much less artelinate was metabolised to dihydroartemisinin, (4.3%) compared with artesunate, (up to 73%), and it is suggested that this accounts for the lower toxicity of artelinate seen in rats since dihydroartemisinin is more toxic to rats than its analogues (Li *et al.*, 1998). For the above reasons, it is hoped that artelinate will prove to be an effective orally active antimalarial.

Due to the problems mentioned above, it has not been possible to use pharmacokinetic data to optimise dose regimens for artemisinin derivatives, but as parasite clearance is very rapid, monitoring of blood parasitaemia may be used to assess and compare drugs. When the effects of artemisinin derivatives on parasite clearance, (in non-immune subjects), are compared with those seen with quinine a number of important differences are observed (White, 1994). These are described below, but it should be remembered that malaria is a complex disease and the following hypothesis may be simplistic.

Quinine is thought to stop development at the mature trophozoite stage of the parasite life cycle so that in patients with ring stage parasites parasitaemia remains unchanged for several hours after treatment has started. When the parasitaemia does fall it is initially the result of sequestration of infected cells into the tissues. If quinine is given to patients with parasites at the mature meront (schizont) stage they are not inhibited and parasitaemia intially increases due to the young ring stages produced. With artemisinin derivatives, this increase in parasitaemia is prevented or reduced because these drugs are able to block development of the very early trophozoites soon after entry into the red blood cell (i.e. these drugs have a broader "window" of activity in the parasites erythrocytic cycle). In addition, the decline in parasitaemia is more rapid with artemisinin derivatives than with quinine, possibly as a result of oxidative damage and removal of the abnormal cells by the spleen; as discussed above, artemisinin may modulate neutrophil function. By comparing parasitaemia-time curves published for quinine or mefloquine with those obtained during artemisinin treatment it has been shown that mean parasite clearance rates for quinine and mefloquine were 5.8%/hour while that for artemisinin was 11.1%/hour (Hien and White, 1993), which is consistent with the rapid antimalarial action of the latter. However, it should be noted that this effect is more obvious where there is quinine resistance such as in south east Asia where the above study was carried out. *In vitro* studies have also shown that artemisinin-like drugs are more effective than quinine in preventing cytoadherence of infected red cells, probably by preventing development to the mature trophozoite stage (White, 1994). Further studies have shown that these drugs also reduce the tendency of infected red cells to form rosettes and to adhere to endothelial cells (Udomsangpetch *et al.*, 1996).

As discussed previously, artemisinin derivatives are initially metabolised to dihydroartemisinin; when [13–^{14}C] labelled dihydroartemisinin was administered intravenously to rats most of the label was recovered in the bile with only about 1% excreted in the urine (Maggs *et al.*, 1997). The principal metabolites were the glucuronides of both 12α- and 12β-dihydroartemisinin; the former is devoid of antiplasmodial activity while the β-glucuronide is about 20-fold less potent than dihydroartemisinin, (Ramu and Baker, 1995), which is consistent with the observation that 12β-ethers are more active than 12α-ethers (Kar *et al.*, 1989). Other metabolites which were identified by liquid chromatography/mass spectroscopy were deoxydihydroartemisinin (analogous to **11**, Scheme 1), and its glucuronide, 3-hydroxydihydroartemisinin glucuronide and the glucuronide of a ring-contracted tetrahydrofuran acetate isomer of dihydroartemisinin (analogous to **17**, Scheme 1), (Maggs *et al.*, 1997). Previous studies on microbial and mammalian metabolism of arteether have shown that hydroxylation of dihydroartemisinin may also give rise to the 2α- and 14-hydroxy analogues as well as the 9α and 9β-hydroxy derivatives (Hufford and Lee, 1990).

Some hydroxylated metabolites of arteether (and which retain the endoperoxide moiety) have *in vitro* antiplasmodial activities higher than that of the parent (Chi *et al.*, 1991) which is thought to be because their lipophilicities, (log P values), are closer to that required for optimum activity than that of the parent (Ramu and Baker, 1995). The latter authors reported that, as expected, the glucuronides of hydroxyarteether had lower lipophilicity and correspondingly lower antiplasmodial activities with the exception of 9β-hydroxyarteether glucuronide which was the least lipophilic but the most active; although its *in vitro* potency was 10-fold less than dihydroartemisinin and 90-fold less than that of artemether itself, it is suggested that this metabolite may contribute to the *in vivo* activity of artemether and may also merit consideration as an antimalarial suitable for intravenous use (Ramu and Baker, 1995).

While a knowledge of the metabolism of artemisinin derivatives is essential in order to understand their pharmacokinetics it also enables predictions to be made concerning the potential for clinically important interactions with other drugs. Using recombinant human P450 liver enzymes, it has been shown that the de-ethylation of arteether to dihydroartemisinin is carried out by the CYP 2B6, 3A4, and 3A5 isozymes, CYP3A4 being the most important (Grace *et al.*, 1998). Patients treated with artemisinin derivatives may also receive, or may have previously received other antimalarial drugs such as quinidine, mefloquine and halofantrine, the latter two of which have very long elimination half-lives, with the possibility of cardiotoxicity due to the inhibition of their metabolism by artemisinin derivatives; conversely, other antimalarials may inhibit the metabolism of artemisinin-like drugs. When tested for their ability to inhibit de-ethylation of arteether by CYP3A4 *in vitro*, halofantrine was found to be the most potent, while mefloquine and quinidine were moderately potent inihibitors but it is estimated that *in vivo* mefloquine and quinidine would inhibit arteether de-ethylation by less that 10% and halofantrine by about 20% which is thought not to be significant (Grace *et al.*, 1998). In this context, it is interesting to note that mefloquine and artemisinin act synergistically in mice infected with *P. berghei*, (Chawira *et al.*, 1987), although potentiation is also observed with *P. falciparum in vitro*. A clinical trial in which artemether was used in combination with mefloquine did not reveal an inter-action (Shwe *et al.*, 1988), but the possibility that artemisinin derivatives may inhibit the metabolism of other antimalarials leading to cardiotoxicity has not been investigated. Drugs such as ketoconazole and erythromycin are more potent CYP3A4 inhibitors than the above, and therefore may have a greater potential for interaction with artemisinin derivatives.

Recently it has been reported that grapefruit juice (which inhibits CYP3A4) increases the oral bioavailability of artemether; peak levels (C-max) increased by more than two-fold while the time taken to reach maximum blood levels decreased when the drug was taken at the same time as grapefruit juice (van Agtmael *et al.*,1999). In contrast, the elimination half-life was unchanged suggesting that intestinal CYP3A4 may be involved in the presystemic metabolism of artemether.

RESISTANCE OF MALARIA PARASITES TO ARTEMISININ-LIKE AGENTS

In the late 1950's resistance of *P. falciparum* to chloroquine was first reported and this is now widespread; furthermore, resistance to other antimalarial drugs including quinine and mefloquine has also emerged and is increasing. In some parts of south east Asia artemisinin derivatives are the only effective agents for the treatment of malaria caused by multi-drug resistant *P. falciparum* and so the development of resistance to these agents would be a very worrying prospect.

Artemisinin-resistant strains of *P. falciparum* have been cultivated in the labora-tory by exposing parasites to mutagenic agents, (Inselburg, 1985), and, more significantly, *P. yoelii* resistant to artemisinin was selected in mice as a result of drug pressure (Chawira *et al.*, 1986). Similarly, *Toxoplasma gondii* (which is normally sensitive to inhibition by artemisinin), became resistant when cultured in the pres-ence of gradually increased concentrations of the drug, and these mutants were about 65-fold less sensitive to artemisinin and exhibited cross-resistance to dihydro-artemisinin and arteether (Berens *et al.*, 1998). In this context it is also worth noting that *Pneumocystis carinii* has also been reported to be inhibited by artemisinin *in vitro* at concentrations close to those achieved in patients (Merali and Meshnick, 1991).

In 1992, Oduola *et al.*, reported the isolation of four strains of *P. falciparum* from malaria patients in Africa which exhibited a transient 7–14 fold decrease in sensitivity to artemisinin *in vitro* during their adaptation to continuous culture. A highly resistant strain of *P. falciparum* was isolated from a traveller to Africa, (Gay *et al.*, 1994), but to date there have been no reports of treatment failure due to artemisinin resistant parasites. In a number of studies, parasites isolated from malaria patients have been tested *in vitro* for their susceptibilities to a number of antimalarial drugs in order to detect patterns of cross-resistance. For example. Alin, (1997), reported that artemsinin was highly effective against 100% of wild Tanzanian isolates of *P. falciparum* including strains resistant to other drugs and no correlation of *in vitro* activities with other drugs was observed i.e. no cross-resistance present.

In a recent study by Pradines *et al.* (1998), artemether was equally effective against both chloroquine sensitive and chloroquine-resistant strains (i.e. no cross-resistance), but showed a positive correlation with mefloquine, quinine, and halofantrine. This shows that strains sensitive to the latter are also sensitive to artemether, but there is also the possibility that there could be cross-resistance between these drugs and artemether. The authors warn that if a similar correlation is seen *in vivo*, then this should be investigated. Peters *et al.* (1999), studied the development of resistance to artemisinin and to a number of other antimalarial endoperoxides in mice infected with chloroquine-sensitive *P. berghei* N or chloroquine resistant *P. yoelii* ssp. NS in an attempt to forecast the risk of resistance developing as a result of the use of artemisinin-like agents in man. Moderate resistance to artemisinin developed in both parasites following drug-selection pressure but only a low level of resistance was seen with the other compounds and resistance was readily lost when drug-selection pressure was withdrawn. The data indicate that stable resistance is less readily developed to artemisinin-like agents than it is to other antimalarials such as atovaquone and mefloquine. Recent work has shown that mutations in the pfmdr1 gene of *P. falciparum* which encodes for the P-glycoprotein homologue 1 protein can confer resistance to mefloquine, quinine and halofantrine and also influence parasite sensitivity to chloroquine and artemisinin (Reed *et al.*, 2000). While chloroquine resistant parasites are generally more sensitive to artemisinin than are chloroquine sensitive strains, resistance to mefloquine and halofantrine is associated with a decreased response to artemisinin and this is consistent with other studies (see above).

The use of combination therapy is likely to prevent, or at least slow the development of drug resistance and there is evidence which suggests that the progression of resistance to mefloquine in the western border region of Thailand has been reduced since the introduction of artesunate plus mefloquine for malaria treatment (van Vugt *et al.*, 1998). The latter authors also showed that a combination of benflumetol (lumefantrine) with artemether was effective in preventing the recrudescence normally seen with doses of artemether greater than those used in the combination.

Chawira *et al.* (1987), used malaria-infected mice to detect antagonism and synergism with a variety of antimalarial drugs in combination with artemisinin. Using *P. berghei* marked synergism was seen with mefloquine, tetracycline and spiramycin but antagonism was observed with antifolates including sulfadoxine. With an artemisinin resistant strain of *P. yoelii* (strain QS), potentiation was demonstrated with mefloquine, primaquine, tetracycline and clindamycin and also with mefloquine against mefloquine-resistant *P. berghei* (strain N/1100). Some synergism was seen with artemisinin and primaquine with drug-sensitive *P. berghei* but a high degree of potentiation was seen with primaquine resistant *P. berghei* (P strain). In the case of chloroquine, artemisinin antagonises the former *in vitro*, but this effect is not seen *in vivo* (Chawira *et al.*, 1986a). Using *P. yoelii* infected mice, Peters and Robinson, (1997), reported that artemisinin in combination with the Chinese antimalarial pyronaridine was additive against chloroquine resistant *P. yoelii* but marked potentiation was seen with both artemisinin resistant (ART strain), and pyronaridine resistant (SPN strain), parasites.

It has been suggested that combinations of artemisinin with mefloquine or with tetracycline may be of value in the treatment of drug-resistant malaria, as the rapid action of artemisinin may enhance the slower actions of the former drugs (Chawira *et al.*, 1987).

Using an *in vitro* simulation of *in vivo* pharmacokinetics, in which cultures of *P. falciparum* were dosed intermittently with mefloquine and artemisinin, Bwijo *et al.* (1998), showed that at lower dose combinations the two drugs acted synergistically while at higher concentrations an additive effect was seen. A number of methoxylated flavones have also been shown to potentiate the action of artemisinin *in vitro* and this raises the possibility that there may be flavones present in *A. annua* which may potentiate the action of artemisinin when extracts of the plant are used in traditional medicine for malaria treatment (Elford *et al.*, 1987).

A reduced response to artemisinin may not necessarily be due to the presence of drug-resistant parasites and there is some evidence that the genetic type of the host erythrocyte may be important. Artemisinin has been shown to have reduced effectiveness against *P. falciparum* infecting α- and β-thallassaemic red blood cells as compared with normal cells (Senok *et al.*, 1997). This effect was however, only observed in old thalassaemic erythrocytes but could be important as it may theoretically increase the risk of development of drug resistant parasites. Radiolabelled dihydroartemisinin has been shown to accumulate preferentially in α-thalassaemic erythrocytes (haemoglobin (Hb) H or HbH/Hb Constant Spring erythrocytes as compared with normal erythrocytes (Vattanaviboon *et al.*, 1998). While some drug was associated with the membrane fraction, most was associated with the cytosol. Hb H bound dihydroartemisinin with a 3-fold higher affinity, (lower *Kd*) and a 2-fold higher capacity than did Hb A. Hb H, which has four β-globin subunits appeared to bind two molecules of drug while Hb A with only two subunits bound only one. However, the increased accumulation of drug in α-thalassaemic cells could not be accounted for solely by the amount of Hb H present indicating that other factors, yet to be identified are also involved. Although only a small proportion of drug bound covalently to isolated haemoglobin it is possible that more covalent binding occurs in intact cells as thalassaemic cells are under greater oxidative stress which could result in increased covalent binding and inactivation of the drug and may explain the decreased sensitivity of *P. falciparum* to artemisinin (Vattanaviboon *et al.*, 1998).

ARTEMISININ DERIVATIVES AND MALARIA TRANSMISSION

Besides being more rapidly acting than other antimalarials and highly effective against multi-drug resistant *P. falciparum*, artemisinin and its derivatives appear to be able to reduce the transmission of malaria. When artesunate plus mefloquine replaced mefloquine alone for the treatment of falciparum malaria in the western border of Thailand in 1994 the incidence of falciparum malaria fell by 50% while there was no effect on the incidence of malaria caused by *Plasmodium vivax*; similarly, the incidence of falciparum malaria in southern Vietnam has declined following the introduction of artemisinin derivatives (Price *et al.*, 1996). The latter authors showed that, compared with mefloquine, artemisinin derivatives reduced the presence of gametocytes (detected in thick blood films) by a factor of eight in primary infections and by a factor of 18.5 in recrudescent infections. Gametocyte carriage was also reduced significantly by artemisinin derivatives compared to treatment with quinine or halofantrine. Thus a decrease in transmission is a likely explanation for the reduced incidence of falciparum malaria.

Although artemisinin derivatives do not kill mature gametocytes, early (stage I-III) gametocytes are inhibited, and their development is prevented because the asexual forms from which they develop are also killed (Kumar and Zheng, 1990). The superior transmission-blocking ability of artemisinin as compared to mefloquine has also been shown in a Chinese study in which gametocytes were counted and their infectivity to *Anopheles dirus* mosquitoes determined using a membrane feeding technique (Chen *et al.*, 1994). Arteether, (as a mixture of α and β isomers), given orally has been shown to possess gametocytocidal activity against *P. cynomolgi B* in rhesus monkeys (Tripathi *et al.*, 1996a), and similarly, sodium β-artelinate given orally or intravenously as a single dose sterilised circulating gametocytes though the latter compound did not possess sporontocidal activity (Tripathi *et al.*, 1996).

CONCLUSION

The artemisinin-like antimalarial drugs are remarkable because of their high potency and low toxicity. The selectivity of these agents appears to be due to their "activation" by haem iron which is present in significant amounts only in red blood cells which are infected with malaria parasites. However, the precise mechanism(s) by which artemisinin kills malaria parasites remains speculative; while the reaction of artemisinin with haem has been shown to yield a variety of reactive products these have been formed in the presence of organic solvents and their relevance to the antimalarial action of artemisinin in man is therefore uncertain.

To date there is little evidence that the neurotoxic effects of artemisinin seen in experimental animals are a problem in man. Of greater concern is the possible development of malaria parasites resistant to these agents and steps are being taken to prevent or at least delay the emergence of resistant strains by recommending that, as far as possible, artemisinin derivatives are used in combination with another effective antimalarial (see chapters 14 and 15). Currently, there is interest in promoting the local cultivation and use of *A. annua* as a herbal antimalarial treatment but while this practice is very attractive it carries with it the risk that the development of resistant parasites may be encouraged especially if patients receive inadequate doses of artemisinin. Recrudescence is a problem even with therapeutic doses of artemisinin-like drugs and is likely to be related to the relatively short elimination half-lives of these compounds. The development of new derivatives with improved pharmacokinetic profiles such as sodium artelinate may help to overcome this problem. Much energy has been put into the synthesis of artemisinin derivatives and also into the preparation of simple analogues based on the trioxane and tetra-oxane ring sytems (see chapter 12).

It is also possible that other naturally occurring peroxides may prove to be lead compounds for novel antimalarial agents. Like *A. annua*, the Chinese herb yingzhao (*Artabotrys unciatus*) is used traditionally in China for malaria treatment and this species contains the terpenoid peroxides yingzhaosu A and C (Zhang *et al.*, 1988). The synthetic antimalarial arteflene has been developed from yinghaosu A and has been evaluated for the treatment of falciparum malaria in Gabonese children (Radloff *et al.*, 1996). Treatment with a single oral dose of 25 mg/kg was not effective in eradicating parasites and this was shown to be due to recrudescence rather than re-infection. More importantly, indications of both R2 and R3 resistance development were seen in parasites from 8 of the 20 patients treated. The development of arteflene as an antimalarial has now been discontinued due to the problem of recrudescence but other analogues are being investigated, (Posner, 1998) and it is to be hoped that the continued investigation of natural products will help to meet the current need for safe and effective antimalarial drugs.

ACKNOWLEDGEMENTS

We thank Mr Godfrey Stell and Dr Richard Wheelhouse for reviewing the manuscript.

REFERENCES

Adams, P.A. and Berman, P.A. (1996) Reaction between ferriprotoporphyrin IX and the anti-malarial endoperoxide artesunate gives an intermediate species with enhanced redox catalytic activity. *J. Pharm. Pharmacol.*, **48**, 183–187.

Alin, M.H. (1997) *In vitro* susceptibility of Tanzanian wild isolates of *Plasmodium falciparum* to artemisinin, chloroquine, sulfadoxine/pyrimethamine and mefloquine. *Parasitol.*, **114**, 503–506.

Alin, M.H., Ashton, M., Kihamia, C.M., Mtey, G.J.B. and Bjorkman, A. (1996) Multiple-dose pharmacokinetics of oral artemisinin and comparison of its efficacy with that of oral artesunate in falciparum-malaria patients. *Trans. Roy. Soc. Trop. Med. Hyg.*, **90**, 61–65.

Anon. (1979) Pharmacological study of qinghaosu. *Xin Yi Yao Xue Za Zhi*, **1**, 23 (CA, 95:108182m); Cited in Chen *et al.*, 1998.

Anon. (1982) China Cooperative Research Group. Chemical studies on qinghaosu (artemisinin). *J. Trad. Chin. Med.*, **2**, 3–8.

Anon. (1982a) China Cooperative Research Group on qinghaosu and its derivatives as antimalarials. *J. Trad. Chin. Med.*, **2**, 31.

Asawamahasakda, W., Benakis, A. and Meshnick, S.R. (1994) The interaction of artemisinin with red cell membranes. *J. Lab. Clin. Med.*, **123**, 757–762.

Asawamahasakda, W., Ittarat, I., Chang, C.C., McElroy, P. and Meshnick, S.R. (1994a) Effects of antimalarials and protease inhibitors on plasmodial hemozoin production. *Mol. Biochem. Parasitol.*, **67**, 183–191.

Asawamahasakda, W., Ittarat, I., Pu, Y.M., Ziffer, H. and Meshnick, S.R. (1994b) Reaction of antimalarial endoperoxides with specific parasite proteins. *Antimicrob. Agents Chemother.*, **38**, 1854–1858.

Augustijns, P., D'Hulst, A., Van Daele, J. and Kinget, R. (1996) Transport of artemisinin and sodium artesunate in caco-2 intestinal epithelial cells. *J. Pharm. Sci.*, **85**, 577–579.

Avery, M.A., Fan, P., Karle, J.M., Miller, R. and Goins, K. (1995) Replacement of the non-peroxidic trioxane oxygen atom of artemisinin by carbon: Total synthesis of (+)-13-carbaartemisinin and related structures. *Tet. Lett.*, **36**, 3965–3968.

Avery, M.A., Fan, P.-C., Karle, J.M., Bonk, J.D., Miller, R. and Goins, D.K. (1996) Structure-activity relationships of the antimalarial agent artemsinin. 3. Total synthesis of (+)-13-carba-artemisinin and relatd tetra and tricyclic structures. *J. Med. Chem.*, **39**, 1885–1887.

Bakhshi, H.B., Gordi, T. and Ashton, M. (1997) In-vitro interaction of artemisinin with intact human erythrocytes, erythrocyte ghosts, haemoglobin and carbonic anhydrase. *J. Pharm. Pharmacol.*, **49**, 223–226.

Beekman, A.C., Woerdenbag, H.J., Kampinga, H.H. and Konings, A.W.T. (1996) Cytotoxicity of artemisinin, a dimer of dihydroartemisinin, artemisitene and eupatoriopicrin as evaluated by the MTT and clonogenic assay. *Phytother. Res.*, **10**, 140–144.

Beekman, A.C., Barentsen, A.R.W., Woerdenbag, H.J., Van Uden, W. and Pras, N. (1997) Stereochemistry-dependent cytotoxicity of some artemisinin derivatives. *J. Nat. Prod.*, **60**, 325–330.

Beekman, A.C., Woerdenbag, H.J., Van Uden, W., Pras, N., Konings, A.W.T. and Wikstrom, H.V. (1997a) Stability of artemisinin in aqueous environments: Impact on its cytotoxic action to Ehrlich ascites tumour cells. *J. Pharm. Pharmacol.*, **49**, 1254–1258.

Berens, R.L., Krug, E.C. and Nash, P.B. (1998) Selaection and characterisation of *Toxoplasma gondii* mutants resistant to artemisinin. *J. Inf. Dis.*, **177**, 1128–1131.

Berman, P.A. and Adams, P.A. (1997) Artemisinin enhances heme-catalysed oxidation of lipid membranes. *Free radical biology and medicine*, **22**, 1283–1288.

Bhisutthibhan, J., Pan, X-Q., Hossler, P.A., Walker, D.J., Yowell, C.A., Carlton, J. *et al.* (1998) The *Plasmodium falciparum* translationally controlled tumor protein homolog and its reaction with the antimalarial drug artemisinin. *J. Biol. Chem.*, **273**, 16192–16198.

Bradley, D.J. (1995) The epidemiology of malaria in the tropics and in travellers. In: Pasvol, G. (Ed.), *Clinical infectious diseases*, **2**, 211–226.

Brewer, T.G., Peggins, J.G., Grate, S.J., Petras, J.M., Levine, B.S., Weina, P.J. *et al.* (1994) Neurotoxicity in animals due to arteether and artemether. *Trans. Roy. Soc. Trop. Med. Hyg.*, **88 Suppl. 1**, 33–36.

Brown, S.B., Hatzikonstantinou, H. and Herries, D.G. (1980) The structure of porphyrins and haems in aqueous solution. *Int. J. Biochem.*, **12**, 701–707.

Bwijo, B., Alin, M.H., Abbas, N., Wernsdorfer, W., and Bjorkman, A. (1997) Efficacy of artemisinin and mefloquine combinations against *Plasmodium falciparum. In vitro* simulation of *in vivo* pharmacokinetics. *Trop. Med. Int. Health*, **2**, 461–467.

Chawira, A.N., Warhurst, D.C. and Peters, W. (1986) Qinghaosu resistance in rodent malaria. *Trans. Roy. Soc. Trop. Med. Hyg.*, **80**, 477–480.

Chawira, A.N., Warhurst, D.C. and Peters, W. (1986a) Artemisinin (qinghaosu) combinations against chloroquine-sensitive and resistant *Plasmodium falciparum in vitro. Trans. Roy. Soc. Trop. Med. Hyg.*, **80**, 335.

Chawira, A.N., Warhurst, D.C., Robinson, B.L. and Peters, W. (1987) The effect of combinations of qinghaosu (artemisinin) with standard antimalarial drugs in the suppressive treatment of malaria in mice. *Trans. Roy. Soc. Trop. Med. Hyg.*, **81**, 554–558.

Chen, P.Q., Li, G.Q., Guo, X.B., He, K.R., Fu, Y.X., Fu, L.C. *et al.* (1994) The infectivity of gametocytes of *Plasmodium falciparum* from patients treated with artemisinin. *Chin. Med. J.*, **107**, 709–711.

Chen, Y., Zhu, S-M., Chen, H-Y. and Li, Y. (1998) Artesunate interaction with hemin. *Bioelectrochem. and Bioenergetics*, **44**, 295–300.

Chi, H.T., Ramu, K., Baker, J.K., Hufford, C.D., Lee, I.-S., Yan-Lin, Z. *et al.* (1991) Identification of the *in vivo* metabolites of the antimalarial arteether by thermospray-high performance liquid chromatography/mass spectrometry. *Biol. Mass. Spectrom.*, **20**, 609–628.

Chou, A., Chevli, R. and Fitch, C.D. (1980) Ferriprotoporphyrin IXX fulfills the criteria for identification as the chloroquine receptor of malaria parasites. *Biochem.*, **19**, 1543–1549.

Cumming, J.N., Polypradith, P. and Posner, G.H. (1997) Antimalarial activity of artemisinin (qinghaosu) and related trioxanes: Mechanism of action. In: August, J.T., Anders, M.W., Murad, F. and Coyle, J.T. Eds. *Adv. in Pharmacol.*, **37**, 253–297.

Dorn, A., Stoffel, R., Matile, H., Bubendorf, A. and Ridley, R.G. (1995) Malarial haemozoin/β-haematin supports haem polymerisation in the absence of protein. *Nature*, **374**, 269–271.

Egan, T.J., Ross, D.C. and Adams, P.A. (1994) Quinoline anti-malarial drugs inhibit spontaneous formation of β-haematin (malaria pigment). *FEBS Lett.*, **352**, 54–57.

Elford, B.C., Roberts, M.F., Phillipson, J.D. and Wilson, R.J.M. (1987) Potentiation of the antimalarial activity of qinghaosu by methoxylated flavones. *Trans. Roy. Soc. Trop. Med. Hyg.*, **81**, 434–436.

Fishwick, J., McLean, W.G., Edwards, G. and Ward, S.A. (1995) The toxicity of artemisinin and related compounds on neuronal and glial cells in culture. *Chem. Biol. Interact.*, **96**, 263–271.

Fishwick, J., Edwards, G., Ward, S.A. and McLean, W.G. (1998) Morphological and immunocytochemical effects of dihydroartemisinin on differentiating NB2a neuroblastoma cells. *Neurotox.*, **19**, 393–403.

Fishwick, J., Edwards, G., Ward, S.A. and McLean, W.G. (1998a) Binding of Dihydroartemisinin to differentiating neuroblastoma cells and rat cortical homogenate. *Neurotox.*, **19**, 405–412.

Gay, F., Ciceron, L., Litaudon, M., Bustos, M.D., Astagneau, P., Diquet, B. *et al.* (1994) *In vitro* resistance of *Plasmodium falciparum* to qinghaosu derivatives in West Africa. *Lancet*, **343**, 850–851.

Genovese, R.F., Newman, D.B., Li, Q., Peggins, J.O. and Brewer, T.G. (1998) Dose-dependent brainstem neuropathology following repeated arteether administration in rats. *Brain Res. Bull.*, 45, 199–202.

Genovese, R.F., Newman, D.B., Petras, J.M. and Brewer, T.G. (1998a) Behavioural and neural toxicity of arteether in rats. *Pharm. Biochem. Behav.*, 60, 449–458.

Grigorov, M., Weber, J., Tronchet, J.M.J., Jefford, C.W., Milhous, W.K. and Maric, D. (1997) A QSAR study of the antimalarial activity of some synthetic 1,2,4-trioxanes. *J. Chem. Inf. Comput. Sci.*, 37, 124–130.

Groves, J.T. and Viski, P. (1990) Asymmetric hydroxylation, epoxidation, and sulfoxidation catalysed by vaulted bisnaphthyl metalloporphyrins. *J. Org. Chem.*, 55, 3628–3634.

Grace, J.M., Aguilar, A.J., Trotman, K.M. and Brewer, T.G. (1998) Metabolism of β-arteether to dihydroqinghaosu by human liver microsomes and recombinant cytochrome P450. *Drug Metab. Disp.*, 26, 313–317.

Halliwell, B. and Gutteridge, J.M.C. (1989) Free raidicals in biology and medicine. 2nd edition. Oxford, Clarendon Press.

Haynes, R.K. and Vonwiller, S.C. (1996) The behavior of qinghaosu (artemisinin) in the presence of heme iron (II) and iron (III). *Tetrahedron. Lett.*, 37, 253–256.

Haynes, R.K. and Vonwiller, S.C. (1996a) The behaviour of qinghaosu (artemsininin) in the presence of non-heme iron(II) and (III). *Tetrahedron. Lett.*, 37, 257–260.

Hien, T.T. and White, N.J. (1993) Qinghaosu. *Lancet*, 341, 603–608.

Hong, Y.L., Yang, Y.Z. and Meshnick, S.R. (1994) The interaction of artemisinin with malarial haemozoin. *Mol. Biochem. Parasitol.*, 63, 121:128.

Hu, Y., Ziffer, H., Li, G. and Yeh, H.J.C. (1992) Microbial oxidation of the antimalarial drug arteether. *Bioorg. Chem.*, 20, 148–154.

Hufford, C.D., Lee, I., ElSohly, H.N., Chi, H.T. and Baker, J.K. (1990) Structure elucidation and thermospray high performance liquid chromatography/mass spectroscopy (HPLC/MS) of the microbial and mammalian metabolites of the antimalarial arteether. *Pharm. Res.*, 7, 923–927.

Inselburg, J. (1985) Induction and isolation of artemisinin resistant mutants of *Plasmodium falciparum. Am. J. Trop. Med. Hyg.*, 34, 417–418.

Jefford, C.W., Vicente, M.G.H., Jaquier, Y., Favarger, F., Mareda, J., Millasson-Schmidt, P. *et al.* (1996) *Helv. Chim. Acta.*, 79, 1475–1487.

Kager, P.A., Schultz, M.J., Zijlstra, E.E., van den Berg, B. and van Boxtel, Ch. J. (1994) Arteether administration in humans: preliminary studies of pharmacokinetics, safety and tolerance. *Trans. Roy. Soc. Trop. Med. Hyg.*, 88 Suppl. 1, 53–54.

Kamchonwongpaisan, S., Paitayatat, S., Thebtaranonth, Y., Wilairat, P. and Yuthavong, Y. (1995) Mechanism-based development of new antimalrials: synthesis of derivatives of artemisinin attached to iron chelators. *J. Med. Chem.*, 38, 2311–2316

Kamchonwongpaisan, S., McKeever, P., Houssier, P., Ziffer, H. and Meshnick, S.R. (1997) Artemisinin neurotoxicity: Neuropathology in rats and mechanistic studies *in vitro. Am. J. Trop. Med. Hyg.*, 56, 7–12.

Kar, K., Nath, A., Bajipai, R., Dutta, G.P. and Vishwakarma, R.A. (1989) Parmacology of α/β arteether – a potential antimalarial drug. *J. Ethnopharmacol.*, 27, 297–305.

Khalifa, S.L., Baker, J.K., Jung, M., McChesney, J.D. and Hufford, C.D. (1995) Microbial and mammalian metabolism studies on the semisynthetic antimalarial. *deoxoartemisinin. Pharmaceut. Res.*, 12, 1493–1498.

Klayman, D.L. (1985) Qinghaosu (artemisinin): an antimalarial drug from China. *Science*, 228, 1049–1055.

Krogstad, D.J., Schlesinger, P.H. and Gluzman, I.Y. (1985) Antimalarials increase vesicle pH in *Plasmodium falciparum*. *J. Cell Biol.*, **101**, 2302–2309.

Krungkrai, S.R. and Yuthavong, Y. (1987) The antimalarial action of qinghaosu and artesunate in combination with agents that modulate oxidant stress. *Trans. Roy. Soc. Trop. Med. Hyg.*, **81**, 710–714.

Kumar, N. and Zheng, H. (1990) Stage-specific gametocytocida effect *in vitro* of the antimalarial drug qinghaosu on *Plasmodium falciparum*. *Parasitol. Res.*, **76**, 214–218.

Lai, H. and Singh, N.P. (1995) Selective cancer cell cytotoxicity from exposure to dihydroartemisinin and holotransferrin. *Cancer Lett.*, **91**, 41–46.

Levander, O.A., Ager, A.L., Morris, V.C. and May, R.G. (1989) Qinghaosu, dietary vitamin E, selenium and cod-liver oil: effects on the susceptibility of mice to the malarial parasite *Plasmosium yoelii*. *Am. J. Clin. Nut.*, **50**, 346–352.

Lee, I.S. and Hufford, C.D. (1990) Metabolism of antimalarial sesquiterpene lactones. *Pharmacol. Ther.*, **48**, 345–355.

Li, L.P., Liu, R.C., Xu, Y.S., Zhou, Z.Y. and Lu, L. (1981) Effect of sodium artesunate on the ultrastructure of erythrocytic form of *Plasmodium knowlesi*. Zong Cao Yao (*Chin. Tradition. Herbal Drugs*), **12**, 175. (Cited in Chen *et al.*, 1998).

Li, Q.-G., Peggins, J.O., Lin, A.J., Masonic, K.J., Trotman, K.M. and Brewer, T.G. (1998) Pharmacology and toxicology of artelinic acid: preclinical investigations on pharmacokinetics, metabolism, protein and red blood cell binding, and acute and anorectic toxicities. *Trans. Roy. Soc. Trop. Med. Hyg.*, **92**, 332–340.

Maeno, Y., Toyoshima, T., Fujioka, H., Yoshihiroito, Meshnick, S.R., Benakis, A. *et al.* (1993) Morphologic effects of artemisinin in *Plasmodium falciparum*. *Am. J. Trop. Med. Hyg.*, **49**, 485.

Maggs, J.L., Madden, S., Bishop, L.P., O'Neill, P.M. and Park, K. (1997) The rat biliary metabolites of dihydroartemisinin, an antimalarial endoperoxide. *Drug Metab. Disp.*, **25**, 1200–1204.

Merali, S. and Meshnick, S.R. (1991) Susceptibility of *Pneumocystis carinii* to artemisinin *in vitro*. *Antimic. Agents Chemother.*, **35**, 1225–1227.

Meshnick, S.R., Tsang, T.W., Lin, F.B., Pan, H.Z., Chankg, C.N., Kuypers, F. *et al.* (1989) Activated oxygen mediates the antimalarial activity of qinghaosu. *Prog. Clin. Biol. Res.*, **313**, 95–104.

Meshnick, S.R., Thomas, A., Ranz, A., Xu, C.M. and Pan, H.Z. (1991) Artemisinin (qinghaosu): The role of intracellular hemin in its mechanism of antimalarial action. *Mol. Biochem. Parasitol.*, **49**, 181–190.

Meshnick, S.R., Yang, Y.Z., Lima, V., Kuypers, F., Kamchonwongpaison, S. and Yuthavong, Y. (1993) Iron dependent free radical generation from the antimalarial agent artemisinin (qinghaosu). *Antimicrob. Agents Chemother.*, **37**, 1108–1114.

Meshnick, S.R. (1994) The mode of action of antimalarial peroxides. *Trans. Roy. Soc. Trop. Med. Hyg.*, **88 Suppl. 11**, 31–32.

Meshnick, S.R., Taylor, T.E. and Kamchnongwongpaisan, S. (1996) Artemisinin and the antimalarial endoperoxides: from herbal remedy to targetted chemotherapy. *Microb. Rev.*, **60**, 301–315.

Miller, L.G. and Panosian, C.B. (1997) Ataxia and slurred speech after artesunate treatment for falciparum malaria. *N. Eng. J. Med.*, **336**, 1328.

Moore, J.C., Lai, H., Li, J.R., Ren, R.L. McDougall, J.A., Singh, N.P. *et al.* (1995) Oral-administration of dihydroartemisinin and ferrous sulphate retarded implanted fibrosarcoma growth in the rat. *Cancer Lett.*, **98**, 83–87.

Muhia, D.H., Thomas, C.G., Ward, S.A., Edwards, G., Mberu, E.K. and Watkins, W.M. (1994) Ferriprotoporphyrin catalysed decomposition of artemether: analytical and pharmacological implications. *Biochem. Pharm.*, **48**, 889–895.

Murphy, J., Clyde, D., Herrington, D., Bagar, S., Davis, J., Palmer, K. *et al.* (1990) Continuation of chloroquine-susceptible *Plasmodium falciparum* parasitemia in volunteers receiving chloroquine. *Antimic. Agents Chemother.*, **34**, 676–679.

Murphy, S.A., Mberu, E., Muhia, D.K., English, M., Crawley, J., Waruiru, C. *et al.* (1997) The disposition of intramuscular artemether in children with cerebral malaria; a preliminary study. *Trans. Roy. Soc. Trop. Med. Hyg.*, **91**, 331–334.

Nontprasert, A., Nosten-Bertrand, M., Pukrittayakamee, S., Vanijanonta, S., Angus, B.J. and White, N.J. (1998) Assessment of the neurotoxicity of parenteral artemisinin derivatives in mice. *Am. J. Trop. Med. Hyg.*, **59**, 519–522.

Oduola, A.M.J., Sowunmi, A., Milhous, W.K., Kyle, D.E., Martin, R.K., Walker, O. *et al.* (1992) Innate resistance to new antimalarial drugs in *Plasmodium falciparum* from Nigeria. *Am. J. Trop. Med. Hyg.*, **86**, 123–126.

Orjih, A.U. (1996) Haemolysis of *Plasmodium falciparum* trophozoite-infected erythrocytes after artemisinin exposure. *Brit. J. Haematol.*, **92**, 324–328.

Paitayatat, S., Tarnchompoo, T., Thebtaranonth, Y and Yuthavong, Y. (1997) Correlation of antimalarial activity of artemisinin derivatives with binding affinity with ferroprotoporphyrin IX. *J. Med. Chem.*, **40**, 633–638.

Parker, F.C., Wesche, D.L. and Brewer, T.G. (1994) Does iron have a role in dihydroartemisinin-induced *in vitro* neurotoxicity?. *Am. J. Trop. Med. Hyg.*, **51**, 260.

Peters, W., Lin, L., Robinson, B.L. and Warhurst, D.C. (1986) The chemotherapy of rodent malaria XL. The action of artemisinin and related sesquiterpenes. *Annal. Trop. Med. Parasit.*, **80**, 483–489.

Peters, W. and Robinson, B.L. (1997) The chemotherapy of rodent malaria. LV. Interactions between pyronaridine and artemisinin. *Ann. Trop. Med. Parasit.*, **91**, 141–145.

Peters, W. and Robinson, B.L. (1999) The chemotherapy of rodent malaria. LVI. Studies on the development of resistance to natural and synthetic endoperoxides. *Ann. Trop. Med. Parasit.*, **93**, 325–339.

Petras, J.M., Kyle, D.E., Gettayacamin, M., Young, G.D., Bauman, R.A., Webster, H.K. *et al.* (1997) Arteether: Risks of two-week administration in *Macaca mulatta*. *Am. J. Trop. Med. Hyg.*, **56**, 390–396.

Posner, G.H. (1998) Antimalarial peroxides in the qinghaosu (artemisinin) and yingzhaosu families. *Exp. Opin. Ther. Pat.*, **8**, 1487–1493.

Posner, G.H. and Oh, C.H., (1992) A regiospecific oxygen-18-labelled 1,2,4-trioxane: A simple chemical model system to probe the mechanism(s) for the antimalarial activity of artemisinin (qinghaosu). *J. Am. Chem. Soc.*, **114**, 8328–8329.

Posner, G.H., Oh, C.H., Wang, D., Gerena, L., Milhouse, W.K., Meshnick, S.R. *et al.* (1994) Mechanism-based design, synthesis, and *in vitro* antimalarial testing of new 4-methylated trioxanes structurally related to artemisinin: The importance of a carbon centred radical for antimalarial activity. *J. Med. Chem.*, **37**, 1256–1258.

Posner, G.H., Cumming, J.N., Ploypradith, P. and Oh, C.H. (1995) Evidence for Fe(IV)=O in the molecular mechanisms of action of the trioxane antimalarial artemisinin. *J. Am. Chem. Soc.*, **117**, 5885–5886.

Posner, G.H., McGarvey, D.J., Oh, C.H., Kumar, N., Mehsnick, S.R. and Asawamahasakda, W. (1995a) Structure-activity relationships of lactone ring-opened analogs of the antimalarial 1,2,4-trioxane artemisinin. *J. Med. Chem.*, **38**, 607–612.

Posner, G.H., Wang, D., Cumming, J.N., Oh, C.H., French, A.N., Bodley, A.L. *et al.* (1995b) Further evidence supporting the importance of, and restrictions on, a carbon-centred radical for high antimalarial activity of 1,2,4-trioxanes like artemisinin. *J. Med. Chem.*, **38**, 2273–2275.

Posner, G.H., Park, S.B., Gonzalez, L., Wang, D., Cumming, J.N., Klinedinst, D. *et al.* (1996) Evidence for the importance of high valent Fe=O and of a diketone in the molecular mechanism of action of antimalarial trioxane analogs of artemisinin. *J. Am. Chem. Soc.*, **118**, 3537–3538.

Pradines, B., Rogier, C., Fusal, T., Tall, C., Trape, J.F. and Doury, J.C. (1998) *In vitro* activity of artemether against african isolates (Senegal) of *Plasmodium falciparum* in comparison with standard antimalarial drugs. *Am. J. Trop. Med. Hyg.*, **58**, 354–357.

Price, R.N., Nosten, F., Luxemberger, C., ter Kuile, F.O., Paiphun, L., Chongsuphajaisiddhi, T. *et al.* (1996) Effects of artemisinin derivatives on malaria transmissibility. *Lancet*, **347**, 1654–1658.

Radloff, P.D., Phillips, J., Nkeyi, M., Stuchler, D., Mittelholzer, M.L. and Kremsner, P.G. (1996) Arteflene compared with mefloquine for treating *Plasmodium falciparum* malaria. *Am. J. Trop. Med. Hyg.*, **55**, 259–262.

Ramu, K. and Baker, J.K. (1995) Synthesis, characterisation, and antimalarial activity of the glucuronides of the hydroxylated metabolites of arteether. *J. Med. Chem.*, **38**, 1911–1921.

Reed, M.B., Saliba, K.J., Caruana, S.R., Kirk, K. and Cowman, A.F. (2000) PghI modulates sensitivity and resistance to multiple antimalarials in *Plasmodium falciparum*. *Nature*, **403**, 906–909.

Robert, A. and Meunier, B. (1997) Characterisation of the first covalent adduct between artemisinin and a heme model. *J. Am. Chem. Soc.*, **119**, 5968–5969.

Senok, A.C., Nelson, E.A.S., Li, K. and Oppenheimer, S.J. (1997) Thalassaemia trait, red blood cell age and oxidant stress: effects on *Plasmodium falciparum* growth and sensitivity to artemisinin. *Trans. Roy. Soc. Trop. Med. Hyg.*, **91**, 585–589.

Scott, M.K., Meshnick, S.R., Williams, R.A., Chiu, D.T.Y., Pan, H., Lubin, B.H. *et al.* (1989) Qinghaosu-mediated oxidation in normal and abnormal erythrocytes. *J. Lab. Clin. Med.*, **114**, 401–406.

Shen, J.X. (Ed.) (1989) Antimalarial drug development in China. Beijing: National Institute of Pharmaceutical Research and Development, pp. 31–95.

Shukla, K.L., Gundi, T.M. and Meshnick, S.R. (1995) Molecular modeling studines of the artemsinin (qinghaosu)-hemin interaction. Docking between the antimalarial agent and its putative receptor. *J. Mol. Graph.*, **13**, 215–222.

Shwe, T., Myint, P.T., Htut, Y., Myint, W. and Soe, L. (1988) The effect of mefloquine-artemether compared with quinine on patients with complicated falciparum malaria. *Trans. Roy. Soc.Trop. Med. Hyg.*, **82**, 665–666.

Smith, S.L., Fishwick, J., McLean, W.G., Edwards, G. and Ward, S.A. (1997) Enhanced *in vitro* neurotoxicity of artemisinin derivatives in the presence of haemin. *Biochem. Pharm.*, **53**, 5–10.

Slater, A.F.G. and Cerami, A. (1992) Inhibition by chloroquine of a novel heam polymerase enzyme activity in malaria trophozoites. *Nature*, **355**, 167–169.

Sullivan, A.D. and Meshnick, S.R. (1996) Haemozoin: Identicication and quantification. *Parasitology Today*, **12**, 161–163.

Teha-Isavadharm, P., Nosten, F., Kyle, D.E., Luxemberger, C., Ter Kuile, F., Peggins, J.O. *et al.* (1996) Comparative bioavailability of oral, rectal, and intramuscular artemether in healthy subjects: use of simultaneous measurement by high performance liquid chromatography and bioassay. *Br. J. Clin. Pharmacol.*, **42**, 599–604.

Titulaer, H.A.C., Zuidema, J., Kager, P.F., Wetsteyn, J.C.F.M., Lugt, C.H.B. and Merkus, F.W.H.M. (1990) The pharmacokinetics of artemisinin after oral, intramuscular, and rectal administration to human volunteers. *J. Pharm. Pharmac.*, **42**, 810–813.

Titulaer, H.A., Eling, W.M. and Zuidema, J. (1993) Pharmacokinetic and pharmacodynamic aspects of artelinic acid in rodents. *J. Pharm. Pharmac.*, **45**, 830–835.

Traylor, T.G. and Miksztal, A.R. (1987) Mechanisms of hemin-catalysed epoxidations: Electron transfer from alkenes. *J. Am. Chem. Soc.*, **114**, 3445–3455.

Tripathi, R., Dutta, G.P. and Vishwakarma, R.A. (1996) Gametocytocidal activity of alpha/beta arteether by the oral route of administration. *Am. J. Trop. Med. Hyg.*, **54**, 652–654.

Tripathi, R., Puri, S.K. and Dutta, G.P. (1996a) Sodium beta-artelinate – a new potential gametocytocide. *Exp. Parasitol.*, **82**, 251–254.

Udomsangpetch, R., Pipitaporn, B., Krishna, S., Angus, B., Pukrittayakamee, S., Bates, I. *et al.* (1996) Antiplasmodial drugs reduce cytoadherence and rosetting of *Plasmosium falciparum. J. Inf. Dis.*, **17**, 691–8.

Ullrich, V. and Brugger, R. (1994) Prostacylin and thromboxane synthase: New aspects of hemethiolate catalysis. *Angew. Chem. Int. Ed. Eng.*, **33**, 1911–1919.

van Agtmael, M.A., Gupta, V., vander Wosten, T.H., Rutten, J.P.B. and van Boxtel, C.J. (1999) Grapefruit juice increases the bioavailability of artemether. *Eur. J. Clin. Pharmacol.*, **55**, 405–410.

van Vugt, M., Brockman, A., Gemperli, B., Luxemberger, C., Gathmann, I., Royce, C. *et al.* (1998) Randomized comparison of artemether-benflumatol and artesunate-mefloquine in treatment of multidrug-resistant falciparum malaria. *Antimic. Agents. Chemother.*, **42**, 135–139.

Vattanaviboon, P., Wilairat, P. and Yuthavong, Y. (1998) Binding of dihydroartemisinin to hemoglobin H: role in drug acumulation and host-induced antimalarial ineffectiveness of α-thalassemic erythrocytes. *Mol. Pharm.*, **53**, 492–496.

Waki, S., Kurihara, R., Nemeto, H. and Suzuki, M. (1993) Effect of recombinant human colony-stimulating factor on the course of parasitaemia in non-lethal rodent malaria. *Parasitol. Res.*, **79**, 703–705.

Wenisch, W., Parschalk, B., Zedwitz-Liebenstein, K., Wernsdorfer, W. and Graninger, W. (1997) The effect of artemisisin on granulocyte function assessed by flow cytometry. *J. Antimicrob. Chemother.*, **39**, 99–101.

White, N.J. (1994) Clinical pharmacokinetics and pharmacodynamics of artemisinin and derivatives. *Trans. Roy. Soc. Trop. Med. Hyg.*, **88 Suppl. 1**, 41–43.

White, N.J. (1998) Why is that antimalarial drug treatments do not always work? *Annal. Trop. Med. Parasit.*, **92**, 449–458.

Wu, W.-M., Yao, Z-J., Wu, Y.-L., Jiang, K., Wang, Y-F., Chen, H-B. *et al.* (1996) Ferrous ion induced cleavage of the peroxy bond in qinghaosu and its derivatives and the DNA damage associated with this process. *Chem. Commun.*, p. 2213–2214.

Wu, W.-M., Wu, Y., Wu, Y.-L., Yao, Z-J., Zhou, C-M., Li., Y. *et al.* (1998) Unified mechanistic framework for the Fe(II)-induced cleavage of qinghaosu and derivatives/analogues. The first spin-trapping evidence for the previously postulated secondary C-4 radical. *J. Am. Chem. Soc.*, **120**, 3316–3325.

Yang , S.D., Ma, J.M., Sub, J.H., Chen, D.X. and Song, Z.Y. (1986) Clinical pharmacokinetics of a new effective antimalarial artesunate, a qinghaosu derivative. *Chinese J. Clin. Pharmacol.*, **1**, 106–109.

Yang, Y-Z., Asawamahasakda, W. and Meshnick, S.R. (1993) Alkylation of human albumin by the antimalarial artemisinin. *Bioch. Pharm.*, **46**, 336–339.

Yang, Y-Z., Little, B. and Meshnick, S.R. (1994) Alkylation of proteins by artemisinin, effects of heme, pH, and drug structure. *Biochem. Pharm.*, **48**, 569–573.

Zhang, F., Gosser, D.K. Jr. and Meshnick, S.R. (1992) Hemin-catalysed decomposition of artemisinin (qinghaosu). *Bioch. Pharm.*, **43**, 1805–1809.

Zhang, L., Zhou, W-S. and Xu, X-X. (1988) A new sesquiterpene peroxide (yninghaosu C) and sesquiterpenol (yinghaosu D) from *Artabotrys unciatus* (L.)Meer. *J. Chem. Soc. Chem. Commun.*, pp. 523.

Zhao, Y., Hanton, W.K. and Lee, K-H. (1986) Antimalarial agents, 2. Artesunate, an inhibitor of cytochrome oxidase activity in *Plasmodium berghei*. *J. Nat. Prod.*, **49**, 139–142.

Zhao, K.C., Chen, Z.X., Lin, B.L., Guo, X.B., Li, G.Q. and Song, Z.H. (1988) Studies on the phase 1 clinical pharmaocokinetics of artesunate and artemether. *Chinese J. Clin. Pharmacol.*, **4**, 76–81.

Zouhiri, F., Desmaele, D., d'Angelo, J., Riche, C., Gay, F. and Cicerone, L. (1998) Artemisinin tricyclic analogs: Role of a methyl group at C-5a. *Tet. Lett.*, **39**, 2969–2972.

14. THE CLINICAL USE OF ARTEMISININ AND ITS DERIVATIVES IN THE TREATMENT OF MALARIA

POLRAT WILAIRATANA AND SORNCHAI LOOAREESUWAN

Hospital for Tropical Diseases, Faculty of Tropical Medicine, Mahidol University, 420/6 Rajvithi Road, Rajthevi, Bangkok 10400, Thailand.

INTRODUCTION

Malaria is one of the world's most important tropical diseases, causing great human suffering and loss of life in the affected countries. There are four species of *Plasmodia* causing human malaria: *P. falciparum, P. vivax, P. ovale, and P. malariae.* Of the four species of *Plasmodia*, it is *P. falciparum* with which causes the most problems. Mosquito control is often not cost-effective in areas where the disease is most severe and where transmission interruption cannot be sustained. Therefore, current malaria control places emphasis on early diagnosis, treatment with effective antimalarials, and selective use of preventive measures including vector control where they can be sustained. The treatment of malaria has changed over the past two decades in response to declining drug sensitivity in *P. falciparum* and a resurgence of the disease in tropical areas. Some other problems are increasing numbers of severe and cerebral malaria cases, the high cost of drugs, side effects and the need for multiple dose regimens. Although chloroquine was recently the drug of choice in Africa, resistance has now spread to all major malarial-endemic areas. For multidrug resistant *P. falciparum* malaria, the choice of treatment is mefloquine, halofantrine, or quinine plus tetracycline. After mefloquine became the drug of choice in some Southeast Asian countries, *P. falciparum* malaria resistant to mefloquine developed. The alternative treatment with oral quinine or quinidine is not well tolerated. Although the combination of quinine with tetracycline or doxycycline remains more than 85 percent effective nearly everywhere (Watt *et al.*, 1992), compliance with seven-day courses of treatment required for resistant *P. falciparum* infections is poor. The second-line drug, a combination of a long-acting sulfonamide (usually sulfadoxine) and pyrimethamine carries a risk of severe side effects and is facing growing resistance.

Traditional remedies derived from plants continue to be a source of interesting biologically active compounds. A highly promising new class of antimalarial drug is derived from qinghaosu – a substance extracted from a plant which the Chinese have used to control fever for at least two thousand years. Artemisinin and its derivatives have been proven to be potent antimalarials – this chapter will concentrate on the clinical aspects of the treatment of falciparum malaria with these agents.

CLINICAL STUDIES

The currently available preparations of artemisinin and its derivatives are artesunate for intravenous injection, oral tablets, and suppositories, artemether in oil for intramuscular injection and as oral capsules, arteether for intramuscular injection, and dihydroartemisinin in tablets (Looareesuwan and Wilairatana, 1997). China and Vietnam are the main producer countries. Artemisinin and its derivatives have been imported into many countries in Southeast-Asia and to a few African countries. The most widely available oral preparations are artemisinin tablets, artemether capsules, artesunate (artesunic acid) tablets, and dihydroartemisinin tablets. Intramuscular preparations are artemether in oil, artesunate as anhydrous powder, and recently a preparation of arteether. The intravenous preparation of artesunate is identical to the preparation for intramuscular injection. Suppository preparations are available for artemisinin, artesunate, and dihydroartemisinin. Table 1 shows the most widely available formulations (WHO, 1998).

Artemisinin and its derivatives have been studied in many countries, e.g., China and Vietnam for many years, some Southeast Asian countries such as Thailand and Myanmar for a few years, and Africa. Artemisinin and its derivatives, artesunate, and artemether give more rapid clearance of fever and malaria parasites than other drugs and survival rates are comparable to those with standard treatment regimens.

The use of artemisinin and its derivatives with other antimalarials seems to be obligatory if a high cure rate is to be obtained. Because of their short half-lives, artemisinin and its derivatives have to be given once or twice daily for 5–7 days. The shorter the duration of drug administration, the higher the rate of recrudescence. A three-day course of these compounds has been associated with about 50% recrudescence. A longer treatment duration, for example of 5–7 days, improves the cure rate to 90–98%, however, patient compliance is poor. The use of antimalarial combinations (i.e. an artemisinin derivative plus mefloquine, tetracycline or doxycycline), may be used for treatment on an out-patient basis because it reduces the duration of drug administration (i.e. 2–3 days *vs*. 5–7 days) and increases patient compliance. Many studies have shown that combinations of these drugs with other antimalarials having a longer half-life gives more advantage. Mefloquine combinations (Karbwang *et al*., 1992a; Looareesuwan *et al*., 1992b,c, 1993, 1994a,b, 1995, 1996b,d; Luxemburger *et al*., 1994; Nosten *et al*., 1994) with either artemisinin rectally, artesunate orally or parenterally or rectally, artemether orally or parenterally have been studied extensively in uncomplicated and severe malaria. Mefloquine allows a short treatment course due to its long-half life. Combination treatment gives a rapid initial therapeutic response and protects the artemisinin compounds from resistance since the other antimalarial drug should eliminate residual parasites.

Uncomplicated Malaria

Although several different artemisinin derivatives have been used for the treatment of uncomplicated malaria in China and South-East Asia, there appears to be no significant difference in efficacy or tolerability between the different derivatives

Table 1   The most widely available formulations of artemisinin and its derivatives.

Oral formulations (country of production)

Artemisinin tablets (Viet Nam)	250 mg
Artesunate tablets (China and Viet Nam)	50 mg
Artesunate tablets (Switzerland)	200 mg
Artesunate tablets (France)	50 mg
Artemether capsules (China)	40 mg
Artemether composite tablets (China)	50 mg
Dihydroartemisinin tablets (China)	20, 60, 80 mg

Parenteral formulations (country of production)
  (i) *Intramuscular administration*

Artemether injection (China and France*)	80 mg/1 ml ampoule
Arteether injection (India)	150 mg/2 ml ampoule

  *also available as 40 mg/ampoule for pediatric use.
  (ii) *Intravenous or intramuscular administration*

Artesunate (China and Viet Nam)	60 mg/1 ml vial

Suppository formulations

Artemisinin (Viet Nam)	100, 200, 300, 400, 500 mg/suppository
Artesunate (China)	100 mg/suppository
Artesunate (Switzerland)	200 mg/rectocap
Dihydroartemisinin (China)	80 mg/suppository

Formulations under development
- Arteether, intramuscular injection-Artecef B.V., Maarssen, The Netherlands. (This is now approved for use in children up to 16 years in the Netherlands with two formulae; 50 mg/vial and 150 mg/vial)
- Co-artemether (artemether plus lumefantrine (benflumetol; Riamet®), oral formulation – Novartis Pharma A.G., Basle, Switzerland.
- Dihydroartemisinin, oral formulation – Artecef B. V., Maarssen, The Netherlands.
- Artelinate intravenous injection – Walter Reed Army Institute of Research, Washington, USA.

when given at the currently recommended doses but comparative data are limited (WHO, 1998).

Despite the limitations of many of the reported clinical studies, artemisinin derivatives consistently produce faster relief of clinical symptoms and clearance of parasites from the blood than other antimalarial drugs. In around 90% of the patients given these drugs, the fevers resolved and the parasitemias cleared within 48 hours of treatment. It has been estimated that they reduce the parasite biomass by a factor of approximately 10^4 for each 36–48 hour asexual cycle of the parasite and by a factor of 10^6–10^8 over a 3-day course of treatment (White, 1997). Despite their short elimination half lives, i.e. 1.6–2.6 hours, artemisinin-like drugs are effective when given daily. Data from 23 trials (with 1891 patients) comparing artemisinin and its derivatives with other antimalarials have recently been reviewed (Hien *et al.*, 1993). The mean shortening of fever clearance time compared with intravenous quinine was 17% (7.7 h) and parasite clearance time was 32% (19.8 h). Artesunate appeared to have more rapid action than other derivatives. No serious toxicity was observed in these trials.

The aim of treating uncomplicated malaria is radical cure. Up to 50% of patients treated with artemisinin and its derivatives alone may recrudesce (become parasitemic again after having cleared the initial parasitemia) within 28 days (Meshnick et al., 1996). The high rate of recrudescence results from the rapid elimination of these drugs and the need for the antimalarial drug to be present in the blood at effective concentrations during four asexual cycles (>6 days) to ensure elimination of the parasite (White, 1997). Higher efficacy can be obtained by 5– and especially 7–day regimens but this is associated with reduced compliance in out-patients. At present, to achieve a 28-day cure with artemisinin drugs alone, treatment courses of five to seven days are required. Since the clinical symptoms of most patients improve rapidly following treatment with these drugs, however, the patients are less likely to continue treatment courses lasting longer than 3 to 4 days (Luxemburger et al., 1994).

To solve the problem of recrudescence following short courses of artemisinin monotherapy, combination chemotherapy, comprising artemisinin (or one of its derivatives) and another antimalarial drug is used. Combinations of antimalarial drugs with different modes of action should be adopted as a basic principle to delay the emergence of resistance. Therefore, the second drug must be chosen carefully. Antimalarial drug interactions should be considered; in vitro synergy has been shown between arteether and both mefloquine and quinine (Ekong et al., 1990) and in vivo synergy between artemisinin and both mefloquine and tetracycline (Chawira et al., 1987). However, tetracycline is contraindicated in pregnancy and in children under 8 years. Research is still required to determine the optimal timing for administration of mefloquine. Compliance in taking mefloquine would be maximised if it were administered on the first day of treatment. However, since mefloquine can induce vomiting this regimen could interfere with absorption and reduce compliance. Although various regimens have been evaluated particularly in Thailand (Table 2); and several are effective, combinations of artemisinin and its derivatives plus mefloquine show great promise for the treatment of malaria (Looareesuwan et al., 1992c; Meshnick et al., 1996). At the present time there is little published data on combination therapy with drugs other than mefloquine. However, a fixed combination of artemether with lumefantrine (Riamet®) is in advanced stages of development by Novartis Pharma A.G (Looareesuwan et al., 1999).

In Africa, oral formulations of artesunate and artemether have been shown to clear parasitemias more effectively than chloroquine and sulfadoxine/pyrimethamine in Nigeria and Tanzania (Alin et al., 1996a,b; Ezedinachi, 1996).

In some parts of Southeast Asia (particularly the eastern and western borders of Thailand) where the failure rates of treatment with high-dose mefloquine alone in falciparum malaria now exceed 40%, oral artesunate given for three to five days in combination with mefloquine still remains highly effective (Looareesuwan et al., 1992c; Nosten et al., 1994; White 1996a). When used alone (for example, for the

Table 2 Clinical trials of artemisinin derivatives either alone or in combination with other antimalarial drugs for adult patients in Thailand.

Reference and drug[a]	Total dose (mg)	Duration (d)	No. of patients	Cure rate (%)
Bunnag et al., 1991a				
Artesunate (o)	1200	5	6	100
Artesunate (o)	600	5	10	90
Artemether (i.m.)	180	5	6	100
Artesunate (i.v.)	300	3	5	20
Artesunate	600	2	5	0
Artesunate	600	1	5	0
Artesunate	650	5	21	95
Artesunate plus chloroquine (o) 1.5 g	200	Single dose	5	0
Artesunate (o) plus Fansidar® (o) 3 tablets single dose	200	Single dose	5	0
Bunnag et al., 1991b				
Artesunate (o)	600	5	46	85
Artesunate (o)	600	7	43	93
Bunnag et al., 1991c				
Artesunate (o)	600	5	25	72
Artesunate (o)	600	5	25	76
Bunnag et al., 1992				
Artemether (i.m.) for uncomplicated malaria	480	5	33	84
Artemether (i.m.)	600	5	28	92
Artemether (i.m.) for severe malaria	480	5	53	65
	600	5	53	76
Karbwang et al., 1992a				
Artemether (i.m.) for severe malaria	640	7	14	93[b]

Table 2 Clinical trials of artemisinin derivatives either alone or in combination with other antimalarial drugs for adult patients in Thailand. (*Continued*)

Reference and drug[a]	Total dose (mg)	Duration (d)	No. of patients	Cure rate (%)
Karbwang *et al.*, 1992b				
Artemether (o)	700	5	34	97
Artemether (o)	500	5	40	74
Looareesuwan *et al.*, 1992b				
Artesunate (o) followed by mefloquine 25 mg/kg divided into 2 doses	600	5	24	100
Looareesuwan *et al.*, 1992c				
Artesunate (o)	600	5	40	88
Artesunate (o) followed by mefloquine 25 mg/kg divided into 2 doses	600	5	39	100
Looareesuwan *et al.*, 1993				
Artesunate (o) followed by mefloquine 15 mg/kg divided into 2 doses	300	2.5	50	90
Looareesuwan *et al.*, 1994b				
Artesunate (200 mg every 12 h) followed by mefloquine 750 mg	800	2	63	92
Looareesuwan *et al.*, 1994c				
Artesunate (o) plus doxycycline 200 mg/d × 7 d	300	2.5	55	80
Looareesuwan *et al.*, 1995				
Artesunate (s) followed by mefloquine 25 mg/kg divided into 2 doses	1600	4	30	92

Table 2 Clinical trials of artemisinin derivatives either alone or in combination with other antimalarial drugs for adult patients in Thailand. (*Continued*)

Reference and drug[a]	Total dose (mg)	Duration (d)	No. of patients	Cure rate (%)
Looareesuwan et al., 1996b				
Artesunate (o): 400 mg every 12 h followed by mefloquine 25 mg/kg divided into 2 doses	800	2	67	84
Artesunate (o): 200 mg every 12 h followed by mefloquine 25 mg/kg divided into 2 doses	800	3	44	100
Looareesuwan et al., 1996e				
Dihydroartemisinin (o)	480	7	53	90
Looareesuwan et al., 1996f				
Ateether (i.m.)	600	4	8	83
Arteether (i.m.)	200	4	8	40
Looareesuwan et al., 1997				
Artemether (o)	750	7	58	98
Artemether (o) followed by mefloquine 25 mg/kg divided into 2 doses	600 mg (200 × 3)	1	53	98
Looareesuwan et al., 1999				
Co-artemether (o)	4 doses of 4 tablets (each containing 20 mg of artemether and 120 mg of lumefantrine)	2	126	69.3

Table 2   Clinical trials of artemisinin derivatives either alone or in combination with other antimalarial drugs for adult patients in Thailand. (*Continued*)

Reference and drug[a]	Total dose (mg)	Duration (d)	No. of patients	Cure rate (%)
Luxemburger *et al.*, 1994				
Artesunate (o)	10 mg/kg (3 doses) plus mefloquine 15 mg/kg single dose	1	270	81
van Vugt *et al.*, 1998a				
Coartemether (o)	4 doses of 4 tablets (each containing 20 mg of artemether and 120 mg of lumefantrine)	2	309	85.2
van Vugt *et al.*, 1998b				
Co-artemether (o)	6 doses of 4 tablets (each containing 20 mg of artemether and 120 mg of lumefantrine)	3	118	96.9
		6	121	99.1

[a] Abbreviations: o = oral, i.m. = intramuscular injection, i.v. = intravenous injection, s = suppository
[b] Survival rate.

treatment of recrudescence after treatment with mefloquine), artemisinin and its derivatives should be given for seven days (White 1996a). Apart from mefloquine, doxycycline has been combined with artesunate (Looareesuwan et al., 1994c), however the cure rate of this combination (80%), is not high.

Although there are many reports of clinical trials on artesunate, artemether and their combinations, there are few clinical reports on arteether (Looareesuwan et al., 1996f), dihydroartemisinin (Looareesuwan et al., 1996e; Wilairatana et al., 1998), and co-artemether (artemether plus lumefantrine (benflumetol)) (Looareesuwan et al., 1996a; van Vugt et al., 1998a; 1998b). However, preliminary studies showed that the latter three drugs rapidly clear the parasites and gave high cure rates (Table 2).

Severe and Complicated Malaria

The goal of treating patients with severe or complicated malaria is survival, not parasitological cure (White 1994). In severe and complicated falciparum malaria (WHO 1990), parenteral antimalarial therapy should be given to the patients. WHO recommended that the use of parenteral artemether or artesunate should be confined to areas where quinine resistance is demonstrable since the limited data available did not show a marked advantage of these drugs over quinine (WHO, 1994). Artesunate may be given by intravenous or intramuscular injection. Artemether is formulated in peanut oil and is given by intramuscular injection. Arteether is a very similar compound to artemether. It is the oil-soluble ethyl ether and is given by intramuscular injection. Artelinic acid is a water-soluble second-generation compound under development (White 1996b). Both arteether and artelinic acid have been developed in the West, but only arteether has undergone clinical trials for severe malaria treatment. In recent studies, rectal suppositories of artemisinin and artesunate have proved as effective as the parenteral drugs (Hien et al., 1991, 1992; Looareesuwan et al., 1995, 1996d). Therefore, effective drug treatment for severe malaria can be given in rural areas where parenteral administration by injection is no possible. These drugs are the most rapidly acting antimalarials known (ter Kuile et al., 1993). They also have a broad window of antimalarial activity during the 48 hour asexual life cycle of the parasite from the ring forms to the mature trophozoites and prevent parasitized erythrocytes from adhering to uninfected cells (rosetting) or to vascular endothelium (cytoadherance) (Udomsangpetch et al., 1996). These drugs produce rapid parasite clearance and appear to be very safe in clinical practice.

Of the currently available artemisinin derivatives, artesunate is the most rapidly acting drug, possibly because it is immediately bioavailable (as dihydroartemisinin) after intravenous injection, and is absorbed rapidly after oral or intramuscular administration. In recent large comparative studies, treatment with intramuscular artemether accelerated parasite clearance but slightly prolonged recovery from coma, and it did not reduce mortality significantly in comparison with quinine (Boel van Hensbrock et al., 1996; Hien et al., 1996). These trials confirm that artemether is an alternative to quinine in severe malaria. Parenteral artemisinin derivatives are, however, easier to use than quinine and do not induce hypoglycaemia.

Mortality associated with cerebral malaria treated with the parenteral artemisinin and its derivatives has been around 13% in countries in both Africa and Asia (Danis *et al.*, 1996; Hien *et al.*, 1996; Wilairatana and Looareesuwan 1995). However, the mortality in quinine-treated groups has consistently been around 20% in countries of Southeast Asia, compared with 15% in others, such as countries in Africa, indicating the decreased efficacy of quinine against some Southeast Asian strains of *P. falciparum* (Looareesuwan *et al.*, 1992a).

Children

Most of the reported information on artemisinin and its derivatives concerns adults, but there are sufficient data to conclude that children also tolerate the drugs very well, that no serious adverse effects have been observed, and that the therapeutic response resembles that of adults with similar levels of immunity (Hien *et al.*, 1991; Taylor *et al.*, 1993; White *et al.*, 1992).

Pregnancy

Non-clinical studies showed no mutagenicity or teratogenicity but the drug caused foetal resorption in rodents even at relatively low doses of 1/200–1/1400 of the LD_{50} (i.e. above 10 mg) when administered after the sixth day of gestation (Qinghaosu Antimalarial Coordinating Research Group, 1979). There are few reports of the use of artemisinin and its derivatives in pregnancy, and they show no abnormalities in children whose mothers were treated with artemisinin or artemether during the second and third trimesters (Li *et al.*, 1990; Shen 1989; Wang 1989). For the management of uncomplicated malaria in pregnancy, artemisinin and its derivatives can be used in the second and third trimester, but their use in the first trimester is not recommended. In severe malaria, artemisinin derivatives are the drug of choice in the second and third trimester. For the treatment of severe malaria in the first trimester, the advantages of artemisisnin-like drugs over quinine, especially the lower risk of hypoglycaemia, must be weighed against the fact that there is still limited documentation on pregnancy outcomes following their use (WHO, 1998).

Resistance

Cross-resistance with mefloquine, quinine, and amodiaquine has been observed following *in vivo* induction of resistance to artemisinin in chloroquine-resistant strains of *P. yoelii* in rodents (Ding 1988; Chawira *et al.*, 1986, 1987). The resistance to artemisinin is thought to be due to a change in the membrane structure which could in part explain resistance to other antimalarials. Multidrug resistant isolates of *P. falciparum* from Southeast Asia appear to be less sensitive *in vitro* to artemisinin and its derivatives than isolates from other geographical areas (WHO 1994). However, there is no good evidence of true resistance to these drugs. Infections recurring after treatment have been attributed to inadequate primary treatment resulting in recrudescence. At present, artemisinin and its derivatives are the only group of antimalarial drugs to which resistance of *P. falciparum* has not yet developed in the field (WHO, 1998).

Overdosage

No reports on overdosage in human have become available.

Adverse Effects

Over two million patients are estimated to have been treated with artemisinin and its derivatives, and no adverse effects have been noted (Hien *et al.*, 1993; WHO 1994). Cardiac and gut toxicity has also occurred in animals, usually with higher doses (White 1996b). Although transient dose-related reductions in reticulocyte counts were noted in some early preclinical toxicology studies, they have not been observed in human clinical trials. Foetal resorption was observed in some animal toxicology studies (China Cooperative Research Group on Qinghaosu and its Derivatives as Antimalarials, 1982). The principal toxicity in animals is a dose-related selective pattern of neurotoxicity affecting brain stem nuclei involved in auditory relays (Brewer *et al.*, 1994). Although there has been one report (Miller *et al.*, 1997), of a patient with ataxia and slurred speech after artesunate treatment, there is no definite proof of a causal relationship in man; ataxia and slurred speech may be due to other causes such as malaria itself.

In some countries, artemisinin and its derivatives can be obtained easily because of wide-spread availability of over-the-counter antimalarial chemotherapy in shops and the market places. Self-treatment, presumptive treatment, and repeated treatment of uncomplicated infections are most frequently performed with oral-dosage forms because these are most convenient for outpatients. Once a drug is released and available, drug toxicity particularly with chronic or repeated administration may become a very important issue (Meshnick *et al.*, 1996).

GUIDELINES ON SELECTION AND USE

Premature and unregulated use of artemisinin and its derivatives, especially oral forms, could accelerate the development of drug resistance, thus compromising the efficacy of these drugs when multidrug resistance develops in Africa. Also, over-prescription of artemisinin and its derivatives particularly in areas without multi-drug-resistance is a matter of concern. Therefore, WHO (1994; 1998) have proposed guidelines for using these drugs. Following are the present guidelines for using the artemisinin group of antimalarials.

Situations Where Multi-Drug Resistant Malaria is Prevalent

Uncomplicated malaria

Artemisinin and its derivatives should be used in areas where quinine and mefloquine resistant *P. falciparum* has been reported (WHO 1994, 1998). Oral artemisinin and its derivatives have been extensively studied in Thailand (Table 2). These drugs should be administered in combination with another effective blood schizontocide to reduce recrudescence and to slow the development of resistance. There are many reports on the efficacy of the use of artemsinin derivatives with mefloquine (15–25 mg base/kg). Tolerance to 25 mg base/kg doses of mefloquine may be further improved by administering 15 mg base/kg on the second or third day with the rest 6–24 hours later. If compliance is a concern, mefloquine can be given on the first day. The dose of mefloquine depends on the local sensitivity of the parasite to mefloquine (WHO, 1998). Recently, a fixed combination of artemether with lumefantrine (benflumatol) has been developed and this appears to have a promising future (Looareesuwan, 1999; van Vugt *et al.*, 1998a,b). Under exceptional circumstances, such as when there is a history of an adverse reaction to the combination agent, artemisinin monotherapy may be indicated, but a 7-day course of treatment is recommended and effort should be made to ensure compliance. In areas where mefloquine is already the first-line drug or a change to mefloquine is being considered, a 3-day course of an artemisinin-like drug should be given in combination with mefloquine (White, 1997; WHO, 1998). Artemisinin derivatives are not indicated for the treatment of malaria due to *P. malariae, P. ovale*, or chloroquine-sensitive *P. vivax*. Possible treatment of chloroquine-resistant *P. vivax* with these drugs requires further study. The use of parenteral preparations of these drugs in patients who can take oral medications is not indicated.

Severe and complicated malaria

Severe falciparum malaria is a serious disease with high mortality that requires intensive care. There is the same urgency to treat this condition with adequate doses of parenteral antimalarials as there is to treat septicaemia with antibacterials. Parenteral artesunate or artemether are effective alternatives to parenteral quinine for the treatment of severe falciparum malaria. Use of these drugs should be confined to areas where there is quinine-resistance. If parenteral formulations are not available, artemisinin or artesunate suppositories may be an effective alternative (Looareesuwan *et al.*, 1996d). Following parenteral treatment, radical cure should be effected by completing treatment with a fully curative dose of an effective antimalarial (e.g., a single 15–25 mg/kg oral dose of mefloquine).

The recommended schedules are shown in Table 3. Although adverse effects with the recommended regimens have rarely been reported, prolonged or higher doses in animals have demonstrated neurotoxic and cardiotoxic effects. Therefore, it is important not to exceed the recommended doses, or to repeat courses of treatment at short intervals until such time as the acute and cumulative risk can be quantitatively evaluated clinically.

Table 3   Treatment of falciparum malaria with artemisinin and its derivatives.

Uncomplicated malaria

(i)   Oral artesunate (50 mg/tablet): In combination with a total of 25 mg/kg mefloquine, give a total of 10–12 mg/kg in divided doses over 3–5 days (e.g., 4 mg/kg daily for 3 days or 4 mg/kg followed by 1.5 mg/kg/day for 4 days). If used alone, the same total dose is given over 7 days (usually 4 mg/kg initially, followed by 2 mg/kg on days 2 and 3 and 1 mg/kg on days 4–7). Recently, Mepha Company (Switzerland) has produced oral artesunate (200 mg/tablet). It has been used effectively for falciparum malaria treatment.

(ii)   Oral artemether (40 mg/capsule): Same regimens as for artesunate.

(iii) Oral dihydroartemisin (20 mg/tablet): 3 mg/kg immediately, then 1.5 mg/kg once daily for 7 days.

<u>Under investigation</u>:
Oral co-artemether (CGP 56697, a combination of artemether 20 mg and lumefantrine (benflumetol) 120 mg/tablet): 4 doses of 4 tablets at 0, 8, 24 and 48 h in 2 days or 6 doses of 4 tablets at 0, 8, 24, 36, 48 and 60 h or at 0, 8, 24, 48, 72 and 96 h.

Severe and complicated malaria

(i)   Artesunate (60 mg/vial); 2.4 mg/kg intravenously or intramuscularly, followed by 1.2 mg/kg at 12 and 24 h, then 1.2 mg/kg daily. Artesunate is dissolved in 0.6 ml of 5% sodium bicarbonate, diluted to 3–5 ml with 5% dextrose, and given by bolus intravenous or intramuscular injection.

(ii)   Artemether (80 mg/ampoule); 3.2 mg/kg intramuscularly, followed by 1.6 mg/kg daily. Not to be given intravenously

(iii) Artesunate suppository (200 mg/rectocap, Mepha Company, Switzerland); 4 mg/kg given twice daily at 0, 4, 8, 12, 24, 36, 48, and 60 h

(iv)   Artemisinin suppository: 10 mg/kg at 0 and 4 h followed by 7 mg/kg at 24, 36, 48, and 60 h.

Important notes:

1.   For monotherapy using an artemisinin derivative as a single antimalarial drug, treatment given for 5–7 days will give a cure rate of over 85%. If the duration of treatment is shorter than 5 days, artemisinin derivatives should be combined with a long acting antimalarial drug, preferably mefloquine 25 mg/kg given orally in two divided doses of 15 mg/kg and 10 mg/kg given 12 hours apart. The two drugs may be given simultaneously on the first day of treatment (e.g. artesunate plus mefloquine) or in sequence, the mefloquine being given immediately following the completion of treatment with the artemisinin derivative.

2.   At present it is recommended that none of the artemisinin derivatives should be used as a single agent (monotherapy). An artemisinin derivative should always be used in combination with a long acting antimalarial in order to reduce recrudescence and slow the development of resistance.

Situations Where Multidrug-Resistant Malaria is Rare or Absent

The use of artemisinin and its derivatives was previously not recommended in areas where multidrug-resistant malaria is rare or absent, e.g. Africa (WHO 1994). The reasons for this were as follows; first, available drugs such as chloroquine, sulfa-pyrimethamine combinations, and mefloquine are still effective for the treatment of uncomplicated malaria. Secondly, quinine is still effective for the treatment of severe and complicated malaria. Thirdly, premature and unregulated use of the artemisinin and its derivatives, particularly oral preparations, might result in the development of resistance and could thereby compromise their efficacy in the future when they may be needed for the treatment of severe and complicated malaria because of decreased efficacy of quinine. Recently, however experts have taken the view that as artemisinin derivatives are potent antimalarials without significant adverse effects they may be used more widely; however, they should be used in combination with another effective blood schizontocide for the treatment of both multidrug-resistant and non-drug-resistant falciparum malaria to reduce recrudescence and to slow the development of resistance (Looareesuwan *et al.*, 1998).

Treatment of Malaria in Pregnancy

Although, information concerning the risk of artemisinin and its derivatives in pregnancy is incomplete, these drugs should not be withheld from pregnant women in areas where these drugs are indicated (White 1996a; WHO 1994). For pregnant women with uncomplicated malaria, it is preferable to avoid the use of these drugs in the first trimester but for those suffering from severe malaria in areas with low quinine efficacy the risk-benefit ratio may favour their use even in the first trimester.

Use Against Species other than *P. falciparum*

P. vivax, P. malariae, and *P. ovale* still respond well to chloroquine and other antimalarials, therefore the use of artemisinin or its derivatives against these malaria species is not recommended. Hypnozoites of *P. vivax* and *P. ovale* (i.e. liver stages) are not eliminated by artemisinin and its derivatives.

Prophylaxis

At the present time, there is a strong consensus that artemisinin and its derivatives should not be used as prophylaxis (WHO 1994). This is based on concerns of current uncertainities on the effect of higher doses and frequently repeated doses, and about continuous drug pressure leading to the emergence of resistance.

Drug Interactions

In the rodent malaria *P. berghei*, artemisinin potentiated the antimalarial effects of mefloquine, primaquine and tetracycline, is additive with chloroquine, and antagonises pyrimethamine, cycloguanil and sulphonamides (Chawira *et al.*, 1987). When mefloquine was given at 6 h after oral artesunate, the mefloquine concentrations were below those observed with mefloquine alone. Later mefloquine administrations had no such effect.

Co-administration of artemisinin derivatives and desferrioxamine B to cerebral-malaria patients was thought to be useful, combining the rapid parasite clearance of the artemisinin and the potential central-nervous system protection of the iron chelator. Although data from *in vitro* studies indicate that artesunate and desferrioxamine B are antagonistic in terms of antiparasitic efficacy (Meshnick *et al.*, 1993), subsequent studies in mice have found no evidence of such antagonism (Meshnick, unpubl. obs). In humans there was no evidence of adverse effects or toxicity resulting from this combination (Looareesuwan *et al.*, 1996c).

ACKNOWLEDGEMENT

We are grateful to Professor N.J. White, Faculty of Tropical Medicine, Mahidol University for reviewing the manuscript.

REFERENCES

Alin, M.H., Ashton, M., Kihamia, C.M., Mtey, G.J. and Bjorkman, A. (1996a) Clinical efficacy and pharmacokinetics of artemisinin monotherapy and in comination with mefloquine in patients with falciparum malaria. *Br. J. Clin. Pharmacol.*, 41, 587–592.

Alin, M.H., Ashton, M., Kihamia, C.M., Mtey, G.J. and Bjorkman, A. (1996b) Multiple dose pharmacokinetics of oral artemsinin and comparison of its efficacy with that of oral artesunate in falciparum malaria patients. *Trans. Roy. Soc. Trop. Med. Hyg.*, 90, 61–65.

Boele van, H.M., Onyiorah, E. and Jaffar, S. (1996) A trial of artemether or quinine in children with cerebral malaria. *N. Engl. J. Med.*, 335, 69–75.

Brewer, T.G., Peggins, J.O. and Grate, S.J. (1994) Neurotoxicity in animals due to arteether and artemether. *Trans. R. Soc. Trop. Med. Hyg.*, 88, (suppl 1):S33-S36.

Bunnag, D., Viravan, C., Looareesuwan, S., Karbwang, J. and Harinasuta, T. (1991a) Clinical trial of artesunate and artemether on multidrug resistant falciparum malaria in Thailand: a preliminary report. *Southeast Asian J. Trop. Med. Public Health*, 22, 380–385.

Bunnag, D., Viravan, C., Looareesuwan, S., Karbwang, J. and Harinasuta, T., (1991b) Double blind randomized clinical trial of two different regimens of oral artesunate in falciparum malaria. *Southeast Asian J. Trop. Med. Public Health*, 22, 534–538.

Bunnag, D., Viravan, C., Looareesuwan, S., Karbwang, J. and Harinasuta, T. (1991c). Double blind randomized clinical trial of oral artesunate at once or twice daily dose in falciparum malaria. *Southeast Asian J. Trop. Med. Public Health*, 22, 539–543.

Bunnag, D., Karbwang, J. and Harinasuta, T. (1992) Artemether in the treatment of multiple drug resistant falciparum malaria. *Southeast Asian J. Trop. Med. Public Health*, 23, 762–768.

Chawira, A.N., Warhurst, D.C. and Peters, W. (1986) Qinghaosu resistance in rodent malaria. *Trans. R. Soc. Trop. Med. Hyg.*, **80**, 447–480.

Chawira, A.N., Warhurst, D.C., Robinson, B.L. and Peters, W. (1987) The effect of combinations of qinghaosu (artemisinin) with standard antimalarial drugs in the suppressive treatment of malaria in mice. *Trans. R. Soc. Trop. Med. Hyg.*, **81**, 554–558.

China Cooperative Research Group on Qinghaosu and its Derivatives as Antimalarials (1982) Studies on the toxicity of quinghaosu and its derivatives. *J. Tradit. Chin. Med.*, **2**, 31–38.

Danis, M., Chandenier, J., Doumbo, O., Kombila, M., Kouame, J., Louis, F. *et al.* (1996) Results obtained with i.m. artemether *versus* i.v. quinine in the treatment of severe malaria in a multicentre study in Africa. *Jpn. J. Trop. Med. Hyg.*, **24**, (suppl 1), 93–96.

Ding, G.S. (1988) Recent studies on antimalarials in China: a review of the literature since 1980. *Int. J. Exp. Clin. Chem.*, **1**, 9–22.

Ezdinachi, E. (1996) *In vivo* efficacy of chloroquine, halofantrine, pyrimethamine-sulfadoxine, and qinghaosu (artesunate) in the treatment of malaria in Calabar, Nigeria. *Central Afr. J. Med.*, **42**, 109–111.

Ekong, R. and Warhurst, D.C. (1990) Synergism between arteether and mefloquine or quinine in a multidrug-resistant strain of *Plasmodium falciparum in vitro. Trans. R. Soc. Trop. Med. Hyg.*, **84**, 757–758.

Hien, T.T., Tam, D.T.H., Cuc, N.T.K. and Arnold, K. (1991) Comparative effectiveness of artemisinin suppositories and oral quinine in children with acute falciparum malaria. *Trans. Roy. Soc. Trop. Med. Hyg.*, **85**, 201–211.

Hien, T.T., Arnold, K., Vinh, H., *et al.* (1992) Comparison of artemisinin suppositories with intravenous artesunate and intravenous quinine in the treatment of cerebral malaria. *Trans R. Soc. Trop. Med. Hyg.*, **86**, 582–583.

Hien, T.T. and White, N.J. (1993) Qinghaosu. *Lancet*, **341**, 603–608.

Hien, T.T., Day, N.P.J., Phu, N.H., *et al.* (1996) A controlled trial of artemether or quinine in Vietnamese adults with severe falciparum malaria. *N. Engl. J. Med.*, **335**, 76–83.

Karbwang, J., Na Bangchang, K., Thanavibula, A., Bunnag, D., Chongsuphajaisiddhi, T. and Harinasuta, T. (1992a) Comparison of artemether and mefloquine in acute uncomplicated falciparum malaria. *Lancet*, **340**, 1245–1248.

Karbwang, J., Sukontason, K., Rimchala, W., Namsiripongpun, W., Tin, T., Auprayoon, P., Tumsupapong, S., Bunnag, D. and Harinasuta, T. (1992b) Preliminary report: a comparative clinical trial of artemether and quinine in severe falciparum malaria. *Southeast Asian J. Trop. Med. Public Health*, **23**, 768–772.

Li, G.Q., Guo, X.B. and Yang, F. (1990) Clinical trials on qinghaosu and its derivatives. Guangzhou College of Traditional Chinese Medicine, Sanya Tropical Medicine Institute, 1, 1–90.

Looareesuwan, S., Wilairatana, P., Vanijanonta, S., Kyle, D. and Webster, K. (1992a). Efficacy of quinine-tetracycline for acute uncomplicated falciparum malaria in Thailand. *Lancet*, 339–369.

Looareesuwan, S., Kyle, D.E., Viravan, C., Vanijanonta, S., Wilairatana, P., Charoenlarp, P. *et al.* (1992b). Treatment of patients with recrudescent falciparum malaria with a sequential combination of artesunate and mefloquine. *Am. J. Trop. Med. Hyg.*, **47**, 794–799.

Looareesuwan, S., Viravan C., Vanijanonta, S. Wilairatana, P., Suntharasamai, P., Charoenlarp, P. *et al.* (1992c). A randomised trial of mefloquine, artesunate and artesunate followed by mefloquine in acute uncomplicated falciparum malaria. *Lancet*, **339**, 821–824.

Looareesuwan, S., Viravan, C., Vanijanonta, S., Wilairatana, P., Charoenlarp, P., Canfield, C. *et al.* (1993) Treatment of acute uncomplicated falciparum malaria with a short course of artesunate followed by mefloquine. *Southeast Asian J. Trop. Med. Public Health*, **24**, 230–234.

Looareesuwan, S. (1994a) Overview of clinical studies on artemisinin derivatives in Thailand. *Trans. R. Soc. Trop. Med. Hyg.*, **88**, Suppl 1,S9–S11.

Looareesuwan, S., Vanijanonta, S., Viravan, C., Wilairatana, P., Charoenlarp, P. and Andrial, M. (1994b) Randomized trial of mefloquine alone and artesunate followed by mefloquine for the treatment of acute uncomplicated falciparum malaria. *Ann. Trop. Med. Parasitol.*, **88**, 131–136.

Looareesuwan, S., Viravan, C., Vanijanonta, S., Wilairatana, P., Charoenlarp, P., Canfield, C. *et al.* (1994c). Randomized trial of mefloquine-doxycycline, and artesunate-doxycycline for treatment of acute uncomplicated falciparum malaria. *Am. J. Trop. Med. Hyg.*, **50**, 784–789.

Looareesuwan, S., Wilairatana P., Vanijanonta, S. and Viravan, C. (1995) Efficacy and tolerability of a sequential, artesunate suppository plus mefloquine, treatment of severe falciparum malaria. *Ann. Trop. Med. Parasitol.*, **89**, 469–475.

Looareesuwan, S. (1996a) A phase III study of CGP 56697 *versus* mefloquine in Thai adults. 5th Western Pacific Congress on Chemotherapy and Infectious Diseases, Singapore, December 1996.

Looareesuwan, S., Viravan, C., Vanijanonta, S., Wilairatana, P., Pitisuttithum, P. and Andrial, M. (1996b) Comparative clinical trial of artesunate followed by mefloquine in the treatment of acute uncomplicated falciparum malaria: two and three-day regimens. *Am. J. Trop. Med. Hyg.*, **54**, 210–213.

Looareesuwan, S., Wilairatana, P., Vannaphan, S., Gordeuk, V.R., Taylor, T.E., Meshnick, S.R. *et al.* (1996c) Co-administration of desferioxamine B with artesunate in malaria: an assessment of safety and tolerance. *Ann. Trop. Med. Parasitol.*, **90**, 551–554.

Looareesuwan, S., Wilairatana, P. and Andrial, M. (1996d) Artesunate suppository for treatment of severe falciparum malaria in Thailand. *Jpn. J. Trop. Med. Hyg.*, **24** (suppl 1), 13–15.

Looareesuwan, S., Wilairatana, P., Vanijanonta, S., Pitisuttithum, P. and Viravan, C. (1996e) Treatment of acute, uncomplicated, falciparum malaria with oral dihydroartemisinin. *Ann. Trop. Med. Parasitol.*, **90**, 21–28.

Looareesuwan, S., Wilairatana, P., Vanijanonta, S., Oosterhuis, B., Peeters, P.A.M., Lugt, C.H.B. *et al.* (1996f) Clinical experiences on β-arteether (Artecef® Adult) in the management of falciparum malaria in Thailand. Abstracts : XIVth International Congress of Tropical Medicine, Nagasaki, Japan, November 1996.

Looareesuwan, S. and Wilairatana, P. (1997) Malaria. In R.E. Rakel, (ed.), *Conn's Current Therapy*, 49th edn., W.B. Saunders, Philadelphia, pp. 104–115.

Looareesuwan, S., Wilairatana, P., Viravan, C., Vanijanonta, S., Pitisuttithum, P. and Kyle, D.E. (1997a) Open randomized trial of oral artemether alone and a sequential combination with mefloquine for acute uncomplicated falciparum malaria. *Am. J. Trop. Med. Hyg.*, **56**, 613–617.

Looreesuwan, S., Wilairatana, P., Chokejindachai, W., Viriyavejakul, P., Krudsood, S. and Singhasavanon, P. (1998) Research on new antimalarial drugs and the use of drugs in combination at the Bangkok Hospital for Tropical Diseases. *Southeast Asian J. Trop. Med. Public Health*, **29**, 344–354.

Looareesuwan, S., Wilairatana, P., Chokejindachai, W., Chalermrut, K., Wernsdorfer, W., Gemperli, B. *et al.* (1999) A randomized, double-blind, comparative trial of a new oral combination of artemether and benflumtol (CGP 56697) with mefloquine in the treatment of acute *Plasmodium falciparum* malaria in thailand. *Am. J. Trop. Med. Hyg.*, **60**, 238–243.

Luxemburger, C., ter Kuile, F., Nosten, F., Dolan, G., Bradol, J.H., Phaipun, L. *et al.* (1994) Single day mefloquine-artesunate combination in the treatment of multidrug resistant falciparum malaria. *Trans. R. Soc. Trop. Med. Hyg.*, **88**, 213–217.

Meshnick, S.R., Taylor, T.E. Kamchonwongpaisan, S. *et al.* (1996) Artemisinin and the antimalarial endoperoxides: from herbal remedy to targeted chemotherapy. *Microbiol. Rev.*, **60**, 301–315.

Meshnick, S.R., Yang, Y.Z., Lima,V., Kuypers, F., Kamchonwongpaisan, S. Yuthavong, Y. *et al.* (1993) *Antimicrobial Agents and Chemotherapy*, **37**, 1108–1114.

Miller, L.G. and Panosian, C.B. 1997. Ataxia and slurred speach after artesunate treatment for falciparum malaria. *N. Eng. J. Med.*, **336**, 1328.

Nosten, F., Luxemburger, C., ter Kuile, F.O., *et al.* (1994) Treatment of multidrug-resistant *Plasmodium falciparum* malaria with 3-day artesunate-mefloquine combination. *J. Infect. Dis.*, **170**, 971–977. [Erratum, *J. Infect. Dis.* 1995, **171**, 519.]

Qinghaosu Antimalarial Coordinating Research Group. Antimalarial studiues on qinghaosu. *Chin. Med. J.*, **92**, 811–816.

Shen, J.X. (1989) *Antimalarial drug development in China.* National Institute of Pharmaceutical Research and Development, Beijing, pp. 31–95.

Taylor, T.T. *et al.* (1993) Rapid coma resolution with artemether in Malawian children with cerebral malaria. *Lancet*, **341**, 661–62

ter Kuile, F., White, N.J., Holloway, P., Pasvol, G. and Krishna, S. (1993) *Plasmodium falciparum: in vitro* studies of the pharmacodynamic properties of drugs uses for the treatment of the severe malaria. *Exp. Parasitol.*, **76**, 85–95.

Udomsangpetch, R., Pipitaporn, B., Krishna, S., Angus, B., Pukrittayakamee, S., Bates, I. *et al.* (1996) Antimalarial drugs reduce cytoadherence and rosetting of *Plasmodium falciparum*. *J. Infect. Dis.*, **173**, 691–698.

van Vugt, M., Brockman, A., Gemperli, B., Luxemburger, C., Gathman, I., Royce, C. *et al.* (1998a) A randomised comparison of artemether-benflumetol and artesunate-mefloquine in the treatment of multidrug resistant falciparum malaria. *Antimicrob Agents Chemother.*, **42**, 135–139.

van Vugt, M., Wilairatana, P., Gemperli, B., Gathmann, I., Phaipun, L., Brockman, A. *et al.* (1998b) Efficacy of six doses of artemether-lumefantrine in the treatment of multi-drug resistant falciparum malaria. *Submitted for publication.*

Wang, T.Y. (1989) Follow-up observations on the therapeutic effects and remote reactions of artemisinin (qinghaosu) and artemether in treating malaria in pregnant women. *J. Trad. Chin. Med.*, **9**, 28–30.

Watt, G., Loesuttivibool, L., Shanks, G.D., Boudreau, E.F., Brown, A.E., Pavanand, K. *et al.* (1992) Quinine with tetracycline for the treatment of drug-resistant falciparum malaria in Thailand. *Am J. Trop. Med. Hyg.*, **47**, 108–11.

White, N.J., Waller, D., Crawley, J., Nosten, F., Chapman, D., Brewster, D. *et al.* (1992) Comparison of arthemether and chloroquine for severe malaria in Gambian children. *Lancet*, **339**, 317–321.

White, N.J. (1994) Artemisinin: current status. *Trans. R. Soc. Trop. Med. Hyg.*, **88**(Suppl 1):S3–S4.

White, N.J. (1996a) The treatment of malaria. *N. Engl. J. Med.*, **335**, 800–806.

White, N.J. (1996b) Malaria. In G.C. Cook, (ed.), *Manson's Tropical Diseases*, W.B. Saunders, London, pp. 1087–1164.

White, N.J. (1997) Assessment of the pharmacodynamic properties of antimalarial drugs *in vivo. Antimicrobial Agents Chemther.*, **41**, 1413–1422.

Wilairatana, P. and Looareesuwan, S. (1995) APACHE II scoring for predicting outcome in cerebral malaria. *J. Trop. Med. Hyg.*, **98**, 256–260.

Wilairatana, P., Chanthavanich, P., Singhasivanon, P., Treeprasertsuk, S., Krudsood, S., Chalermrut, K. *et al.* (1998) A comparison of three different dihydroartemisinin formulations for the treatment of acute uncomplicated falciparum malaria in Thailand. *Int. J. Parasitol.*, **28**, 1213–1218.

WHO, Division of Control of Tropical Diseases. (1990) Severe and complicated malaria. *Trans R. Soc. Trop. Med. Hyg.*, **84**, (Suppl 2),1–65.

WHO (1994) The role of artemisinin and its derivatives in the current treatment of malaria (1994–1995): Report of an informal consultation: 27–29 September 1993. WHO/MAL 94.1067, Geneva, pp. 1–49.

WHO (1998) The use of artemisinin & its derivatives as anti-malarial drugs: Report of an informal consultation: 10–12 June 1998. WHO/MAL 98.1086, Geneva, pp. 1–28.

15. REGULATION OF THE QUALITY AND USE OF ARTEMISININ AND ITS DERIVATIVES

PENELOPE PHILLIPS-HOWARD

Malaria Control Unit, Division of Control of Tropical Diseases, World Health Organization, Geneva, Switzerland.

Current address: CDC/KEMRI,
Vector Biology and Research Center, Kenya Medical Research Institute, PO Box 1578, Kisumu, Kenya, and Division of Parasitic Diseases, National Centers for Infectious Diseases, Centers for Disease Control and Prevention, Atlanta, Georgia.

SUMMARY

Artemisinin and its derivatives have an essential role to play in the treatment of severe and complicated malaria, and for uncomplicated multidrug-resistant falciparum malaria. Their rapid action and lack of perceived side effects have prompted widespread use. To promote rational use and curtail misuse, drug regulations and policies were formulated through WHO and a multi-disciplinary team of experts. Regulatory policies included prescription and marketing of the artemisinin derivatives only in areas with multidrug-resistant falciparum malaria, for the purpose of treatment (not prophylaxis); prescription in combination with another effective antimalarial, preferably mefloquine; avoidance of drug promotion which would cause consumer misuse, and development of post-marketing surveillance strategies to ensure maximum quality, efficacy, and safety. To promote regulatory standards, guidelines formulated by the World Health Organization were distributed to the national drug regulatory centers of each country, to their ministries of health and to international organizations.

INTRODUCTION

Over the last decades, there has been a continuing deterioration of the malaria situation in many parts of the world. This is particularly so in areas where *Plasmodium falciparum* has developed resistance to chloroquine, sulfa-pyrimethamine combinations and, to some extent, to quinine which previously has been effective in the treatment of severe and complicated disease. In some areas, such as on the Thai/Cambodian and Thai/Myanmar borders, high levels of resistance to mefloquine, the only alternative drug available in control programmes, has also developed.

The urgent need for novel and rapidly acting drugs for the treatment of multidrug-resistant falciparum malaria was met by the Chinese discovery of artemisinin, a constituent of *Artemisia Annua* (Qinghaosu Antimalarial Coordinating Research Group, 1979). Now formulated as artemisinin and as water soluble and oil soluble derivatives, this group of drugs has been shown to elicit a rapid parasitological and clinical response in falciparum malaria, including severe and complicated cases and multidrug-resistant infections (Li, 1990; Looareesuwan, 1994). Artemisinin and its derivatives have been used for several years in China and Vietnam, and have been registered in other countries of South-East Asia, Latin America, and Africa in recent years.

A growing number of pharmaceutical companies produce and market artemisinin and its derivatives. Artemisinin formulations are manufactured in Vietnam (several establishments in Hanoi and Ho Chi Minh City), artesunate in Vietnam (Central Pharmaceutical Enterprise II in Hanoi) and in China (Guilin Pharmaceutical Works), and artemether in China (Kunming Pharmaceutical Factory). Neither the parenteral nor oral formulation of artesunate has yet been produced according to Good Manufacturing Practice (GMP) standards. The French company Rhone-Poulenc Rorer Doma, in association with the Kunming Pharmaceutical Factory, has upgraded the facility for production of artemether to GMP standard. WHO has assisted in the development of quality control mechanisms in countries of south-east Asia and reference centers are planned to strengthen national capabilities. A Belgian company, Profarma, imports raw materials from Vietnam and then manufactures oral and intramuscular artemether, without local registration, for export to developing countries with limited regulatory procedures.

The weak regulatory systems in many countries of the world are one of the greatest obstacles to the correct deployment and rational use of artemisinin and its derivatives. Even where regulatory mechanisms are well established, it is possible that the system maybe by-passed by illegal black market activities, including sub-standard and counterfeit drugs.

Control programme and infectious disease scientists have feared that widespread and indiscriminative use of the artemisinin compounds of varying quality, particularly through self-treatment with the oral formulations, would result in a rapid decline in their efficacy (White, 1994). This prompted WHO to bring together a multi-disciplinary team of experts to determine the role and limitations for use of these drugs, and to develop guidelines on regulation of quality and use which were then broadly circulated to relevant bodies (WHO, 1994). Since these recommendations were made, further experience with the use of artemisinin and its derivatives has been gained, particularly from south-east Asia, and resistance to chloroquine and sulfadoxine-pyrimethamine has intensified in parts of Africa. WHO convened a second informal consultation on the use of artemisinin and its derivatives in June 1998 to update WHO policies accordingly (WHO 1998). The following article summarizes these guidelines. A full copy of the meeting is obtainable through the Malaria Control Unit, Division of Control of Tropical Diseases, World Health Organization, Geneva, Switzerland.

Combination therapy, using the artemisinin derivatives, has recently been proposed as an approach to preventing the emergence of multiple-drug resistance (White 1998). The benefits of combination therapy will need to be measured against the disadvantages of increased direct treatment costs, poor compliance, logistical difficulties in drug supplies and some potential, but as yet unknown, pharmaco-kinetic and toxicity issues.

DRUG POLICY AND REGULATIONS

Unregulated use of artemisinin and its derivatives may result from a variety of factors. These include government failure to enforce regulations to control their importation and distribution, lack of a national drug policy for antimalarial drugs, uncontrolled importation and distribution, ignorance or lack of cooperation of pre-scribers on appropriate use of the drugs, lack of cooperation of pharmacists, and unwarranted demand on the part of the public. Irresponsible promotion and adver-tising by the technical and popular media may influence prescribers, pharmacists and patients (many of the latter having fevers other than malaria) in favour of the use of the artemisinin group of drugs in situations where they are not needed. Strategies recommended at the two WHO meetings to enhance and strengthen the manufacture and marketing, registration and use, and surveillance of the artemisinin and its derivatives are discussed below.

Registration and Marketing Authorization

Registration and marketing authorization of artemisinin and its derivatives should be restricted to areas where multidrug-resistance is prevalent, i.e., where chloro-quine and, sulfa-pyrimethamine combinations and quinine are ineffective. Although these drugs may not be more effective than quinine, clinical experience suggests that they may be preferred to quinine because of their fewer side effects and ease of administration. At present, such areas are found in Brazil, Cambodia, China, Lao Peoples' Democratic Republic, Myanmar, Thailand and Vietnam.

Each artemisinin product should be registered nationally as a prerequisite for routine use (Phillips-Howard, 1994). Drug licencing should only occur when it is deemed essential for use, with a description for its correct use in the national drug policy. All drugs should be registered according to strict guidelines and require a product licence (WHO, 1990).

The WHO Guidelines for small national drug regulatory authorities describe the licensing process as the administrative process assuring the safety, efficacy and quality of marketed products. For imported products the licensing process can be simplified by efficient use of the WHO Certification Scheme of the Quality of Pharmaceutical Products Moving in International Commerce (WHO, 1992a). This provides the means for an importing country to obtain information on the regula-tory status of the product in the country of export and on compliance of the manu-facturer with Good Manufacturing Practice (GMP) standards. If a product has been registered and produced in a country having a reputable drug regulatory authority, a WHO-type certificate might provide sufficient basis for issuing a marketing licence. WHO has offered to advise national authorities regarding local manufacturing facil-ities (WHO, 1994).

A product licence establishes, on legal and scientific grounds, the detailed composition, formulation, intended use of the drug, its efficacy, safety and quality (Table 1). Formulation of the licence by the Ministry of Health requires a precise definition of the drugs and the category of the licence holders, the content and format of licences, the criteria used to assess licence applications, and the guidance needed for the manufacturers, importers and distributors on the content and format of licence application, circumstances for renewal, extension or variation in the licence. As with other drugs, the decision to licence these drugs in a given country should be based on a documented need for them.

Where registration has already proceeded but, according to available evidence, the introduction of these drugs would not lead to any improvement in the treatment of malaria, ministries of health can choose not to give marketing authorization for use. Countries then have the option to approve marketing authorization at a future date when need arises, or for a limited time period during which close scrutiny of quality, efficacy, safety, and use by indication, can be monitored. Countries which have approved a marketing authorization, or anticipate doing so for artemisinin or one of its derivatives, need to evaluate and control the distribution and use of these drugs based on the need for them within the country. Approved marketing authorization may be for a limited period during which time additional data to ensure quality, efficacy and safety may be reviewed. Such authorization must include adoption of adequate product information and dissemination of accurate information, including clear guidelines on conditions in which the drugs should be used. Only when these steps have been taken should distribution and promotion commence.

Licences require routine review to monitor any breaches in conduct. Changes in the quality, safety or efficacy of the drug, or in the manufacturing, importing, distribution practices or changing the labelling of the drug may require amendment or revoking of the licence. Whilst post-licence inspection is one of the most valuable tools to ensure correct drug use, it is in practice the most difficult to maintain and is most rudimentary in countries where counterfeiting and illegal drug use is greatest.

Regional collaboration is essential in respect of product marketing and registration, because tight regulations in one country and not in its neighbours will result in the development of black market smuggling which inevitably leads to sales of sub-

Table 1   Required information for product licence.

☐   name (generic or brand)
☐   name/address of manufacturers and importers
☐   description of the dosage form
☐   description of composition (all ingredients)
☐   the therapeutic class
☐   indications for use
☐   a copy of all labelling including shelf life
☐   copies of all relevant certificates and warranties
☐   state serial number
☐   state date/issue of licence, expiry, special conditions

Source: World Health Organization (1994).

standard or counterfeit products. These problems are particularly applicable in South-East Asia where artemisinin and its derivatives are manufactured to varying marketing quality standards, and which are used in the countries of origin and exported.

In spite of previous recommendations, artemisinin derivatives are widely registered and available on the market in Africa south of the Sahara (WHO 1998). Injectable artemether and oral artesunate have been registered in 25 and 13 countries in Africa, respectively. Weak regulatory systems in many countries of the world are one of the greatest obstacles to correct deployment and rational use of antimalarial drugs, including artemisinin and its derivatives. Even where regulatory mechanisms are well established, it is possible that the system may be by-passed by illegal black market activities, including sub-standard and counterfeit drugs.

The most important constraints in controlling the use of the artemisinin drugs are;

- lack of policies for the selection of essential drugs, and for drug registration;
- inability to enforce regulations controlling the importation and distribution of drugs;
- lack of scientific data to develop an appropriate national antimalarial drug policy;
- ignorance and lack of cooperation of prescribers on the rational use of anti-malarial drugs;
- lack of information, education and communication (IEC) materials on appropriate treatment seeking behaviour leading to unnecessary patient demands for these drugs;
- commercial market forces making antimalarial drugs more widely available.

Control of Quality of Drugs

Given the strategic importance of artemisinin and its derivatives in the management of multidrug-resistant malaria, the quality of supplies should be assured to reach internationally recognized standards (WHO, 1992b). The licensing of quality drugs can be assured by using the WHO Certification Scheme (WHO, 1992a). This provides the means for an importing country to obtain information on the regulatory status of the product in the country of export and on compliance of the manufacturer with GMP standards. Importation of artemisinin and its derivatives should be controlled by national governments through a governmental health authority, a central drug procurement agency or by a limited number of licensed importers. In the absence of a national regulatory system or authority for drug registration, minimum requirements should include the use of the WHO Certification Scheme to obtain information on the regulatory status of the exporting country.

Drug regulatory authorities should be informed that most artemisinin group drugs are not produced to GMP standards, and that in some countries drugs of substandard quality are being promoted and used. Governments should take care in the selection of products for registration, assessing the need and assuring quality, efficacy and safety. It is recommended that regulatory authorities in exporting countries provide information to regulatory authorities in importing countries in respect of the quality and regulatory status of these drugs.

Where these drugs are used, routine sampling of artemisinin and its derivatives should be carried out to monitor their quality and the quality assurance procedure should be verified by independent experts. Screening should be routinely performed at port of entry (if imported) or at the factory, and again at the point of distribution to identify drug degradation following transit and storage. Withdrawal of the whole drug batch is mandatory for drugs not reaching predefined standards, and if the quality of drugs is consistently sub-optimal, the licence would be revoked. Where laboratory facilities are able to monitor drug quality, information on quality control specifications from the pharmaceutical company manufacturing the drugs is required. The specifications file includes the drugs chemical formulation, the manufacturing process, assay techniques, and quality assurance procedures.

There are currently three methods of assessment of drug quality:

- Thin-layer chromatography (TLC) identifies the parent compound as well as related substances and can provide an approximate estimation of the content of the drug ingredient;
- High Performance Liquid Chromatography (HPLC) coupled to an appropriate detector (see chapter 2), allows for the accurate quantitation of the parent drug substance but also the relative quantitation of diluents and breakdown products;
- *In vitro* dissolution profiles are being established for tablets, capsules and suppository formulations.

WHO is presently involved in developing and evaluating quality specifications of the drug substance and formulations for artemisinin, artemether, arteeether, artesunate and dihydroartemisinin that are intended for publication in The International Pharmacopoeia.

WHO has established collaboration for the quality control of antimalarial drugs with the University Sains Malaysia, Penang, Malaysia which is accepting samples of any antimalarial drug including the artemisinin drugs but this facility is insufficient and there is need for the establishment of additional international/regional reference centres as well as for the strengthening of national capacities.

Assistance through WHO would be available to standardize quality assurance mechanisms, with support from industrialized countries to strengthen local capabilities for this activity. Quality assurance may best be established as regional or interregional reference laboratories to serve all countries in the region. Information on potential institutes able to take on this activity is required.

Post-Marketing Surveillance (PMS)

The principal objectives of post-marketing surveillance are to identify problems related to drug quality, stability, and efficacy, adverse reactions that may be of a delayed nature, and the development of drug resistance. Such problems may not have been recognized in the initial trials of the drugs, or were not communicated to the manufacturer. Ideally, these should be carefully monitored by each national malaria control programme on a general basis and possibly through more detailed monitoring at sentinel sites.

There is serious concern that the indiscriminate use and misuse of these drugs may encourage the development of parasite resistance. Monitoring of resistance is a responsibility of national malaria control programmes and should be conducted at sentinel sites on a regular repetitive schedule. It is advisable that the clinical progress of patients treated in hospitals be carefully documented in order to detect, as early as possible, any drug failure that might be an early sign of resistance. Reliable data are accumulating on the safety of artemisinin derivatives (Price *et al.*, 1999). While post-marketing surveillance reports of use in over 4600 patients in Thailand, and over 10,000 treatments during clinical trials suggest the risk of severe adverse reactions to the artemisinin derivatives appears low, they indicate the need for continued vigilance for PMS in all countries where these drugs are used. Further data on the effects of repeated treatments, particularly among pregnant women are needed. Exhaustive enquiries into possible drug-induced deaths should be undertaken.

In practice, post-marketing surveillance methodologies in malaria endemic countries are not adequately developed and many countries do not have the capacity or resources to perform this (Phillips-Howard, 1994). No standardized methods exist yet for monitoring therapeutic efficacy or resistance to this group of drugs in the field and these urgently need to be established. Countries which have initiated activities, such as Thailand (a joint activity by the Thai Food and Drug Administration and the Thai Malaria Control Programme) are potential models for other countries (Ministry of Public Health, Thailand, 1996). Assistance at regional level to correlate information of common interest to neighboring countries should be promoted.

Control of Use

Since artemisinin and its derivatives are among the most effective antimalarials known and resistance to them has not yet been observed outside models, use in each country must be regulated. Country-specific policies to control the importation, promotion, distribution and use of this group of drugs to protect against misuse and exploitation are necessary. National governments need to evaluate carefully the necessity for oral forms of artemisinin and its derivatives and target their use to high-risk populations since national registration of oral formulations brings special problems. These include lack of quality assurance, difficulties in patient compliance with an effective radical cure requiring prolonged dosage, difficulties in controlling distribution and sales, and difficulties in controlling importation and over-the-counter use resulting in over-use and misuse. An analysis of the cost-effectiveness of switching from a less effective antimalarial to one of the artemisinin compounds should be conducted in advance (Phillips & Phillips-Howard, 1996).

Clear guidelines on the use of artemisinin and its derivatives should be produced by national authorities and distributed to all prescribers (Annex: Treatment Guidelines). This includes national drug formulary and pharmacy guidelines; information should be supplied to health workers to ensure correct prescribing in respect of clinical indication and dosage. Countries in which multidrug-resistance does not exist and which have already registered a drug belonging to the artemisinin group need to seriously reconsider their position, or limit the promotion, marketing and use to the maximum extent possible.

Storage and distribution of dugs should be in accordance with national policies and guidelines, bearing in mind that these drugs must be carefully conserved and their use regulated. Subject to national policies and regulations, these drugs should only be available on prescription, and not over the counter. It is important to elicit the support of pharmacists and practitioners in peripheral health areas to ensure proper control of distribution and prevent use of counterfeit drugs (Table 2); supplies through aid agencies and non-governmental organizations should all be in accordance with governmental regulations and procedures. Such agencies should be discouraged from promoting or importing drugs except through officially authorized channels in places where there is a documented need for them.

To avoid consumer misuse, it is important that the general public as potential or actual patients be educated in the correct use, in particular the importance of compliance, of artemisinin and its derivatives. Care should be taken not to promote the inappropriate use of the drugs. National regulatory agencies should be in a position to ban inaccurate or irregular promotion. Strong education, advertising and government controls are required to restrict their distribution. Promotional campaigns need to be monitored to prevent inaccurate and misleading statements during advertising. A series of mechanisms can facilitate control of the use of these drugs, including:

- selecting the most effective compounds and placing them in appropriate WHO publications, i.e. Model Prescribing Information (WHO, 1995) detailing where and when they should be used;
- recommending restrictions in registration and importation;

Table 2   Approaches to reduce black market and counterfeit drugs.

☐	Define licensed drug list
☐	Enforce regulations through penal sanctions
☐	Use WHO Certification Scheme
☐	Designate drug importation points with customs
☐	Routine inspection of imported drugs
☐	Customs officials destroy suspect drugs
☐	Inspectors monitor drug distribution network
☐	Flood the market with information
☐	Set up quality assurance laboratory
☐	Stock reputable drugs in pharmacies
☐	Devise anti-counterfeit measures (local companies)
☐	Minimize drug distribution through intermediaries

Source: World Health Organization (1994).

- advising that retail and distribution outlets be limited to licensed pharmacies and health facilities, making the drugs "prescription only", and eliciting cooperation of retail pharmacists not to sell these drugs without a prescription;
- provision of national guidelines for management of malaria including training of prescribers on the rational use of antimalarials and consumer education.

RESEARCH

Further studies are needed especially in the following areas:

1 Basic Research

Studies on the mode of action and mechanism of resistance.
Identification of genetic markers of resistance.
Studies on the mechanism of selective toxicity.

2 Treatment of Uncomplicated Malaria

Clinical evaluation of combinations of artemisinin derivatives with other antimalarial drugs, such as sulfadoxine/pyrimethamine, amodiaquine, chloroquine, pyronaridine, doxycycline and tetracycline to determine pharmacokinetic interactions, efficacy and tolerability and their impact on the development of resistance to the components. These studies, particularly with sulfadoxine/pyrimethamine, should be given the highest priority.

Determination of the optimum regimens for the oral administration of dihydroartemisinin and artemether.

Studies on the potential role of artemisinin derivatives in the treatment of chloroquine-resistant vivax malaria.

3 Compliance Issues

Studies to determine whether the regimens of artemisinin derivatives can be shortened, without loss of efficacy, when used in combination with other drugs.

Studies on presentation and packaging of artemisinin-based drugs to improve compliance.

4 Treatment of Severe Malaria

Determination of optimal regimens for parenteral and rectal administration of artemisinin derivatives.

Determination of the speed and reliability of absorption of intramuscular and rectal administration of artemisinin derivatives.

Determination of the clinical significance of delayed recovery from coma following administration of artemisinin derivatives.

5 Use in Pregnancy

Determination of the safety of artemisinin drugs in the first trimester of pregnancy by follow up of patients inadvertently given drugs during the early stages of undisclosed pregnancies.

6 Anti-Gametocytogenesis Activity

Determination of the importance of the activity of artemisinin drugs on gametocytogenesis in different epidemiological situations, with particular reference to malaria incidence and transmission.

7 Monitoring of Adverse Reactions

Expansion of post marketing surveillance, currently only carried out in Thailand, to other countries where artemisinin and its derivatives are widely used.

Continued surveillance of pregnant women given artemisinin drugs and all patients receiving repeated treatments, with special attention to the temporary suppression of reticulocyte response and neurotoxicity (see Price *et al.*, 1999).

8 Monitoring of Drug Resistance and Therapeutic Efficacy

Evaluation of recently developed protocols for monitoring therapeutic efficacy of antimalarial drugs for their applicability to artemisinin drugs.

Further evaluation of *in vitro* tests for susceptibility of *P. falciparum* to artemisinin and its derivatives.

Establishment of sentinel site monitoring of therapeutic efficacy testing.

9 Antimalarial Drug Policy

Determination of the cost implications of the use of artemisinin derivatives in combination with other appropriate drugs.

Evaluation of the impact of global recommendations on the use of artemisinin and its derivatives on drug policies and their use in the public and private sector.

10 Drug Development and Quality Assurance

Development of improved formulations and new derivatives of artemisinin.

Development of paediatric formulations for oral administration.

Technology transfer to improve standards of manufacture of all formulations to GMP.

Improvement of shelf life of formulations.

Strengthening of national and regional regulations and capacities for quality assurance of artemisinin and its derivatives.

ANNEX: TREATMENT GUIDELINES

The role of artemisinin and its derivatives are best defined within the context of the national drug policy and should take into account drug recommendations, and the concerns of, neighboring countries within a region. The information below has been published in Model Prescribing Information, through WHO (WHO, 1995, 1998). Care must be taken not to exceed recommended dosages, or to repeat courses of treatment at short intervals, to reduce the risk of adverse drug effects.

1 Treatment of Uncomplicated Malaria

Combination therapy (Oral)

Artemisinin and its derivatives should be administered in combination with another effective blood schizontocide to reduce recrudescences and to slow the development of resistance. At present data only support the operational use of the combination with mefloquine (15–25 mg base/kg) but a fixed combination of artemether with lumefrantrine is at an advanced state of development and research on other combinations is also being carried out. Administration of mefloquine on the second or third day considerably reduces the risk of vomiting once the clinical condition has been improved. Tolerance to the 25 mg base/kg doses of mefloquine may be further improved by administering 15 mg base/kg on the second or third day with the rest 6–24 hrs later. If compliance is a concern, mefloquine can be given on the first day. The dose of mefloquine depends on the local sensitivity of the parasite to mefloquine.

The following regimens are recommended:

- Artemisinin: 20 mg/kg as a divided loading dose on the first day, followed by 10 mg/kg once a day for a further 2 days, plus mefloquine (15–25 mg base/kg) as a single or split dose on the second or third day.
- Artesunate: 4 mg/kg once a day for 3 days, plus mefloquine (15–25 mg base/kg) as a single or split dose on the second or third day.
- Artemether: 4 mg/kg once a day for 3 days, plus mefloquine (15–25 mg base/kg) as a single or split dose on the second or third day.

The combination of dihydroartemisinin with mefloquine and other drugs is still being evaluated in clinical trials.

Monotherapy

In those situations where the use of artemisinin combinations is impossible, for example because of patient intolerance to mefloquine, monotherapy with artemisinin drugs may be used in regimens of 7 days with every effort being made to ensure compliance. Administration of shorter regimens to non-immune patients leads to unacceptably high levels of recrudescences.

The following regimens are recommended:

- Artemisinin: 20 mg/kg in a divided loading dose on the first day, followed by 10 mg/kg once a day for 6 days.
- Artesunate: 4 mg/kg in a divided loading dose on the first day, followed by 2 mg/kg once a day for 6 days.
- Artemether: 4 mg/kg in a divided loading dose on the first day, followed by 2 mg/kg once a day for 6 days.

There are limited data on dihydroartemisinin and further research is required to determine optimal dosage regimens.

2 Treatment of Severe and Complicated Malaria

The following schedules are recommended for adults and children over 6 months.

Intramuscular artemether

3.2 mg/kg as a loading dose on the first day, followed by 1.6 mg/kg daily for a minimum of 3 days until the patient can take oral therapy of an effective antimalarial. The daily dose of artemether can be given as one single injection. In children, the use of a 1 ml tuberculin syringe is advisable since the injection volumes will be small. A formulation (40 mg/1ml) that is more easily used in children is available from one manufacturer.

Intravenous artesunate

2.4 mg/kg as a loading dose on the first day, followed by 1.2 mg/kg daily for a minimum of 3 days until the patient can take oral therapy of an effective antimalarial.

The anhydrous acid contents are dissolved in 0.6ml 5% (w/v) sodium hydrogen carbonate. The solution should be prepared just before use, because of the instability of the acid, and be diluted with 5.4 ml of 5% (w/v) dextrose solution or dextrose in normal saline.

3 Use in Pregnancy

In view of the serious health implications of malaria in pregnancy, artemisinin and related compounds should not be withheld from pregnant women in areas where these drugs are indicated. For the management of uncomplicated malaria in pregnancy, artemisinin and its derivatives can be used in the second and third trimester, but their use in the first trimester is not recommended. For the treatment of severe malaria in the first trimester, the advantages of artemisinin drugs over quinine, especially the lower risk of hypoglycaemia, must be weighed against the fact that there is still limited documentation on pregnancy outcomes following their use. The inadequacy of current knowledge on the use of these drugs during pregnancy should be understood by care providers, and if possible, all pregnancies exposed to these drugs should be monitored. Reports of all clinical outcomes, both successful and adverse events should be made to regulatory authorities.

REFERENCES

Li Guo Qiao (1990) *Clinical Trials on Qinghaosu and its derivatives*, **Vol. 1**. Guangzhou College of Traditional Chinese Medicine, Sanya Tropical Medicine Institute.

Looareesuwan, S., (1994) Overview of clinical studies on artemisinin derivatives in Thailand. *Transactions of the Royal Society of Tropical Medicine and Hygiene*, Suppl., 88, 9–12.

Phillips-Howard, P.A., (1994) Regulation of drug use and post-registration surveillance. *Transactions of the Royal Society of Tropical Medicine and Hygiene*, Suppl., 88, 59–62.

Phillips, M.A., and Phillips-Howard PA (1996). Economic implications of resistance to anti-malarial drugs. *PharmacoEconomics*, 10, 225–238.

Price, R., van Vugt, M., Phaipun, L., Luxemburger, C., Simpson, J. *et al.* (1999) Adverse effects in patients with acute falciparum malaria treated with artemisinin derivatives. *Am. J. Trop Med Hyg.*, 60, 547–555.

Qinghaosu Antimalarial Coordinating Research Group (1979) Antimalarial studies on qing-haosu. *Chinese Medical Journal*, 92, 811–16.

White, N.J., (1994) Summary of discussions and conclusions. *Transactions of the Royal Society of Tropical Medicine and Hygiene*, Suppl., 88, 63–65.

White, N.J., (1998) Preventing antimalarial drug resistance through combinations. *Drug Resistance Updates*, 1, 3–9.

World Health Organization (1990) *Guidelines for small national drug regulatory authorities*. Technical Report Series, N° 790. Geneva, World Health Organization.

World Health Organization (1992a) *Certification Scheme of the Quality of Pharmaceutical Products Moving in International Commerce*. WHO Technical Report Series, N° 823. Geneva, World Health Organization.

World Health Organization (1992b) Good manufacturing practice for pharmaceutical prod-ucts. In: *WHO Expert Committee on Specifications for Pharmaceutical Preparations*. 32nd Report. Technical Report Series, N° 823, Annex 1. Geneva, World Health Organization.

World Health Organization (1994) *The role of artemisinin and its derivatives in the current treatment of malaria (1994–1995)*. WHO/MAL/94.1067. Geneva, World Health Organization.

World Health Organization (1995) *WHO Model Prescribing Information – Drugs used in Parasitic Diseases*. World Health Organization.

World Health Organisation (1998) The use of artemisinin and its derivatives as antimalarial drugs. Report of a joint CTD/DMP/TDR consultation. Geneva, 10–12 June 1998. WHO/MAL98.1086.

INDEX

Chemical structures are indicated by page numbers in *italics*.

Printed and bound by CPI Group (UK) Ltd, Croydon, CR0 4YY

23/10/2024

01778254-0006